Plant Biology: Concepts and Applications

Plant Biology: Concepts and Applications

Editor: Geoffrey Watkins

CALLISTO
REFERENCE

www.callistoreference.com

Callisto Reference,
118-35 Queens Blvd., Suite 400,
Forest Hills, NY 11375, USA

Visit us on the World Wide Web at:
www.callistoreference.com

ISBN: 978-1-64116-162-6 (Hardback)

Trademark Notice: Registered trademark of products or corporate names are used only for explanation and identification without intent to infringe.

Cataloging-in-Publication Data

Plant biology : concepts and applications / edited by Geoffrey Watkins.
    p. cm.
Includes bibliographical references and index.
ISBN 978-1-64116-162-6
1. Botany. 2. Plants. I. Watkins, Geoffrey.
QK47 .P53 2019
580--dc23

Table of Contents

Preface

Plant biology is a natural science that is concerned with the study of various aspects of plants. It is classified into several branches within the three principal groups of core topics, applied topics and organismic topics. The core topics of biology such as cytology, phenology, plant anatomy, plant morphology, plant systematics, etc. address the varied aspects of processes, diversity and classification of plants. Agronomy, biotechnology, forestry and horticulture, among others are the various applied branches of botany that are concerned with the economic application of plants. The organismal domains of orchidology, lichenology, mycology, etc. are the disciplines that specialize in the study of specific plant groups. This book provides significant information of this discipline to help develop a good understanding of botany and related fields. From theories to research to practical applications, case studies related to all contemporary topics of relevance to this discipline have been included in this book. Researchers and students in this field will be assisted by this book.

All of the data presented henceforth, was collaborated in the wake of recent advancements in the field. The aim of this book is to present the diversified developments from across the globe in a comprehensible manner. The opinions expressed in each chapter belong solely to the contributing authors. Their interpretations of the topics are the integral part of this book, which I have carefully compiled for a better understanding of the readers.

At the end, I would like to thank all those who dedicated their time and efforts for the successful completion of this book. I also wish to convey my gratitude towards my friends and family who supported me at every step.

Editor

Relative importance of phenotypic trait matching and species' abundances in determining plant–avian seed dispersal interactions in a small insular community

Aarón González-Castro[1,2,3*], Suann Yang[2,4], Manuel Nogales[1] and Tomás A. Carlo[2]

[1] Island Ecology and Evolution Research Group (CSIC-IPNA), C/Astrofísico Francisco Sánchez n° 3, 38206, La Laguna, Tenerife, Canary Islands, Spain
[2] Department of Biology, Pennsylvania State University, 208 Mueller Laboratory, University Park, PA 16802, USA
[3] Present address: Instituto de Ciencia Innovación Tecnología y Saberes Universidad Nacional de Chimborazo, Avenida Antonio José de Sucre, Riobamba, Ecuador
[4] Present address: Biology Department, Presbyterian College, 503 South Broad Street, Clinton, SC 29325, USA

Guest Editor: Donald Drake

Abstract. Network theory has provided a general way to understand mutualistic plant–animal interactions at the community level. However, the mechanisms responsible for interaction patterns remain controversial. In this study we use a combination of statistical models and probability matrices to evaluate the relative importance of species morphological and nutritional (phenotypic) traits and species abundance in determining interactions between fleshy-fruited plants and birds that disperse their seeds. The models included variables associated with species abundance, a suite of variables associated with phenotypic traits (fruit diameter, bird bill width, fruit nutrient compounds), and the species identity of the avian disperser. Results show that both phenotypic traits and species abundance are important determinants of pairwise interactions. However, when considered separately, fruit diameter and bill width were more important in determining seed dispersal interactions. The effect of fruit compounds was less substantial and only important when considered together with abundance-related variables and/or the factor 'animal species'.

Keywords: Dispersal; frugivory; mutualistic networks; oceanic islands; probability matrices.

Introduction

A ubiquitous mutualistic plant–animal interaction is that between fleshy-fruited plants and the fruit-eating animals that disperse their seeds (Jordano 2000). Seed dispersal interactions are complex because they involve multiple species of animals and plants, the temporal and spatial variability of such interactions (Yang *et al.* 2014) and the influence of frugivore behaviour and physiology, as well as the chemistry of fruits (Jordano 2000; Carlo

* Corresponding author's e-mail address: arongcastro@gmail.com

and Yang 2011). Network theory has emerged as a useful tool to deal with such complexity and to search for organizational and coevolutionary patterns in community-wide plant–frugivore interactions (e.g. Bascompte et al. 2003, 2006; Jordano et al. 2003; Bascompte and Jordano 2007; Rezende et al. 2007; González et al. 2010; Mello et al. 2011; Aizen et al. 2012; Stouffer et al. 2012). However, the mechanisms responsible for interaction patterns in such networks (e.g. nestedness, modularity, interaction asymmetry, degree distribution) remain unclear.

Two hypotheses are available to explain how mutualistic interactions influence the structure of mutualistic networks. The first is the neutrality hypothesis (so-called abundance hypothesis), which states that observed patterns within a community are due to random species interactions. According to neutrality, probabilities of observing a plant–disperser interaction chiefly depend on the abundance of species. For example, observing both common and rare frugivores feeding on common fruiting plant species is more likely than on rare ones. This implies that abundance will be positively correlated with the level of generalization in the mutualistic interactions, i.e. highly abundant species could artificially appear as generalists that are highly connected in the mutualistic network, and rare species as more specialized (e.g. Dupont et al. 2003; Vázquez 2005; Vázquez et al. 2007; Schleuning et al. 2011).

On the other hand, the phenotypic traits hypothesis postulates that interaction patterns result from morphological, physiological, behavioural or evolutionary constraints that condition interaction probabilities between potential mutualistic partners (Jordano et al. 2003; Rezende et al. 2007; Santamaría and Rodríguez-Gironés 2007; Dupont and Olesen 2009; Mello et al. 2011; Olesen et al. 2011). Among phenotypic traits, the most commonly used in analyses of seed dispersal networks are the disperser bill width and fruit diameter (i.e. this will determine whether or not a seed can be swallowed and dispersed), as well as accessibility restrictions by frugivores (e.g. Rezende et al. 2007; Olesen et al. 2011; Burns 2013). However, although it has been shown that the chemical compounds of fruit can be important in determining frugivory and seed dispersal interactions (Jordano 2000 and references therein), such traits have not been used previously in network analyses. In this study we incorporate, for the first time, fruit nutritional compounds into the analysis of a frugivory network.

Some studies have demonstrated that both mechanisms (abundance and phenotypic traits) can work hand-in-hand to shape network structure (e.g. Stang et al. 2006; Bascompte and Jordano 2007). In this vein, Verdú and Valiente-Banuet (2011) demonstrated that facilitative interactions between plants were better explained by a combination of abundance and phylogenetic relationships than by these variables separately. Still, both hypotheses are not necessarily mutually exclusive and ecologists are beginning to examine their relative importance. Because interaction networks can be presented as adjacency matrices, Vázquez et al. (2009a) proposed using probability matrices (derived from species abundance and their spatial–temporal overlap) to assess the relative importance of abundance and phenotypic traits to determine the observed patterns of mutualistic interactions. This approach is useful to predict aggregate network parameters, but not effective in predicting pairwise interactions (Vázquez et al. 2009a).

To improve this approach, here we model pairwise interactions between fleshy-fruited plants and their avian dispersers, as a response to species' phenotypic traits, as well as to the species' abundance. Then we use interaction frequencies predicted by the best statistical model to build a probability matrix as proposed by Vázquez et al. (2009a) to assess the ability of that model to predict aggregate network parameters. Solving these questions will help us to understand evolutionary and ecological forces driving the assemblage of interactions in mutualistic communities (Bascompte and Jordano 2007).

Methods

Study area

The study was carried out in Los Adernos, a Mediterranean scrubland habitat site located in the northwest region of Tenerife (Canary Islands, UTM: 28R 317523 E/3138253 N, 220 m above sea level (a.s.l.)). The climate is Mediterranean, with mean annual rainfall ranging between 200 and 400 mm and mean temperature between 16 and 19 °C. The fleshy-fruited plant community is mainly composed of Asparagus plocamoides Webb ex Svent., Jasminum odoratissimum L., Rubia fruticosa Aiton, Rhamnus crenulata Aiton and Heberdenia excelsa (Aiton) Banks ex DC. There are four avian disperser species: Sylvia atricapilla, S. melanocephala, Turdus merula and Erithacus rubecula. The study was conducted in two sampling periods encompassing two whole years: from June 2008 through May 2009, and from January 2010 through December 2010.

Seed dispersal interactions

In order to characterize the set of interactions between fleshy-fruited plants and avian dispersers, we focussed on seeds recovered from the faeces of birds captured with mist nets that were opened from dawn until dusk 2–3 days per month. We computed mist-netting effort by multiplying the mist-net length by the number of hours they were operated. Faecal samples were analysed with a dissecting scope for seeds, which were counted and identified to species level. In order to take into account interspecific differences on the number of seeds produced per fruit, we divided the number of seeds dispersed by the mean number of seeds produced per fruit. Doing so gave us a better estimation of the number of times a given disperser visit a plant species for fruits. With these data we constructed an interaction matrix based on the interaction frequency between fleshy-fruited plants and avian dispersers **[see Supporting Information]**.

Network theory has been usually applied for large and complex communities, whereas the community in this study is small (four animal and nine plant species). Small communities, however, are less prone to sample bias than large ones (Blüthgen 2010), and the reliability of studies will be greater when the more accurate is the sampling of interactions. We used an accumulation curve to prove the robustness of our sampled interactions, with a curve slope lower than 0.03 after all our 54 mist-netting sessions **[see Supporting Information]**.

Explanatory variables

We considered eight explanatory variables associated with phenotypic traits (six variables) and abundance hypotheses (two variables) in order to explain interaction frequency between plants and animals (i.e. number of dispersed seeds). Explanatory variables for phenotypic traits hypothesis were fruit organic compounds (fibre, lipids, sugars and proteins), the identity of bird species and size overlap between fruit diameter and bill width of birds (hereafter size overlap). Although a wide variety of fruit chemical compounds may influence the choice of fruits by birds (Jordano 2000 and references therein), we selected sugars, fibre, proteins and lipids based on a study on Mediterranean avian-dispersed fruits (Herrera 1987). We decided to include the factor 'animal species' involved in each plant–animal interaction because species identity is important to predict animal interaction patterns (Carlo et al. 2003; Carnicer et al. 2009).

The two explanatory variables used to test for the abundance hypothesis were the product of abundance of interacting species (hereafter abundance) and temporal overlap of species phenophase length (time length which plants display fruits and bird species are present at the study site). Although phenophase length and hence temporal overlap are, to some extent, species-specific traits, they can also be considered as metrics of abundance because a species can be abundant either by producing high fruit densities and/or by being available over long time periods (Vázquez et al. 2009a). Moreover, the phenophase length of fruiting plants can also be affected by external factors to the plant such as weather conditions or the depletion of fruit crops.

Fruit nutrient compounds. Chemical analyses of fruits were performed by Canagrosa Laboratories (http://www.canagrosa.com/). Amount of compounds was calculated as percentage of dry mass by different methods: Kjeldahl method for proteins, gravimetric plus digestion with acid-detergent solution for fibre and Soxhlet extraction with hexane for lipids. The amount of sugars was calculated based on the remaining organic material following the equation:

$$\text{sugars} = \frac{\text{NFES} \times 100}{100 - \text{RH}}, \tag{1}$$

where RH is the relative humidity of the sample and NFES (nitrogen-free extractive substances) is calculated as follows:

$$\text{NFES} = 100 - \text{RH} - \text{proteins} - \text{lipids} - \text{fibre}. \tag{2}$$

Animal species. We accounted for the animal species identity as a factor to explain the interaction frequency. Although quantification of animal traits exists for the species studied here (Herrera 1984; Jordano 1987), our system is unfortunately too small (36 interactions) to support models with these additional explanatory variables. We decided to give priority to use fruit nutrient compounds because animal traits have been previously used in some extent on network analyses, whereas fruit nutrients have not yet.

Size overlap. To account for individual variability of fruit and bill size, we decided to use the range (mean ± SD) in both fruit diameter and bird gape width instead of just comparing their mean values. For each pair of species, we calculated the percentage of range of fruit diameter that was equal or smaller than the maximum value of the bird gape width. For example, if the diameter of a fruit species ranges from 7.0 to 8.0 mm (variation range = 1.0 mm), then the resulting overlap with the gape of a bird that ranges between 9.0 and 10.0 mm would be 100 %. However, for a bird with a gape width of 6.0–7.5 mm, the size overlap is 50 %, because only half of the fruit variation range (0.5 out of 1 mm) could be 'swallowed' by this second bird. For those interactions with pairs of species which size overlap was 0 % we arbitrarily establish a size overlap of 1×10^{-5}.

Abundance and temporal overlap. The variable abundance is the product of abundance of the interacting species. To assess fruit abundance we used 20 plots of 5 m^2 randomly placed. We visited every plot monthly and estimated the number of fruits per metre square for every plant species (visual counting method, Chapman et al.

1992). We estimated the cumulative abundance after the two study years and then calculated the relative fruit abundance for every plant species as the percentage of fruits of each species from the total community-wide fruit crop. Bird abundance was estimated by using a simple mixed effect regression analysis for every 100 h of sampling: [Individuals $\times$ m^{-2} = 2.15 + 4.117 $\times$ (100 $\times$ C); $P = 0.001$; $N = 152$], where C is the number of captured birds per unit of effort. To build this regression we used unpublished data (A.G.-C.) from the same study area. Censuses were performed twice per month. Therefore, we considered the date as a random effect factor to avoid an effect of temporal pseudo-replication. To ensure bird detectability in censuses, a band 25 m wide was surveyed, where all individuals (seen or heard) were counted. We consider that all disperser birds had equivalent capture probability in mist nets because shrubs mostly dominate the study site, and frugivorous birds have same movement patterns, between shrubs, where mist nets where placed.

Every 15 days we recorded the presence of species of fruits and birds to obtain the length of species phenophase. The temporal overlap is defined as the percentage of days with respect to the whole study period (i.e. 730 days) that pairs of species coincided in the study area. For example, if a plant species fruited for 60 days, and a bird species was present in the study area for 60 days, but fruit and bird species coincided for only 30 days, then they had 4.11 % of temporal overlap (30 out of 370 days).

Modelling interaction frequency between pairs of species

We modelled the log-transformation of interaction frequency (estimated as explained above) as a response to the explanatory variables by using a generalized least squares (GLS) model in the *nlme* package (Pinheiro *et al.* 2009) implemented in R 2.11 (R Development Core Team 2015). The variance was not homogeneous and changed with the predictor 'size overlap', thus the GLS model allowed us to work with a normal error distribution and taking into account the variance structure by using the function 'varFixed', implemented in the *nlme* package (Zuur *et al.* 2009).

As our study system was small, with only 36 plant-frugivore interactions, we had to build different models with different subsets of explanatory variables to avoid model over fitting (*sensu* Burnham and Anderson 2002). Therefore, we used different combinations of phenotypic traits and abundance variables in different models **[see Supporting Information]**. By doing so, each statistical model included only a maximum of five explanatory variables (including main effects of variables and/or their statistical). In models that we could not include all fruit

compounds together, we separated them into two different sets: one included 'non-energetic' compounds (fibre and proteins) and the other included 'highly energetic' compounds (sugars and lipids).

We ranked models according to the AIC value and computed the Akaike's weight as an estimation of the probability of a given model to be the best candidate model explaining the observed interactions (Burnham and Anderson 2002). To evaluate the importance of different explanatory variables we used a multi-model inference, based on the sum of Akaike's weight of each model where each explanatory variable appeared (Burnham and Anderson 2002). Different variables appeared in a very different number of models (e.g. size overlap appeared in three models, whereas 'animal species' appeared in 14 models), based on the natural history of the fruiting plants and animals rather than their statistical importance. For such a reason, we averaged the sum of Akaike's weight by dividing it by the total number of models where each variable appeared.

Prediction of aggregate network parameters

With the interaction frequency predicted by the best statistical model (that showed the lowest AIC value) we created an expected interaction matrix. Subsequently, we normalized this matrix by dividing their elements by their total number of predicted interactions to obtain a probability matrix. With this probability matrix, we used simulations based on the approach of Vázquez *et al.* (2009*a*) to assess the capacity of our best model to predict different aggregate network parameters: connectance (proportion of realized interactions respect to total cells in the interaction matrix), interaction evenness (which is a Shannon index proposed by Tylianakis *et al.* 2007; the higher the index, the more evenly distributed are interactions in the matrix), nestedness (the degree to which specialists interact with proper subsets of the species that generalists interact with) and interaction asymmetry for fruits and for dispersers (Vázquez *et al.* 2007, 2009*a*). We performed 1000 randomizations for the model (i.e. the probability matrix), calculated the mean and 95 % confidence interval of each parameter and assessed if the observed value for each parameter in the interaction matrix recorded at Los Adernos fell within such confident interval.

We had problems in simulating nestedness values with the temperature algorithm proposed by Rodríguez-Gironés and Santamaría (2006), perhaps due to small size of our interaction matrix. Therefore, we used the Nestedness metric based on Overlap and Decreasing Fill (NODF; Almeida-Neto *et al.* 2008) and 'weighted nestedness' (Galeano *et al.* 2009) algorithms. All the analyses were run with the R-code provided by Vázquez *et al.*

(2009*a*) but with modifications to include NODF and weighted nestedness measures as implemented in the *bipartite* (Dormann *et al.* 2009) package of R statistical software (R Development Core Team 2015).

Results

Statistical models

We created 24 statistical models, containing each explanatory variable separately, as well as different combinations of them **[see Supporting Information]**. The best model to explain plant–bird interactions combined both species phenotypic traits and species abundance. This model (AIC = −15.00) included the factor 'animal species' and variables related with species matching (size overlap between fruit diameter and bird bill width, temporal overlap of species phenophase length and the product of species abundance). The second best model (AIC = −4.57) included some fruit compounds (fibre and proteins), the factor 'animal species' and both abundance-related variables (temporal overlap and species abundance) **[see Supporting Information]**. The null model (that with only the intercept) was better than three models based only on species abundance, size overlap and 'size overlap × animal species', respectively **[see Supporting Information]**.

Considering the averaged sum of Akaike's weight, the most important variable was size overlap between fruit diameter and bird bill width, followed by species temporal overlap, species abundance and the factor 'animal species' (Table 1). However, the difference between temporal overlap and species abundance was not significant (the order of magnitude of that difference was 1×10^{-15}). According to the Akaike's weight, fruit nutrient compounds were less important variables (Table 1). However, models that combined both fruit compounds with the identity of animal species and/or variables related with abundance were among the best fitted **[see Supporting Information]**.

In general, birds dispersed more frequently plant species with a lower amount of sugars and lipids in their fruits, with the exception of *T. merula*, which tend to select those fruits with higher sugar content (Fig. 1). Birds, with the exception again of *T. merula*, more often dispersed fruits with higher protein content. Curiously, fibre-rich fruits, which are of low digestibility, also had, in general, a high dispersal frequency (Fig. 1). Interaction frequency was also higher for pairs of species with a higher product of their abundance and higher temporal overlap, with the exception of *S. melanocephala* (Fig. 2).

With respect to aggregate network parameters, the probability matrix based on the best statistical model was able to predict only nestedness, based on both 'NODF' and 'weighted nestedness' algorithms (Table 2).

Table 1. Relative importance of each explanatory variable to determine fruit–avian disperser interactions in Los Adernos (Northwest of Tenerife Island). Variables are ranked from most important to less important according to the averaged sum of Akaike's weight, w_i (Burnham and Anderson 2002). The higher the value, the more important the explanatory variable is. As the number of models in which each variable appears is established by our knowledge of fruit and bird natural history, and not because of statistical reasons, we decided to use the averaged sum of Akaike's weight.

Explanatory variable	Number of models in which variable appears	Averaged sum of w_i
Size overlap	3	0.333333333
Temporal overlap	7	0.143707147
Species abundance	7	0.143707147
Animal species	14	0.072078903
Fibre	5	0.001181644
Proteins	5	0.001181644
Sugars	5	0.000568973
Lipids	5	0.000568973

Discussion

Our study shows that the best way to understand pairwise interactions of the plant–frugivore network in the scrublands of Tenerife is using both the phenotypic traits and the abundance of species. Previous studies in other sites and mutualistic interactions (e.g. pollination) have reached similar conclusions (Stang *et al.* 2006; Bascompte and Jordano 2007; Verdú and Valiente-Banuet 2011), but our study stands out in showing that matching of two phenotypic traits (fruit diameter and bird bill width) is a stronger determinant of mutualistic interactions than species' abundances (Table 1). Our best model was able to predict only one network parameter, nestedness (NODF and 'weighted nestedness'), which is in contrast to previous studies able to predict more parameters (Vázquez *et al.* 2009*a*; Verdú and Valiente-Banuet 2011). However, it is important because nestedness has been proposed to be an important structural feature determining species coexistence and diffuse coevolution (Bascompte *et al.* 2003). Therefore, our results support that nestedness may be strongly influenced by both fruit-bill matching, as well as by species abundance as previously proposed (e.g. Vázquez 2005; Rezende *et al.* 2007).

Previous studies have demonstrated that an important trait related to seed dispersal frequency is the overlap between fruit diameter and bird gape width (e.g. Rezende *et al.* 2007; Olesen *et al.* 2011), which is in accord with our best model. Increasing the size overlap between the

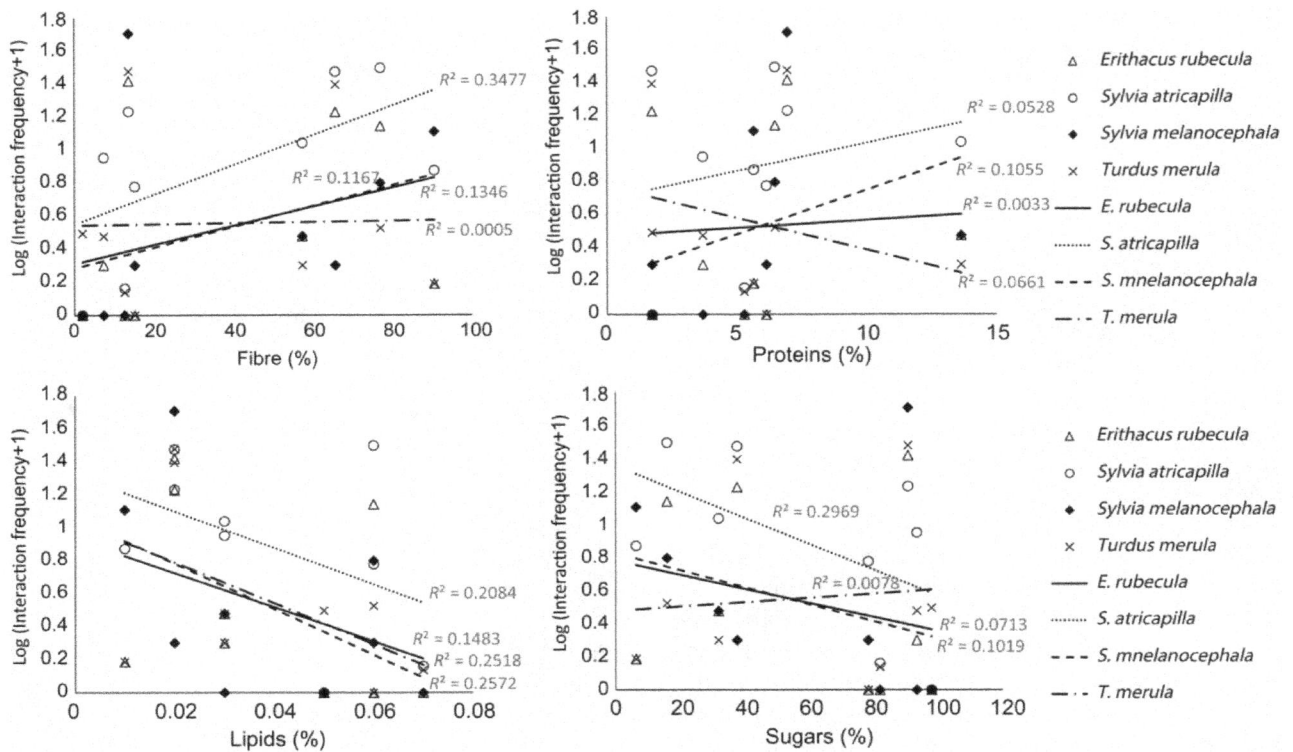

Figure 1. Relationship between the content of different fruit nutrient compounds and interaction frequency in Los Adernos. The figure shows the relationship with different avian disperser species. For each species (different lines) the fit (R^2) is shown.

bill widths and fruit diameters increases the probability of a successful seed dispersal interaction between a bird–plant species pair (e.g. Pratt and Stiles 1985; Wheelwright 1985; Jordano 1987; Levey 1987). Size restriction could explain why in this community, some small bird species depended heavily on smaller fruits of low digestibility (e.g. *S. melanocephala*—*R. crenulata*) than on more profitable but larger fruits (e.g. *Tamus edulis*, with a 97.87 % of sugars). This size restriction could also explain why the smallest passerine (*S. melanocephala*) has very few interactions with the large-fruited *H. excelsa* (Fig. 2), despite these two species having a high temporal overlap and high abundances.

Although the importance of fruit chemistry in mediating plant–frugivore interactions has been amply demonstrated (Jordano 2000), this study is the first to include them as part of structural analyses of a mutualistic network. We found some relationship between fruit nutrient amount and interaction frequency (Fig. 1). However, fruit compounds were weak predictors of fruit–bird interaction frequency when considered independently (Table 1; and see low values of R^2 in Fig. 1). On the other hand, six out of the seven best models included combinations of fruit compounds with the identity of animal species and/or species abundance **[see Supporting Information]**. This suggests that importance of fruit

compounds in determining fruit–bird interactions should be considered in a global context of additional ecological factors.

In general, small birds dispersed plant species with fruits of low digestibility and profitability (i.e. low content of sugars and lipids and high content of fibre) more frequently, whereas *T. merula* tended to show the opposite pattern (Fig. 1). This result makes sense if we consider that small disperser birds of the thermophilous scrublands (*E. rubecula*, *S. atricapilla* and *S. melanocephala*) are characterized by having fruit-dominated diets and short gut-passage times (see Herrera 1984 for some bird species). Thus, when birds with short gut-passage times consume fruits of low digestibility, they would need to increase the rates of fruit intake to maintain their energy and nutrient assimilation balance (Barboza et al. 2009), whereas birds with long gut-passage times (*T. merula* in our case) would not need to increase the intake rate because fruit pulp remains longer in the gut (Barboza et al. 2009).

Our results also confirm the abundance hypothesis because of the positive relationship between the frequency of seed dispersal and the product of species abundances and their temporal overlap, with the above-mentioned exception of the *S. melanocephala*–*H. excelsa* interaction (Fig. 2). According to the averaged sum of Akaike's

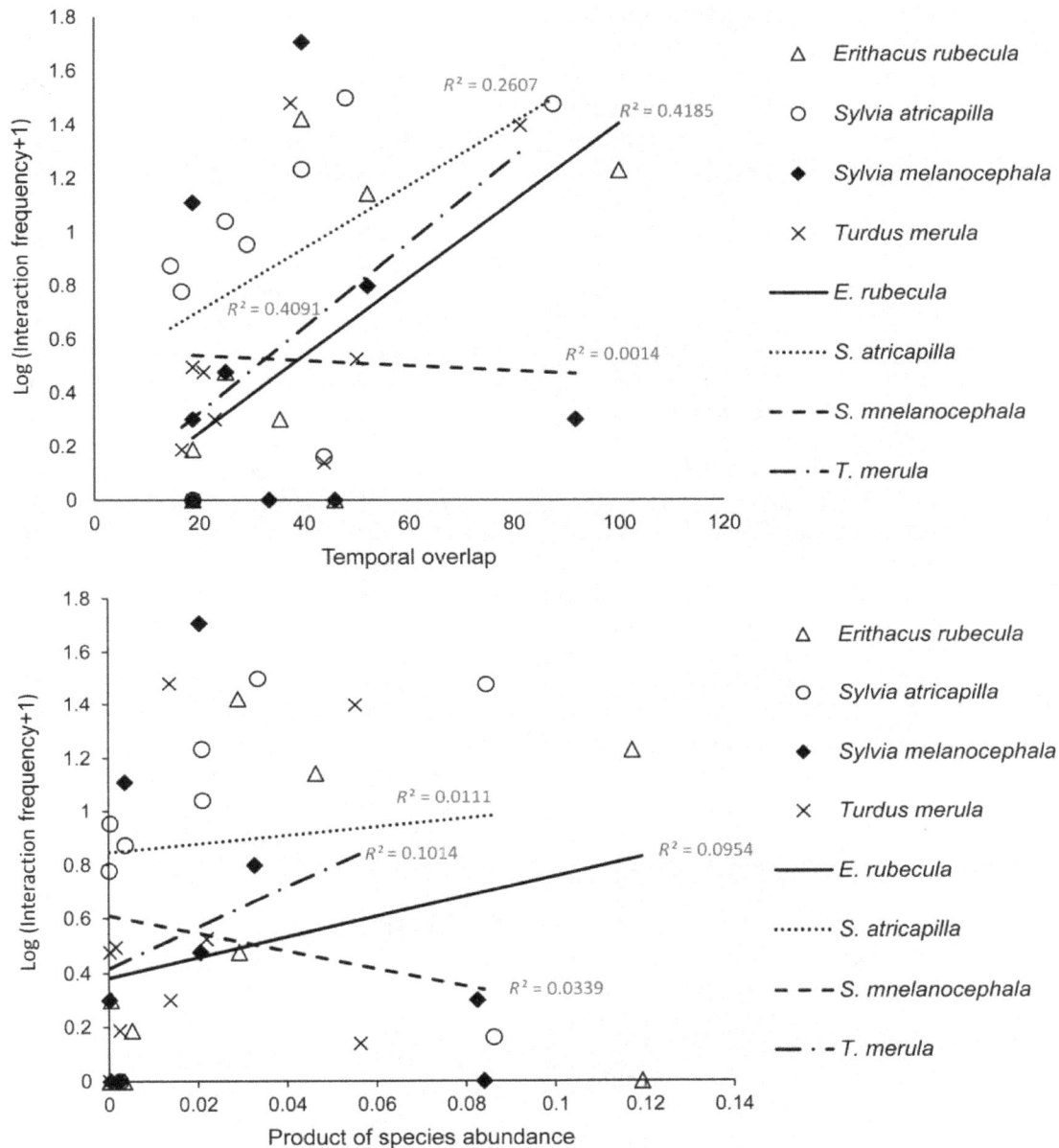

Figure 2. Relationship of species phenophase temporal overlap, and product of species abundance with interaction frequency in Los Adernos. The figure shows the relationship with different avian disperser species. For each species (different lines), the fit (R^2) is shown.

Table 2. Observed and simulated (mean and 95 % confident interval) values of six network parameters. Network parameters with observed value that coincide with confidence interval of simulation are in bold.

Network parameter	Observed value	Mean value	Lower limit	Upper limit
Connectance	0.7777778	0.9170556	0.8611111	0.9444444
Nestedness (NODF)	**41.6666667**	**30.0674603**	**16.6666667**	**48.8095238**
Weighted nestedness	**0.3467869**	**0.1300832**	**−0.1311452**	**0.4306871**
Interaction evenness	0.8341048	0.8738392	0.8471696	0.899341
Interaction asymmetry for birds	−0.2723253	−0.1723619	−0.2142811	−0.1448604
Interaction asymmetry for plants	0.1602762	0.148106	0.1422437	0.15625

weights, temporal overlap and species abundance were more important than fruit chemical compounds (Table 1). Thus, as the product of species abundances and/or their temporal overlap increases, the more likely it is that species will interact with each other. According to previous studies (e.g. Dupont *et al.* 2003; Vázquez 2005; Vázquez *et al.* 2007; Schleuning *et al.* 2011), species abundance (as number of individuals) should be sufficient by itself to explain fruit–bird interactions. However, like for phenotypic traits, a greater importance of species abundance (and temporal overlap) emerged from four models where abundance-related variables appear jointly with other phenotypic traits **[see Supporting Information]**.

Although the importance of temporal overlap and species abundance was not significantly different (Table 1), the model solely based on temporal overlap of species' phenophase fit better than the model only based on the product of species' abundance **[see Supporting Information]** (Fig. 2). This finding is in accordance with previous studies on mutualistic networks (Olesen *et al.* 2008; González-Castro *et al.* 2012*b*). We have to note that although we consider temporal overlap as an abundance-related variable, species' phenophase length is, in some extent, a species-specific trait (Olesen *et al.* 2008; González-Castro *et al.* 2012*b*). Therefore, the importance of species' temporal overlap for plant–disperser interactions might be considered as an influence of phylogeny of species.

We can think of at least two explanations for the relatively weak effect of species abundance on plant–disperser interactions in our community, when compared with the effect of temporal overlap. One could be due to the small size of this community (four animal and nine plant species). Larger communities invariably have a higher potential number of interactions, which makes them more difficult to sample appropriately (Blüthgen 2010). In contrast, in small communities sampling community-wide interactions is more precise. For example, comparing two closely related Mediterranean habitats, González-Castro *et al.* (2012*a*) found that the effect of abundance on interaction asymmetry was lower in the small-sized community than in the large one. Another possibility is that the method used to estimate species abundances affects the outcome of the models (Vázquez *et al.* 2009*b*). Previous studies have used interaction frequency as a measure of abundance (e.g. Vázquez *et al.* 2007; Schleuning *et al.* 2011). Thus, there is an obvious lack of independence between the response (interactions) and the predictor variable (abundance). But, in this study, we measured species abundance independently of the animal–plant interaction using captures and censuses. In this sense, our abundance estimates are uncorrelated to our interaction data, and thus more appropriate than those used by previous studies. We suggest that future studies should use independent estimators of species' abundances when trying to assess the effects of abundance on interaction patterns.

The first six models in our ranking included the factor 'animal species' **[see Supporting Information]**. The inclusion of this factor, and/or its interaction term with fruit compounds or abundance-related variables, generally improved the equivalent models in which 'animal species' was excluded **[see Supporting Information]**. Importance of interaction terms between fruit compounds and 'animal species' reveals the relevance of some animal phenotypic traits in determining fruit-disperser interactions, and suggests that different animal species might respond to fruit nutrients in different ways (Jordano 2000). Therefore, our results support the assertion of Jordano (2000) that: 'the profitability of a given fruit should be examined in the context of an interaction with a particular frugivore species'.

The model including the statistical interaction 'animal species × species abundance' slightly improves the fit of the model with respect to the model only based on species abundance (AIC = 375.124 and 376.489, respectively) **[see Supporting Information]** (see also different bird responses in Fig. 2). This result is consistent with interspecific differences in the capacity of birds to respond to changing fruit abundances (Carnicer *et al.* 2009). This species-specific response of birds to fruit abundance makes it more difficult for abundance to determine community-wide interactions by itself.

Conclusions

Although fruit-bill size overlap seems to be the most important variable when considered independently, both species abundance and phenotypic traits were important in determining fruit–bird interactions. The small community size (36 interactions) constrained us to use different subsets of explanatory variables in different models. However, species abundance and phenotypic traits are inseparable in a community. Therefore, the approach we used will be useful to examine more diverse communities, but using more realistic models (i.e. including all explanatory variables together). Although obtaining detailed phenotypic data in larger communities is challenging, it will allow a better understanding of ecological organization and coevolutionary processes shaping mutualistic plant–animal communities.

Sources of Funding

A.G.-C. benefited from a JAE-PRE fellowship from the Consejo Superior de Investigaciones Científicas (Spain). The National Science Foundation under a grant awarded in 2008 funded S.Y. This work was financed by the

Spanish Ministry of Science and Education project (CGL2007-61165/BOS) and is also framed within project GGL2010-18759/BOS, supported by FEDER funds from the European Union.

Contributions by the Authors

T.A.C. and A.G.-C. conceived the idea. M.N. and A.G.-C. carried out the fieldwork, S.Y. and A.G.-C. analysed the data. S.Y., M.N., T.A.C. and A.G.-C. wrote the paper.

Acknowledgements

We thank Benito Pérez, Daniel González, Yurena Gavilán, David Padilla and Juan Carlos Illera for their help during fieldwork. We also thank Diego P. Vázquez for useful comments on an early version of the manuscript. CANAGROSA Laboratories performed chemical analyses of fruits. The Cabildo (insular council) of Tenerife, Parque Rural de Teno and the landowner Teobaldo Méndez provided permission to work at the study site. The Biology Department and the Eberly College of Science at Penn State University provided support to T.A.C., S.Y. and A.G.-C. during the Spring and Summer of 2011.

Supporting Information

The following additional information is available in the online version of this article –

Table S1. Interaction frequency between plants (columns) and animals (rows) recorded at the study site.

Figure S1. Accumulative curve of interactions recorded at the study site against the sampling effort (mist-netting sessions).

Table S2. Candidate statistical models, ranked according to their AIC value.

Literature Cited

Aizen MA, Sabatino M, Tylianakis JM. 2012. Specialization and rarity predict nonrandom loss of interactions from mutualist networks. *Science* **335**:1486–1489.

Almeida-Neto M, Guimarães P, Guimarães PR, Loyola RD, Ulrich W. 2008. A consistent metric for nestedness analysis in ecological systems: reconciling concept and measurement. *Oikos* **117**: 1227–1239.

Barboza PS, Parker KL, Hume ID. 2009. *Integrative wildlife nutrition*. Berlin: Springer.

Bascompte J, Jordano P. 2007. Plant-animal mutualistic networks: the architecture of biodiversity. *Annual Review of Ecology, Evolution, and Systematics* **38**:567–593.

Bascompte J, Jordano P, Melian CJ, Olesen JM. 2003. The nested assembly of plant-animal mutualistic networks. *Proceedings of the National Academy of Sciences of the USA* **100**:9383–9387.

Bascompte J, Jordano P, Olesen JM. 2006. Asymmetric coevolutionary networks facilitate biodiversity maintenance. *Science* **312**: 431–433.

Blüthgen N. 2010. Why network analysis is often disconnected from community ecology: a critique and an ecologist's guide. *Basic and Applied Ecology* **11**:185–195.

Burnham KP, Anderson DR. 2002. *Model selection and multi-model inference: a practical information-theoretic approach*. New York: Springer.

Burns KC. 2013. What causes size coupling in fruit–frugivore interaction webs? *Ecology* **94**:295–300.

Carlo TA, Yang S. 2011. Network models of frugivory and seed dispersal: challenges and opportunities. *Acta Oecologica* **37**: 619–624.

Carlo TA, Collazo JA, Groom MJ. 2003. Avian fruit preferences across a Puerto Rican forested landscape: pattern consistency and implications for seed removal. *Oecologia* **134**:119–131.

Carnicer J, Jordano P, Melián CJ. 2009. The temporal dynamics of resource use by frugivorous birds: a network approach. *Ecology* **90**: 1958–1970.

Chapman CA, Chapman LJ, Wangham R, Hunt K, Gebo D, Gardner L. 1992. Estimators of fruit abundance of tropical trees. *Biotropica* **24**:527–531.

Dormann CF, Fründ J, Blüthgen N, Gruber B. 2009. Indices, graphs and null models: analyzing bipartite ecological networks. *The Open Ecology Journal* **2**:7–24.

Dupont YL, Olesen JM. 2009. Ecological modules and roles of species in heathland plant–insect flower visitor networks. *Journal of Animal Ecology* **78**:346–353.

Dupont YL, Hansen DM, Olesen JM. 2003. Structure of a plant–flower-visitor network in the high-altitude sub-alpine desert of Tenerife, Canary Islands. *Ecography* **26**:301–310.

Galeano J, Pastor JM, Iriondo JM. 2009. Weighted-Interaction Nestedness Estimator (WINE): a new estimator to calculate over frequency matrices. *Environmental Modelling and Software* **24**: 1342–1346.

González AMM, Dalsgaard B, Olesen JM. 2010. Centrality measures and the importance of generalist species in pollination networks. *Ecological Complexity* **7**:36–43.

González-Castro A, Traveset A, Nogales M. 2012a. Seed dispersal interactions in the Mediterranean Region: contrasting patterns between islands and mainland. *Journal of Biogeography* **39**: 1938–1947.

González-Castro A, Yang S, Nogales M, Carlo TA. 2012b. What determines the temporal changes of species degree and strength in an oceanic island plant-disperser network? *PLoS One* **7**:e41385.

Herrera CM. 1984. Adaptation to frugivory of Mediterranean avian seed dispersers. *Ecology* **65**:609–617.

Herrera CM. 1987. Vertebrate-dispersed plants of the Iberian Peninsula: a study of fruit characteristics. *Ecological Monographs* **57**: 305–331.

Jordano P. 1987. Frugivory, external morphology and digestive system in Mediterranean sylviid warblers *Sylvia* spp. *Ibis* **129**: 175–189.

Jordano P. 2000. Fruits and frugivory. In: Fenner M, ed. *Seeds: the ecology of regeneration in plant communities*. Wallingford: CAB International, 125–165.

Jordano P, Bascompte J, Olesen JM. 2003. Invariant properties in coevolutionary networks of plant-animal interactions. *Ecology Letters* **6**:69–81.

Mello MAR, Marquitti FM, Guimarães PR, Kalko EK, Jordano P, de Aguiar MA. 2011. The modularity of seed dispersal: differences in structure and robustness between bat- and bird-fruit networks. *Oecologia* **167**:131–140.

Olesen JM, Bascompte J, Elberling H, Jordano P. 2008. Temporal dynamics in a pollination network. *Ecology* **89**:1573–1582.

Olesen JM, Bascompte J, Dupont YL, Elberling H, Rasmussen C, Jordano P. 2011. Missing and forbidden links in mutualistic networks. *Proceedings of the Royal Society B: Biological Sciences* **278**:725–732.

Pinheiro J, Bates D, DebRoy S, Sarkar D, The R Core Team. 2009. *nlme: linear and nonlinear mixed effects models*. R package version 3.1-96.

Pratt TK, Stiles EW. 1985. The influence of fruit size and structure on composition of frugivore assemblages in New Guinea. *Biotropica* **17**:314–321.

R Development Core Team. 2015. *R: a language and environment for statistical computing*. Vienna: R Foundation for Statistical Computing. http://www.r-project.org.

Rezende EL, Jordano P, Bascompte J. 2007. Effects of phenotypic complementarity and phylogeny on the nested structure of mutualistic networks. *Oikos* **116**:1919–1929.

Rodríguez-Gironés MA, Santamaría L. 2006. A new algorithm to calculate the nestedness temperature of presence-absence matrices. *Journal of Biogeography* **33**:924–935.

Santamaría L, Rodríguez-Gironés MA. 2007. Linkage rules for plant-pollinator networks: trait complementarity or exploitation barriers? *PLoS Biology* **5**:e31.

Schleuning M, Blüthgen N, Flörchinger M, Braun J, Schaefer HM, Böhning-Gaese K. 2011. Specialization and interaction strength in a tropical plant-frugivore network differ among forest strata. *Ecology* **92**:26–36.

Stang M, Klinkhamer PGL, van der Meijden E. 2006. Size constraints and flower abundance determine the number of interactions in a plant-flower visitor web. *Oikos* **112**:111–121.

Stouffer DB, Sales-Pardo M, Sirer MI, Bascompte J. 2012. Evolutionary conservation of species' roles in food webs. *Science* **335**:1489–1492.

Tylianakis JM, Tscharntke T, Lewis OT. 2007. Habitat modification alters the structure of tropical host–parasitoid food webs. *Nature* **445**:202–205.

Vázquez DP. 2005. Degree distribution in plant-animal mutualistic networks: forbidden links or random interactions? *Oikos* **108**:421–426.

Vázquez DP, Melián CJ, Williams NM, Blüthgen N, Krasnov BR, Poulin R. 2007. Species abundance and asymmetric interaction strength in ecological networks. *Oikos* **116**:1120–1127.

Vázquez DP, Chacoff NP, Cagnolo L. 2009a. Evaluating multiple determinants of the structure of plant-animal mutualistic networks. *Ecology* **90**:2039–2046.

Vázquez DP, Blüthgen N, Cagnolo L, Chacoff NP. 2009b. Uniting pattern and process in plant-animal mutualistic networks: a review. *Annals of Botany* **103**:1445–1457.

Verdú M, Valiente-Banuet A. 2011. The relative contribution of abundance and phylogeny to the structure of plant facilitation networks. *Oikos* **120**:1351–1356.

Wheelwright NT. 1985. Fruit-size, gape width, and the diets of fruit-eating birds. *Ecology* **66**:808–818.

Yang S, Albert R, Carlo TA. 2014. Transience and constancy of interactions in a plant-frugivore network. *Ecosphere* **4**:147.

Zuur AF, Ieno EN, Walker NJ, Savaliev AA, Smith GM. 2009. *Mixed effects models and extensions in ecology with R*. New York: Springer.

2

Long-distance plant dispersal to North Atlantic islands: colonization routes and founder effect

Inger Greve Alsos[1]*, Dorothee Ehrich[2], Pernille Bronken Eidesen[3], Heidi Solstad[4], Kristine Bakke Westergaard[5], Peter Schönswetter[6], Andreas Tribsch[7], Siri Birkeland[3,8], Reidar Elven[9] and Christian Brochmann[9]

[1] Tromsø Museum, University of Tromsø, NO-9037 Tromsø, Norway
[2] Department of Arctic and Marine Biology, Faculty of Biosciences, Fisheries and Economics, University of Tromsø, NO-9037 Tromsø, Norway
[3] The University Centre in Svalbard, PO Box 156, NO-9171 Longyearbyen, Norway
[4] Museum of Natural History and Archaeology, Norwegian University of Science and Technology, NO-7491 Trondheim, Norway
[5] Norwegian Institute for Nature Research, PO Box 5685 Sluppen, NO-7485 Trondheim, Norway
[6] Institute of Botany, University of Innsbruck, Sternwartestraße 15, A-6020 Innsbruck, Austria
[7] Department of Organismic Biology, University of Salzburg, Hellbrunnerstraße 34, A-5020 Salzburg, Austria
[8] Centre for Ecological and Evolutionary Synthesis, Department of Biosciences, University of Oslo, PO Box 1066 Blindern, NO-0316 Oslo, Norway
[9] National Centre for Biosystematics, Natural History Museum, University of Oslo, PO Box 1172 Blindern, NO-0318 Oslo, Norway

Guest Editor: Jose Maria Fernandez-Palacios

Abstract. Long-distance dispersal (LDD) processes influence the founder effect on islands. We use genetic data for 25 Atlantic species and similarities among regional floras to analyse colonization, and test whether the genetic founder effect on five islands is associated with dispersal distance, island size and species traits. Most species colonized postglacially via multiple dispersal events from several source regions situated 280 to >3000 km away, and often not from the closest ones. A strong founder effect was observed for insect-pollinated mixed maters, and it increased with dispersal distance and decreased with island size in accordance with the theory of island biogeography. Only a minor founder effect was observed for wind-pollinated outcrossing species. Colonization patterns were largely congruent, indicating that despite the importance of stochasticity, LDD is mainly determined by common factors, probably dispersal vectors. Our findings caution against a priori assuming a single, close source region in biogeographic analyses.

Keywords: Amplified fragment length polymorphism (AFLP); dispersal vector; founder effect; genetic diversity; islands; long-distance dispersal (LDD); postglacial; species traits.

* Corresponding author's e-mail address: inger.g.alsos@uit.no

Introduction

Long-distance dispersal (LDD) of plants is a complex process (Higgins *et al.* 2003). Direct observations of LDD are rare (Ridley 1930); therefore, it is usually inferred from the geographical distribution of species or genes (de Queiroz 2005). Effective LDD (also termed long-distance colonization) involves seed release, dispersal by one or several vectors, arrival in a favourable microhabitat, germination and successful establishment of a new population (Chambers and MacMahon 1994; Nathan 2006). Several factors may influence each of these components. Dispersal routes and frequencies may depend on historical factors such as past climate shifts and geographical distributions (Taberlet *et al.* 1998), as well as on dispersal vectors such as birds, sea currents and wind (Gillespie *et al.* 2012). Local establishment depends on the number of arriving propagules, adaptation of the newcomer to local ecological conditions, including abiotic factors and relationship to pollinators or mycorrhiza partners (Chambers and MacMahon 1994; Hegland *et al.* 2009). Because LDD of plants is rarely directly observed, quantifying its relationship to potential determining factors is challenging (Nathan 2006). The relative importance of deterministic versus stochastic processes in shaping LDD patterns is not clear (Higgins *et al.* 2003; Nathan 2006; Vargas *et al.* 2012).

Oceanic islands represent good models to study LDD, as every species (or its ancestor) on such islands must have arrived by LDD. According to the equilibrium theory of island biogeography, the number of species on an island increases with island size and decreases with distance to source regions (MacArthur and Wilson 1967), although other factors such as species traits, sea current, past and present climate, and habitat heterogeneity may also play a role (Triantis *et al.* 2012; Weigelt and Kreft 2013). Similarly, the amount of genetic diversity in island populations is expected to be positively correlated with island size, but typically to be lower than in continental populations as only a limited number of genotypes from the source populations are expected to disperse to the recipient region causing a genetic founder effects (Jaenike 1973; Frankham 1997). As the frequency of dispersal events decreases with distance (Nathan 2006; Nathan *et al.* 2008), the initial founder effect and restriction of subsequent immigration, both leading to genetic depauperation of island populations, may increase with distance to the source region (Bialozyt *et al.* 2006; Dlugosch and Parker 2008; Pauls *et al.* 2013). Genetic diversity in plant populations is not only a result of population history but also related to species traits such as pollination mode, breeding system, growth form and morphological adaptations to dispersal (Hamrick and Godt 1996; Thiel-Egenter *et al.* 2009), all factors that may affect the intensity of founder effects. If species diversity and genetic diversity on islands are shaped by the same deterministic colonization processes, relative levels of genetic diversity should be related to the levels of species diversity. Moreover, floristic and genetic similarities should point to the same source regions for island colonization.

The role of LDD in shaping the current northern flora, which contains species that typically are widely distributed across a naturally fragmented biome, is debated (Löve and Löve 1963; Brochmann *et al.* 2013). In the Arctic, efficient LDD may be frequent due to open landscapes, strong winds and numerous migrating birds, a prediction supported by genetic data for the isolated archipelago of Svalbard (Alsos *et al.* 2007). Sea ice may also facilitate dispersal, as a 'bridge' or as a rafting vector (Johansen and Hytteborn 2001). Nevertheless, floristic analyses have indicated that most Arctic islands are not saturated with species (Hoffmann 2012). Similarly, analyses of plant species diversity in the Arctic mainland indicate that species distributions are limited by dispersal and/or establishment conditions (Lenoir *et al.* 2012).

The potentially strongest barrier to plant dispersal in the circumpolar region is the North Atlantic Ocean. For more than 100 years it has been debated whether plants were able to cross it via LDD after the last glaciation, or whether they depended on surviving the last (or several) glaciation(s) in local ice-free refugia in different Atlantic regions (Löve and Löve 1963; Brochmann *et al.* 2003). Molecular evidence clearly shows that trans-Atlantic LDD has occurred recently in many species (Brochmann *et al.* 2003). The Atlantic Ocean (and the Greenlandic ice sheet) is nevertheless a stronger barrier against dispersal than continuous Arctic landmasses, as shown in a recent circumpolar analysis of genetic variation in 17 vascular plant species (Eidesen *et al.* 2013). Even though the current floras in various Atlantic regions mainly have established following postglacial colonization, genetic data for a few species indicate *in situ* glacial persistence (Westergaard *et al.* 2011; see also Parducci *et al.* 2012).

To gain a better understanding of the factors determining LDD, we here analysed genetic structure in 25 plant species in five islands and adjacent mainland regions in the North Atlantic, as well as similarities in species composition among regional floras. We ask whether genetic data (i) support the prevalent hypothesis of postglacial long-distance colonization or, alternatively, local glacial survival, (ii) determine the source areas for postglacial island colonization in the North Atlantic region, (iii) quantify the intensity of the genetic founder effect and investigate how it relates to distance, island size and plant species traits and (iv) compare genetic and floristic relationships among regions.

Methods

Geographical regions

We selected five recipient islands/archipelagos: East Greenland (182 440 km^2 as delimited by Elven *et al.* 2011), Iceland (103 000 km^2), Svalbard (24 453 km^2 of non-glaciated area), the Faroe Islands (1396 km^2) and Jan Mayen (377 km^2). Although East Greenland is only part of an island, we treated it as an island because only narrow strips of land disrupted by glaciers connect it to North and South Greenland, and the Inland Ice Sheet forms a firm dispersal barrier to West Greenland (Eidesen *et al.* 2013). All surrounding land masses in north-eastern North America and Europe were selected as potential source regions. Minimum distances between recipient island and source regions (coast to coast) were estimated using Google Earth version 6.2.0.5905 (beta).

All recipient islands were mainly glaciated during the last glacial maximum (LGM, ~20 000 cal. BP, Ehlers and Gibbard 2004) although minor ice-free areas existed (reviewed in Brochmann *et al.* 2003). Pollen and macro-fossil studies show that a flora including many of the species we analysed for genetic variation existed on East Greenland from 12 800 to 12 300 cal. BP (Bennike 1999; Bennike *et al.* 1999), on Iceland from 13 000 to 10 800 cal. BP (Rundgren 1998; Rundgren and Ingólfsson 1999; Caseldine *et al.* 2003), on Svalbard from 9000 cal. BP (Birks 1991) and on the Faroe Islands from 13 100 cal. BP (Hannon *et al.* 2010). No late glacial or early Holocene palaeobotanical studies exist from Jan Mayen. Iceland, the Faroe Islands and Jan Mayen are true oceanic islands, whereas Svalbard and Greenland are continental islands. However, due to the previous heavy glaciation also of the latter two, they may be viewed as mainly oceanic islands in terms colonization processes.

Genetic data

We assembled amplified fragment length polymorphism (AFLP) datasets for Arctic and north-boreal species of vascular plants present in the five recipient islands. Most data originate from published studies [see Supporting Information—Table S1]. We included only AFLP datasets of high quality, e.g. with error rates estimated from random replicates, test of many primers before selection of final primer set (see Alsos *et al.* 2007, 2012 for details) and based on extensive sampling in the North Atlantic area. Our final dataset comprised 25 species, 1110 local populations, 8932 individual plant samples and 3653 polymorphic markers [see Supporting Information—Table S1]. Details on the AFLP analyses of 24 of the 25 species have been published elsewhere [see Supporting Information—Table S1, for *Sibbaldia procumbens*].

Species traits

We expected four species traits to be most important in determining the intensity of the genetic founder effect: mode of pollination (insect or wind), breeding system (outcrossing, selfing or mixed mating), growth form (woody or herbaceous) and dispersal adaptation (long-distance or short-distance). Dispersal adaptation was defined as 'long-distance-dispersed' if morphologically adapted to wind- or animal-dispersal, even though the regular dispersal distance in such species may be moderate rather than long (Higgins *et al.* 2003; Tamme *et al.* 2014); otherwise as 'short-distance-dispersed'. Only 10 species in the North Atlantic region have adaptations for dispersal by sea current (Löve 1963); as none of them were analysed here, this category was not included. However, a large proportion of the species have seeds that might float (Thiel and Gutow 2005). Higher levels of genetic diversity are typically found in wind-pollinated, outcrossing and woody species than in insect-pollinated, selfing and herbaceous species (Hamrick and Godt 1996; Thiel-Egenter *et al.* 2009). Information on these traits for the 25 species in the genetic dataset was compiled from the literature, following the criteria outlined in Alsos *et al.* (2012); [see Supporting Information—Table S1]. The founder effect has been shown to be related to adaptation to local climate (Alsos *et al.* 2007), but the observed reduction in genetic diversity might be explained by a bottleneck due to cooler climate on Svalbard during the last 2000 years causing a decrease in distribution of, for example, *Betula nana* and *Salix herbaceae* (Birks 1991; Alsos *et al.* 2002). However, as most species are not at their climatic limit on most recipient islands at present (except some on Svalbard, Elven *et al.* 2011), we did not include that factor here.

Genetic data analyses

For each species, the sampled area was divided into regions according to geographically consistent genetic groups identified (cf. Alsos *et al.* 2007; Eidesen *et al.* 2013) [see Supporting Information]. The geographic distribution of the main genetic groups and subgroups for each species are shown in Fig. 1.

The source region for the populations on each of the five recipient islands was inferred by looking at shared genetic groups among regions and by carrying out a multilocus assignment test in AFLPOP (Duchesne and Bernatchez 2002). We used a log-likelihood difference of one as threshold for allocation (i.e. for a genotype to be assigned to one particular source region it should be 10 times as likely assigned to that source region than to any other source regions; [see also Supporting Information]). For each recipient island and each species,

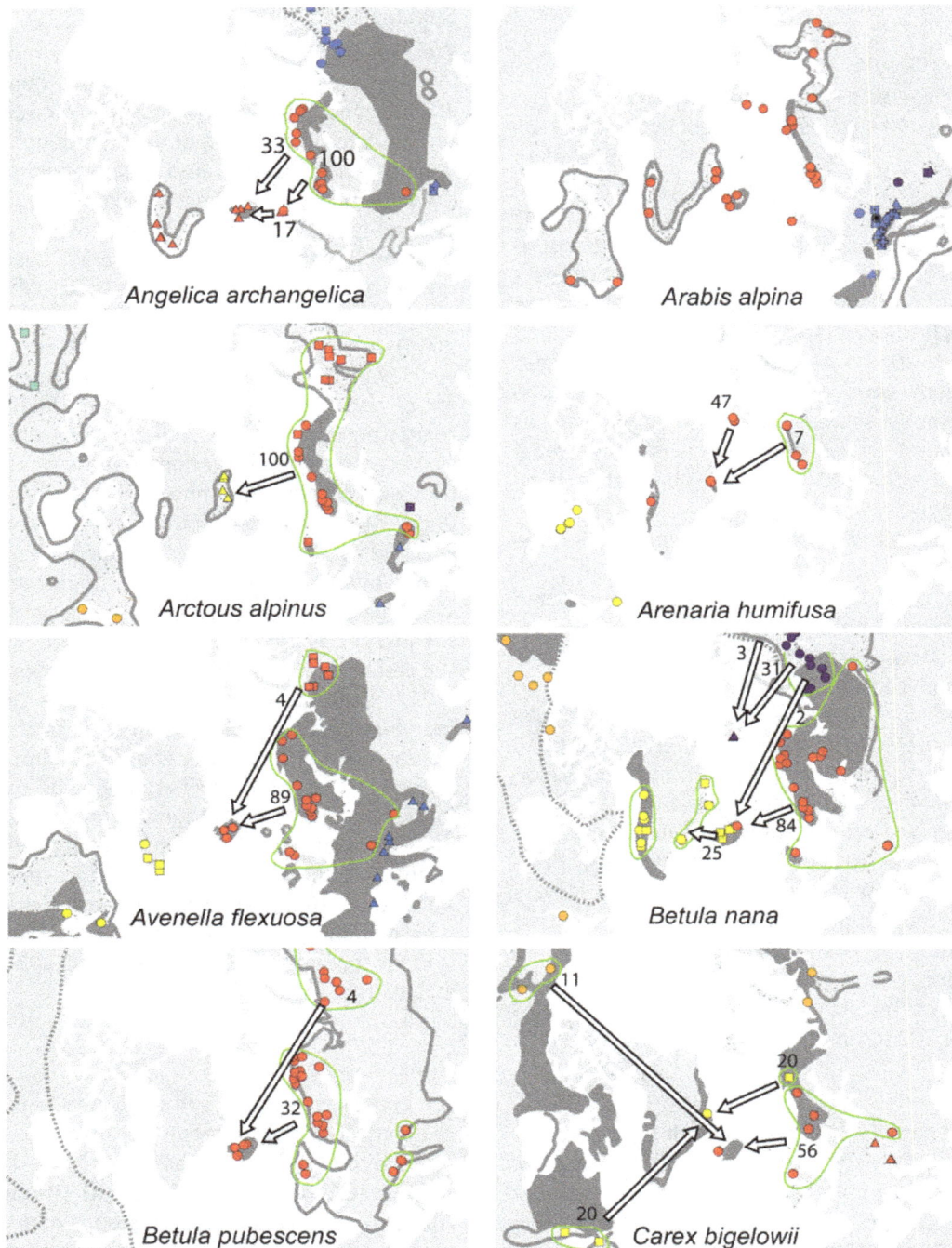

Figure 1. Maps showing the genetic structuring of the 25 species analysed for AFLPs. Colours identify the main genetic groups according to Bayesian clustering analyses run with STRUCTURE and other methods (see text); symbol shapes identify subgroups within main groups. The present distribution of the species is given according to Hultén and Fries (1986; dark grey area, dots and outline whereas stippled lines show vicariant taxa). Arrows represent dispersal routes inferred from assignment tests to geographical regions. Numbers show the proportion (%) of plants allocated to each source region. Due to lack of genetic variation, no assignment test was performed for *Arabis alpina* and *Carex rufina*. No assignment test was performed for *Dryas octopetala* in East Greenland as our sampling from that region only comprises assumed hybrids with *D. integrifolia* (Skrede *et al.* 2006). For *Micranthes stellaris*, Icelandic plants allocated to the combined regions Faroe Island, Scotland and Scandinavia, whereas Faroe Island plants allocated to the Scandinavian-Scottish subgroups (red dots on the map).

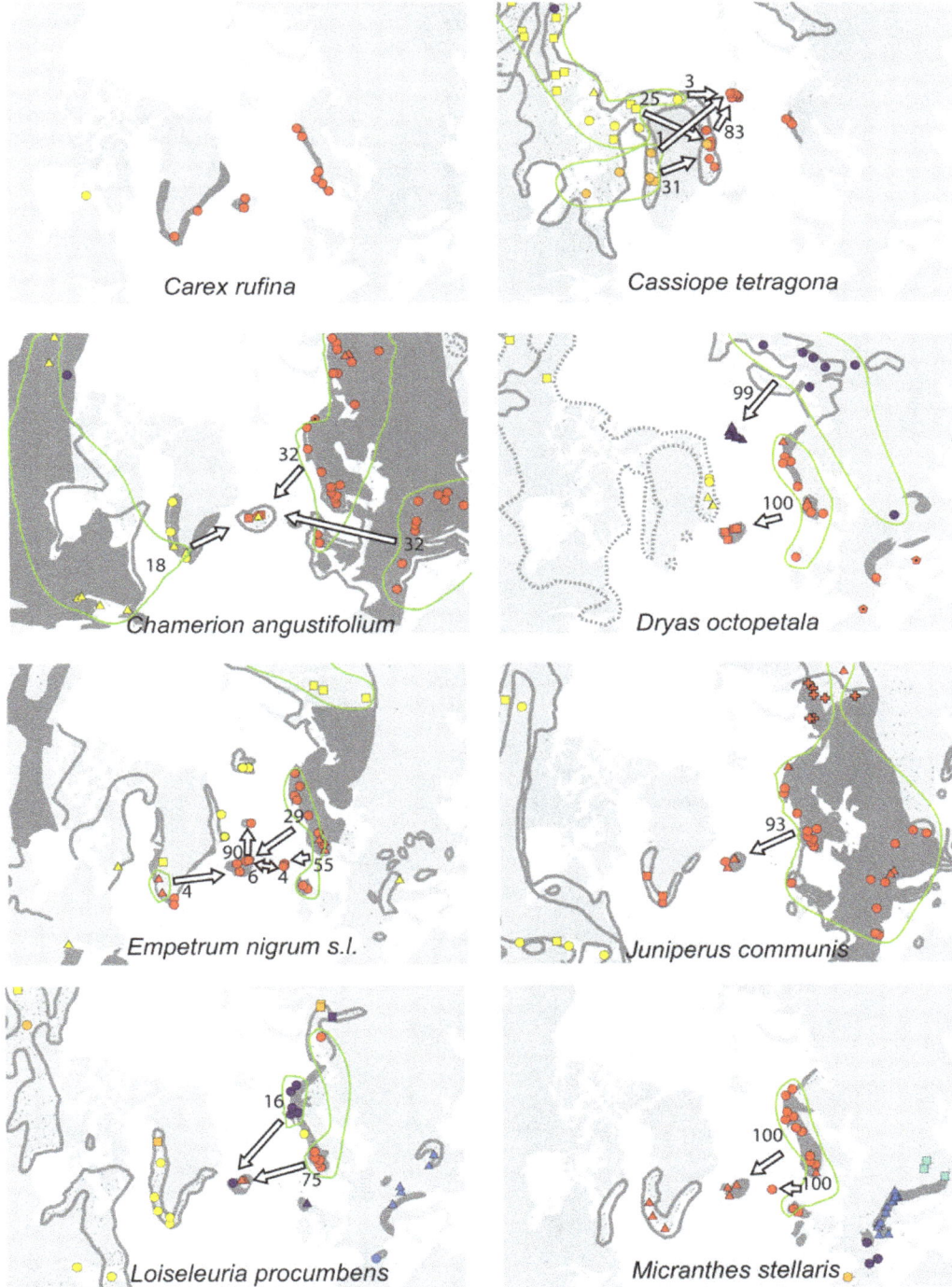

Figure 1. (Continued).

we calculated the proportion of individuals allocated to each source region according to the assignment test (excluding individuals that were not assigned with a log-likelihood difference of 1) resulting in 51 recipient islands × species combinations. The source region with the highest proportion of allocation was considered the main source region. As we are addressing historical colonization and not present day immigration, the results of the assignment test should not be interpreted as revealing individual immigrant individuals. Nevertheless, as the number of generations since colonization is small in evolutionary time for these mainly long-lived plant species (Alsos *et al.* 2002; de Witte and Stöcklin 2010), we are confident that our analyses reveal the

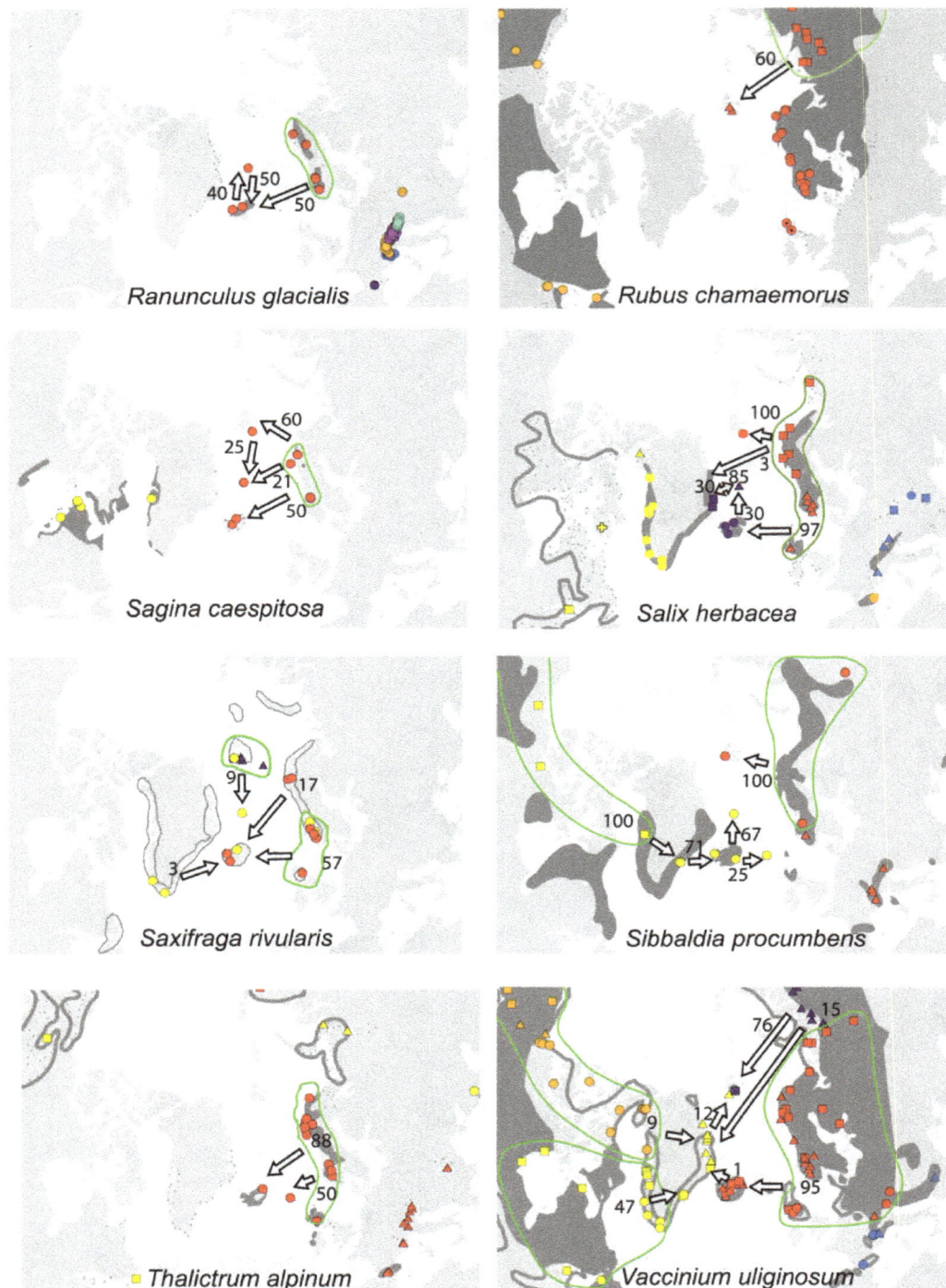

Figure 1. (Continued).

main colonization directions despite possible drift in the founded population.

As in Alsos *et al.* (2007), we quantified the genetic founder effect using six different measures: (i) a minimum number of dispersed propagules that resulted in successful colonization (*propagules*; estimated as the smallest number of individual genotypes in the main source region necessary to bring all observed markers to the recipient island, [**see Supporting Information**]), (ii) proportion of intrapopulation genetic diversity observed in the recipient island relative to that in the main source region (*population diversity*; estimated as the mean of the population

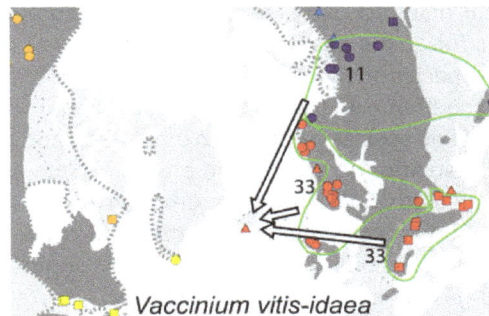

Vaccinium vitis-idaea

Figure 1. (Continued).

averages of number of pair-wise differences among individuals), (iii) proportion of regional (total) genetic diversity in the recipient island relative to that in the main source region (*regional diversity*; estimated across all individuals in the region), (iv) proportion of AFLP markers observed in the recipient island relative to those in the main source region (*markers*), (v) number of source regions inferred in order to account for all markers observed in the recipient island (*sources markers*) and (vi) number of source regions estimated from the assignment test (*sources allocation*). All six measures are also influenced by potential effective dispersal events occurring after initial colonization; in the following we therefore use the term 'founder effect' to encompass the overall reduction in genetic variability in a population in a colonized area as compared with its source. Correlations among the measures and differences in intensity of the founder effect among species and recipient islands were investigated using principle component analysis (PCA) as implemented in the R-package ade4 (http://cran.r-project.org/web/packages/ade4/) and R version 3.01 (R Core Team 2013).

We tested for correlations among the independent variables (species traits, island size and dispersal distance). As growth form and dispersal type were significantly correlated, we chose to use dispersal type because we assume it to be directly relevant to the founder effect (Alsos *et al.* 2012). We also found a significant correlation between pollination mode and breeding system [**see Supporting Information—Tables S1 and S2**], and chose the predictor variable with fewest categories (pollination mode). To determine to what extent distance between source and recipient island, island size, dispersal and pollination mode were correlated with the intensity of the founder effect, we carried out a Principal Component Analyses with Instrumental Variable (PCAIV; function pcaiv in ade4, Thioulouse and Dray 2007). To further test for significant associations, we carried out a linear mixed model (LMM) analysis with regional diversity as a response variable. We chose regional diversity as an

estimate of founder effect here because it was most correlated with the first axis in the PCA. Species was included as a random effect in all models. The explanatory variables included as fixed effects were distance, area, pollination mode and dispersal type. These variables were assembled into 17 candidate models comprising a constant response as the simplest model, each explanatory variable alone, as well as all possible combinations of two variables with or without interactions between them. Models were fitted using the function lmer in the R package lme4 (Bates *et al.* 2014). Maximum likelihood (ML) was used as an optimization criterion to fit models for model selection, whereas restricted ML (REML) was used to obtain parameter estimates (Pinheiro and Bates 2000). The best models were selected based on Akaike's Information Criterion corrected for small sample size (AICc, Burnham and Anderson 2007) using AICcmodavg (Mazerolle 2011) in R. Models with a difference in AICc of <2 were considered equally appropriate. The selected models were checked graphically for constant variance of the residuals, presence of outliers and approximate normality of the random effects.

The likelihood that a species immigrated to the recipient island postglacially (rather than survived the last glaciation *in situ*) was evaluated based on the amount of genetic diversity in the recipient island relative to that in potential source regions, as well as on the number of private markers (markers restricted to one geographical region and thus likely represent local mutations) found in the recipient island (Westergaard *et al.* 2011).

Floristic data

To compile data on species occurrences, we used the Pan Arctic Flora checklist (Elven *et al.* 2011) for those of our regions that are included there, otherwise Hultén and Fries (1986), various regional sources and personal observations [see complete taxon list per region in **Supporting Information—Table S3**]. Taxa closely associated with human activity and agricultural lands, including pasture lands, were assumed to have been introduced by humans to a region (Elven *et al.* 2011; Wasowicz *et al.* 2013, Alsos *et al.* 2015) and therefore omitted. Since the occurrence of some taxa is uncertain [see **Supporting Information—Table S3**], we calculated floristic similarities as the minimum and maximum proportion of recipient island species that also occurred in each potential source region, and used the mean proportion in further analyses.

The number of years between each successful colonization events was estimated as the time since first postglacial palaeorecord/(total number of species on the island × proportion of species assumed to colonize postglacially × average number of propagules per species). Although these numbers contain uncertainties, they provide a rough estimate useful for comparison with other islands. For Jan Mayen, where no palaeobotanical records existed, we assumed a time period of 12 800 years, similar to East Greenland (12 700) and Iceland (13 000).

Results

Genetic data

In most cases, we observed less genetic diversity in the recipient islands than in the source regions both at the population and regional levels and in terms of number of markers, reflecting a founder effect (Table 1). We observed only few private markers in the recipient islands. *Sagina caespitosa* had relatively high numbers of private markers (5) in a recipient island, but this was not combined with high levels of genetic diversity, and thus not interpreted as indicating *in situ* glacial survival. Only *Arenaria humifusa* and *Saxifraga rivularis* showed a genetic pattern consistent with glacial survival on Svalbard [see **Supporting Information**]. Thus, 92 % of the species analysed were assumed to have colonized postglacially.

Of the 12 species analysed from East Greenland, the populations of five species belonged to amphi-Atlantic genetic groups, three to West-Atlantic groups, three to Greenlandic-Icelandic groups, *Arctous alpina* had unique groups and *Cassiope tetragona* had both a western and an eastern genetic group (Fig. 1). Overall, the highest proportions of genetic groups were shared with West Greenland (80 %, Fig. 2).

Among the 21 species analysed from Iceland, the populations of the majority of species belong to eastern (11) or amphi-Atlantic/East-Atlantic (7) genetic groups. Both eastern and western genetic groups were observed in *Betula nana* and *Chamerion angustifolium* (Fig. 1). All genetic groups were shared with Jan Mayen and Faroe Islands, whereas 75 and 84 % also were found in Great Britain and Norway, respectively (Fig. 2).

Among the 11 species analysed from Svalbard, the populations of five species belonged to amphi-Atlantic genetic groups (*Saxifraga rivularis* having a unique group in addition), four to East-Atlantic groups, one to West-Atlantic groups and *Vaccinium uliginosum* had both a western and an eastern genetic group (Fig. 1). The highest proportion of genetic groups was shared with Ural (88 %) followed by Norway (50 %) and East and West Greenland (42–43 %, Fig. 2).

In four of the five species analysed from the Faroe Islands, the populations belong to main genetic groups that were found both east and west of the archipelago although sometimes with different subgroups. Only *Sibbaldia procumbens* belong to a strict western genetic group. The high percentage of genetic groups shared among the regions is an effect of this but should be interpreted with caution due to the general low sample sizes.

All six species analysed from Jan Mayen fell into genetic groups together with individuals from Iceland. The populations of two species belonged to amphi-Atlantic groups, two to East-Atlantic groups and one to a West-Atlantic group, whereas *Salix herbacea* grouped together with Greenland and Iceland with separate subgroups in each of the three regions (Fig. 1). The highest proportion of genetic groups was shared with Iceland, whereas less than five genetic groups were shared with most other regions.

For 23 of the 25 species, populations from the recipient islands were successfully assigned to one ($n = 22$), two ($n = 17$) or more ($n = 7$) source regions (Fig. 1 and Table 1) [see also **Supporting Information**]. Assignment of *Arabis alpina* and *Carex rufina* was not possible because of lack of genetic diversity, and five recipient island × species combinations had to be excluded because the direction could not be determined [see **Supporting Information**]. On average, two source regions had to be inferred to account for all markers observed in the recipient island (Table 1). Only for the Faroe Islands, the most important source region (Scandinavia/Great Britain, 285 km away) was the geographically closest one (Fig. 2). Iceland is only 280 km away from East Greenland, but populations were allocated mainly to Northwest Europe, with Shetland (775 km) and Norway (965 km) being closest. Jan Mayen is 100 km closer to East Greenland than Iceland, where most populations allocated to. Despite the large geographic distance, Northwest Russia was the single most important source region for both East Greenland and Svalbard, although western source regions were also important (Table 1 and Fig. 2).

In the PCA of the six measures of the founder effect, all measures were more or less correlated with axis 1 (horizontal axis), which explained 47.1 % of the variation (Fig. 3A–D). The proportion of regional genetic diversity was most strongly correlated with axis 1, whereas the other five measures were also correlated with axis 2 (vertical axis, 19.9 % of the variation), positively or negatively so. The five recipient islands were placed along the first axis according to their size, although with considerable overlap (Fig. 3C). In the PCAIV, 30.5 % of the variation was explained by the four independent variables (Fig. 3E and F). Island size was mainly related to the first axis, showing the strongest founder effect in small recipient islands. Pollination mode was strongly related to both the first and second axes, with wind-pollinated species being characterized in particular by a higher number of propagules and sources for markers. Dispersal distance was correlated with the second axis. It was negatively correlated with the proportion of markers in

Table 1. Dispersal distance, number of private markers and six measures of the genetic founder effect for each species in each target island. The measures of founder effect are (i) minimum number of colonizing propagules (*Propagules*), (ii) proportion of intrapopulation genetic diversity in target relative to source (*Population diversity*), (iii) proportion of total genetic diversity in target relative to source (*Regional diversity*), (iv) proportion of AFLP markers in target relative to source (*Markers*), (v) number of source regions inferred from AFLP markers (*Sources markers*) and (vi) number of source regions inferred from assignment tests (*Sources allocation*). Target and main source regions are the Faroe Islands (FAROE), Iceland (ICE), East Canada (ECAN), East Greenland (EGRE), Jan Mayen (JM), North Canada (NCAN), mainland Norway (NOR), Northwest Europe (NWEUR), Russia (RUS), Southwest Greenland (SWGRE) and Svalbard (SVALB). Mean $\pm$ standard deviation values for each target island and overall mean are given.

Species	Target island	Main source region	Distance (km)	Number of private markers	Founder effect					
					Propagules	Population diversity	Regional diversity	Markers	Sources markers	Sources allocation
Angelica archangelica	FAROE	NWEUR	570	1	9	0.82	0.79	0.76	2	1
Angelica archangelica	ICE	NWEUR	775	1	8	0.82	0.78	0.81	2	2
Arctous alpinus	EGRE	NWEUR	1270	1	13	0.86	0.76	0.57	3	1
Arenaria humifusa	EGRE	SVALB	570	1	4	1.00	0.82	0.94	2	2
Avenella flexuosa	ICE	NWEUR	775	1	11	0.96	0.94	0.79	2	2
Betula nana	EGRE	ICE	280	0	13	0.73	0.94	0.82	3	1
Betula nana	ICE	NWEUR	775	1	14	1.11	1.00	0.81	3	4
Betula nana	SVALB	RUS	1000	1	7	0.68	0.70	0.59	2	3
Betula pubescens	ICE	NWEUR	775	2	12	1.10	1.06	0.77	2	2
Carex bigelowii	EGRE	ECAN	880	1	12	1.08	0.84	0.75	3	2
Carex bigelowii	ICE	NWEUR	775	3	16	1.58	1.16	0.76	3	2
Cassiope tetragona	EGRE	WGRE	360	1	14	1.00	1.06	1.04	4	2
Cassiope tetragona	SVALB	EGRE	570	1	11	0.94	0.97	0.91	3	3
Chamerion angustifolium	ICE	NWEUR	775	0	14	0.29	1.02	0.66	2	2
Dryas octopetala	ICE	NWEUR	775	0	5	0.63	0.54	0.55	2	1
Dryas octopetala	SVALB	RUS	1000	0	22	0.72	0.86	0.81	4	1
Empetrum nigrum	FAROE	NWEUR	285	0	3	0.72	0.58	0.53	1	2
Empetrum nigrum	ICE	NWEUR	775	0	6	0.75	0.65	0.63	1	3
Empetrum nigrum	JM	ICE	555	0	3	0.30	0.20	0.64	1	1
Juniperus communis	ICE	NWEUR	775	1	11	0.93	0.81	0.73	2	1
Loiseleuria procumbens	ICE	NWEUR	775	1	9	0.93	0.97	0.93	3	2
Micranthes stellaris	FAROE	NWEUR	285	0	1	1.38	1.05	0.71	1	1
Micranthes stellaris	ICE	NWEUR	775	3	2	1.12	0.79	0.71	1	1
Ranunculus glacialis	ICE	NWEUR	775	0	4	0.00	1.20	1.03	3	2

Continued

Table 1. Continued

Species	Target island	Main source region	Distance (km)	Number of private markers	Founder effect				Sources markers	Sources allocation
					Propagules	Population diversity	Regional diversity	Markers		
Ranunculus glacialis	JM	ICE	555	0	1	1.00	0.14	0.86	1	1
Rubus chamaemorus	SVALB	RUS	1000	0	5	0.47	0.51	0.62	1	1
Sagina caespitosa	ICE	NWEUR	965	5	9	0.65	0.81	0.76	2	1
Sagina caespitosa	JM	SVALB	875	1	11	0.35	0.30	0.68	2	2
Sagina caespitosa	SVALB	NOR	640	1	3	0.26	0.22	0.64	2	1
Salix herbacea	EGRE	JM	450	0	14	0.73	0.85	1.03	3	2
Salix herbacea	ICE	NWEUR	775	1	16	0.74	0.67	0.64	5	3
Salix herbacea	JM	ICE	555	0	13	0.97	0.92	0.79	3	2
Salix herbacea	SVALB	NWEUR	1000	1	12	0.67	0.70	0.57	3	1
Saxifraga rivularis	ICE	NWEUR	775	0	7	1.09	0.83	0.81	2	3
Saxifraga rivularis	JM	SVALB	945	0	4	0.26	0.21	0.55	1	1
Sibbaldia procumbens	EGRE	ECAN	360	0	5	0.77	0.69	0.80	2	2
Sibbaldia procumbens	FAROE	ICE	425	0	3	0.91	0.75	0.82	2	1
Sibbaldia procumbens	ICE	EGRE	280	1	5	0.55	0.70	0.98	2	1
Sibbaldia procumbens	JM	ICE	555	0	2	0.55	0.46	0.77	1	1
Sibbaldia procumbens	SVALB	NWEUR	640	0	2	0.00	0.00	0.75	1	1
Thalictrum alpinum	FAROE	NWEUR	285	0	14	1.12	0.94	0.68	3	1
Thalictrum alpinum	ICE	NWEUR	775	0	12	1.18	0.99	0.68	2	1
Vaccinium uliginosum	EGRE	WGRE	360	0	10	1.15	1.12	1.10	3	4
Vaccinium uliginosum	ICE	NWEUR	775	0	11	1.05	0.90	0.75	2	3
Vaccinium uliginosum	SVALB	RUS	1000	0	9	0.33	0.84	0.67	2	2
Vaccinium vitis-idaea	ICE	NWEUR	775	0	6	1.07	0.89	0.73	2	3
East Greenland			566 ± 342	0.5 ± 0.5	11.0 ± 4.0	0.92 ± 0.16	0.89 ± 0.15	0.88 ± 0.18	2.9 ± 0.6	2.0 ± 0.9
Iceland			758 ± 124	1.1 ± 1.4	9.4 ± 4.1	0.88 ± 0.35	0.88 ± 0.17	0.76 ± 0.12	2.3 ± 0.9	2.1 ± 0.9
Svalbard			856 ± 200	0.5 ± 0.5	8.9 ± 6.4	0.51 ± 0.03	0.60 ± 0.34	0.70 ± 0.12	2.3 ± 1.0	1.6 ± 0.9
Faroe Islands			370 ± 127	0.2 ± 0.6	6.0 ± 5.4	0.99 ± 0.26	0.82 ± 0.18	0.70 ± 0.11	1.8 ± 0.8	1.0 ± 0.0
Jan Mayen			673 ± 185	0.2 ± 0.4	5.7 ± 5.1	0.57 ± 0.34	0.37 ± 0.29	0.72 ± 0.11	1.5 ± 0.8	1.3 ± 0.5
Overall mean			693 ± 233	0.7 ± 1.0	8.6 ± 4.9	0.79 ± 0.34	0.76 ± 0.28	0.76 ± 0.14	2.2 ± 0.9	1.7 ± 0.8

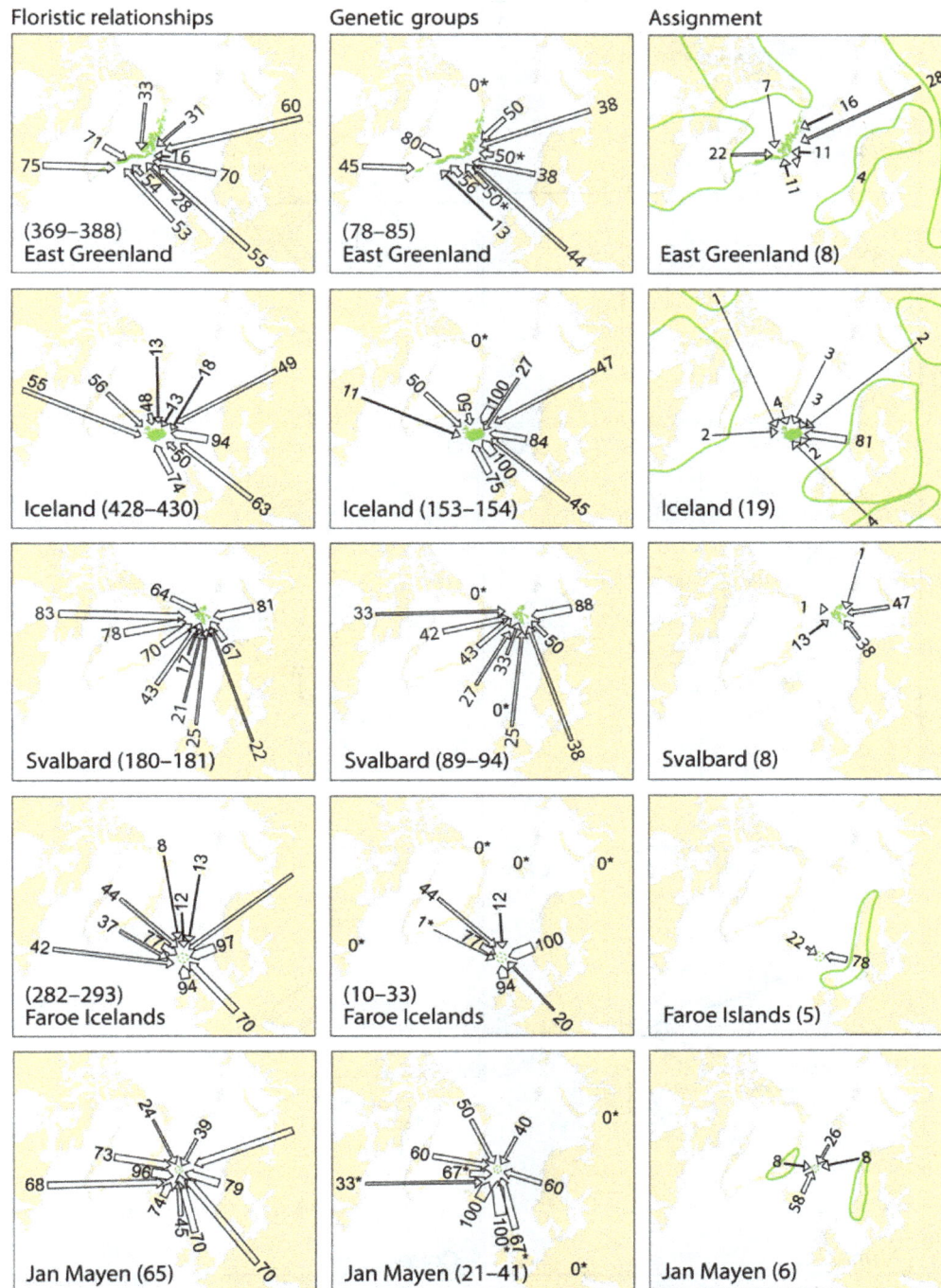

Figure 2. Floristic and genetic relationships between the five recipient islands (East Greenland, Iceland, Svalbard, the Faroe Islands and Jan Mayen), and potential source regions. Floristic relationships are expressed as the proportion (%) of all species occurring in each recipient island that also occur in each potential source region. Genetic relationships are expressed in two ways. First, as the proportions of genetic main groups that also are found in the source region (only counting source regions where populations have been analysed, stars denote regions where less than five comparisons were possible across species). Second, as the mean proportion (%) of plant individuals that were allocated to each source region in assignment tests (excluding individuals that were not assigned). Source regions for floristic and genetic group comparisons are as defined in **Supporting Information—Table S3**. Approximate delimitations of source regions of the assignment test are summarized across species (Fig. 1) and encircled in green. Number of species found (floristic data), genetic groups found (upper range is total number of genetic groups, lower range is number of genetic groups where observations for minimum five species are available) or species assigned (assignment data) in each recipient island are given in parentheses.

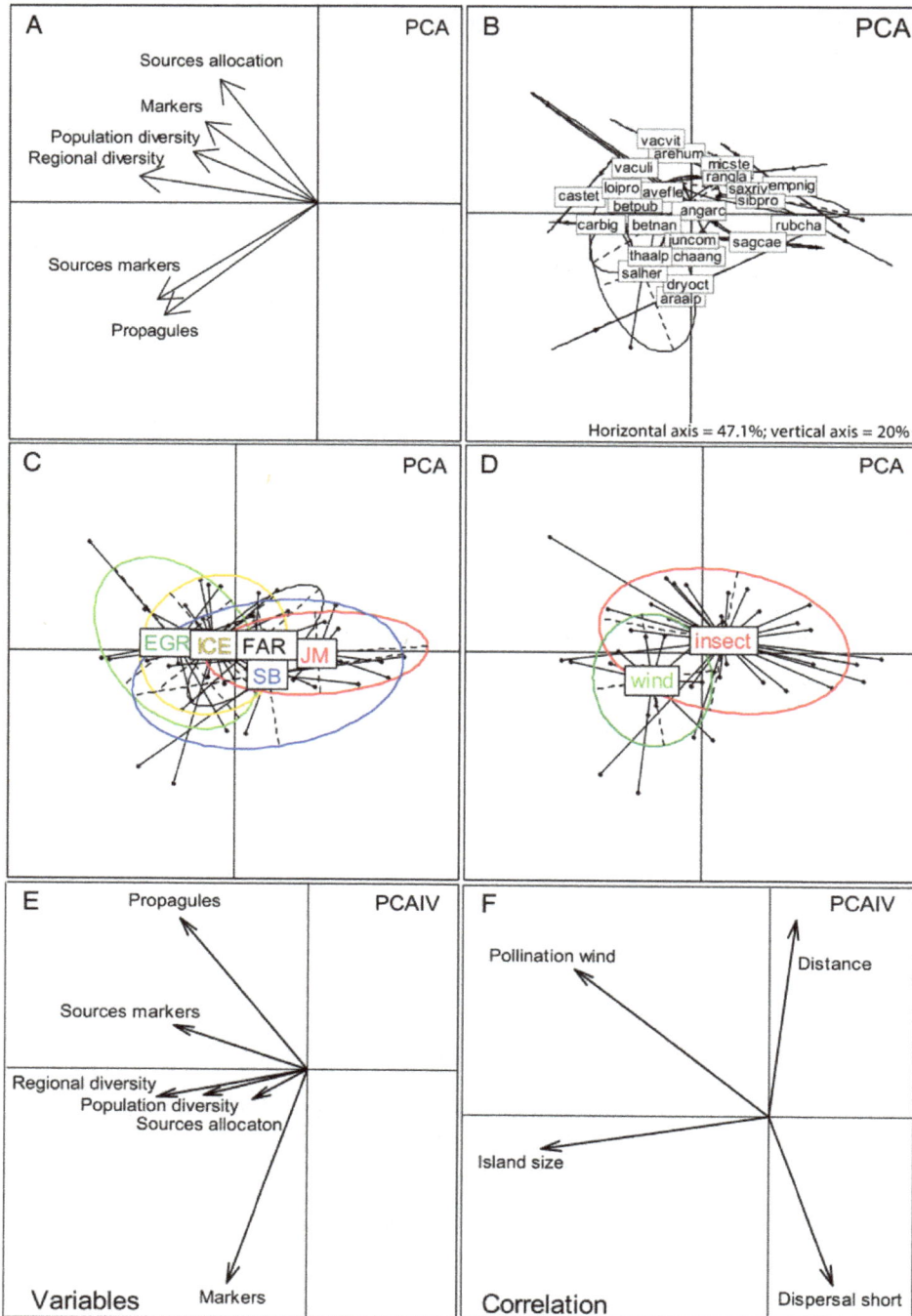

Figure 3. Principal component analyses (PCAs) of six variables expressing the genetic founder effect (see Methods) in the five recipient islands East Greenland (EGR), Iceland (ICE), Svalbard (SB), the Faroe Islands (FAR) and Jan Mayen (JM) relative to the source regions ($n = 38$). The founder effect is increasing from left to right in all panels. (A–D) Principal component analyses showing correlation among the variables (A) and differences in intensity of the founder effect among species (B, some names slightly moved for visibility), recipient islands (C) and pollination mode (D). (E and F) Principal component analyses of instrumental variables (PCAIV), which show to what extent the distance between source and recipient island, size of the recipient island, dispersal adaptation (long- or short-distance) and mode of pollination were correlated with the intensity of the founder effect (ordination taking into account the effect of independent variables).

the recipient island, but positively with the number of propagules. Dispersal type was also correlated to the second axis, but in opposite direction, partly reflecting

that distance was on average somewhat shorter for species lacking adaptations to wind- or animal-dispersal (641 km, SD = 223) than those possessing such adaptations

(750 km, SD = 253, difference not significant, **see Supporting Information—Table S2**). Lack of adaptations to dispersal was indeed associated with a higher proportion of markers but fewer propagules (Fig. 3F).

Variation in the proportion of regional diversity in the recipient island was equally well explained by (i) a model including island size and pollination mode with (best model) or without interaction (AICc difference to the best model 1.05) or (ii) a model with island size, distance from source as well as their interaction (AICc difference 0.68, **Supporting Information—Table S4**) as fixed effects. For insect-pollinated species, the proportion of regional diversity decreased by 0.076 (SE = 0.018, 95 % CI = 0.043–0.113) for reduction by one of the natural logarithm of island size (for instance from 1096 to 403 km^2 or from 162 755 to 59 874 km^2), but this was not the case for wind-pollinated species (marginally significant interaction; Fig. 4A, Table 2). According to the second model, the proportion of regional diversity in the recipient island decreased with distance to source region for smaller islands, but less so for the largest ones (slope: −0.072 per 100 km for an island of 1000 km^2 versus −0.001 for an island of 150 000 km^2; marginally significant interaction, Fig. 4B and Table 2).

Floristic data

The number of species occurring in the five recipient islands were 369–388 in East Greenland, 428–430 in Iceland, 180–181 in Svalbard, 282–293 in the Faroe Islands and 65 in Jan Mayen [**see Supporting Information—Table S3**]. There were 1, 7, 3, 2 and 0 endemic species in the recipient islands, respectively. Thus, species diversity showed the same overall pattern as the genetic data, with highest species diversity and weakest founder effect in larger islands, and lowest species diversity and strongest founder effect in smaller islands (Fig. 3C, Table 1).

Assuming that 92 % of the species colonized postglacially, and using the average number of propagules per species arriving to each recipient island from Table 1, we estimated the following number of years between each successful colonization event: East Greenland 3.3, Iceland 3.5, Svalbard 6.1, Faroe Islands 9.2 and Jan Mayen 42.0. Similarly, using the number of species per island (Fig. 2) and assuming that 92 % colonized postglacially, the years between each successful species establishment were: East Greenland 36.5, Iceland 32.9, Svalbard 54.2, Faroe Islands 49.5 and Jan Mayen 214.0.

All recipient islands showed high floristic similarity with several potential source regions, but with a clear east–west pattern (Fig. 2). For Iceland and the Faroe Islands, the potential source regions showing highest floristic similarities, Fennoscandia and Great Britain, were also identified as source regions by the genetic data. Svalbard had high floristic similarities with both East Canada and Russia, whereas Russia was identified as a major source region by the genetic data. For Jan Mayen, highest floristic similarity was with East Greenland whereas the genetic data identified Iceland as the main source region. For East Greenland, we found highest floristic similarities to West Greenland, Canada and Scandinavia, highest proportion of shared genetic groups with West Greenland, Svalbard and Iceland, whereas the

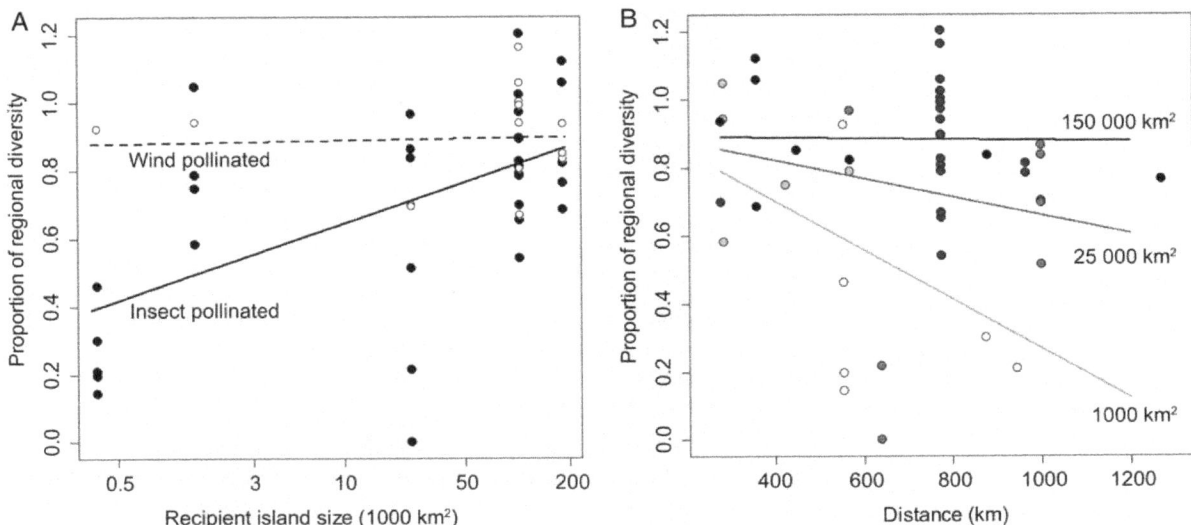

Figure 4. Proportion of regional genetic diversity found on islands relative to source region in relation to (A) size of the recipient island and mode of pollination (open circles represent wind pollinated and black circles indicate insect pollinated) and (B) distance to source region and size of the recipient island (increasing darkness reflects increasing island size).

Table 2. Parameter estimates for the two most suitable LMMs with REML explaining the proportion of regional genetic diversity in the target island compared with that in the source region ($n = 46$) in function of mode of pollination (insect or wind; reference level is insect), the natural logarithm of island size in square kilometres and distance to main source region (parameter estimates per 100 km increase in distance). Species was included as random effect and variance $\pm$ standard deviation is given.

	Fixed effect	Estimate	CI	SE
(1)				
Regional diversity ~ pollination × island size	Intercept	−0.052		0.185
Species 0	Log(size)	0.076	0.039 to 0.112	0.018
	Pollination wind	0.912	0.050 to 1.754	0.416
	Pollination wind × log(size)	−0.073	−0.157 to 0.005	0.039
(2)				
Regional diversity ~ island size × distance	Intercept	1.129		0.459
Species 0.004 ± 0.062	Log(size)	−0.020	−1.049 to 0.077	0.045
	Distance (100 km)	−0.170	−0.318 to −0.018	0.072
	Distance × size	0.014	−0.0004 to 0.028	0.007

assignment test suggest highest colonization rates from the North Russia followed by West Greenland/Canada (Fig. 2).

Discussion

We have presented the first comprehensive study of LDD to oceanic islands based on combined population genetic and floristic similarity analyses. We show that the relative intensity of the founder effect is similar at the level of species and genes, and broadly corresponds to the predictions of the Island Equilibrium Theory (MacArthur and Wilson 1967). This indicates that species and genetic diversity on islands are shaped by the same processes. Compared with the floristic data, the genetic data give more detailed information particularly as it allows identification of source regions and estimating the number of colonization events. The genetic data also allowed us to quantify the founder effect in relation to island size, distance to source region and species traits.

Source regions and colonization patterns

We were able to identify postglacial dispersal routes for most species (Figs 1 and 2), [see also **Supporting Information**], and only find indications of *in situ* glacial persistence in 2 of the 48 combinations of species and recipient islands we analysed here (Table 1; **Supporting Information**, Westergaard *et al.* 2010, 2011). It is still possible that glacial survivor populations did exist but remained undetected in our analyses because they were swamped by postglacial immigrants; however, this scenario would also involve postglacial LDD. Also, our revision of the flora confirmed earlier analyses that the number of

endemic species is low on these islands (Brochmann *et al.* 2003), indicating a young age of the local floras.

The differences between our floristic and genetic analyses with respect to relative importance of source regions (Fig. 2) may have been affected by the selection of species for the genetic analyses, and in the case of East Greenland, also by the delimitation of this region as the proportion of eastern and western species varies (Elven *et al.* 2011). The genetic data were nevertheless most informative; in cases where the same species occurred in many regions, we could identify with reasonable certainty which and how many of them actually served as sources. We therefore give priority to the inferences based on the genetic data in our summary of dispersal routes in the amphi-Atlantic region (Fig. 5).

If LDD is a mainly stochastic process, we would expect that the closest potential source region served as the most important one, as the probability of a dispersal event decreases with distance (Nathan 2006; Nathan *et al.* 2008). On the contrary, with a single exception (the Faroe Islands) we found that the closest potential source region was not the most important one in the North Atlantic. The most extreme case was Iceland, where the main source regions Scandinavia/Great Britain are situated 2.8–4.5 times further away than East Greenland. In addition, gene-based inference of the same main source region was made for 18 of the 19 species analysed, consistent with the floristic similarities (Fig. 2). Thus, although the distance to source region has been found previously to be the second most important factor in determining species diversity on oceanic islands based on stochasticity (after island size, Weigelt and Kreft 2013), our results suggest that other, deterministic

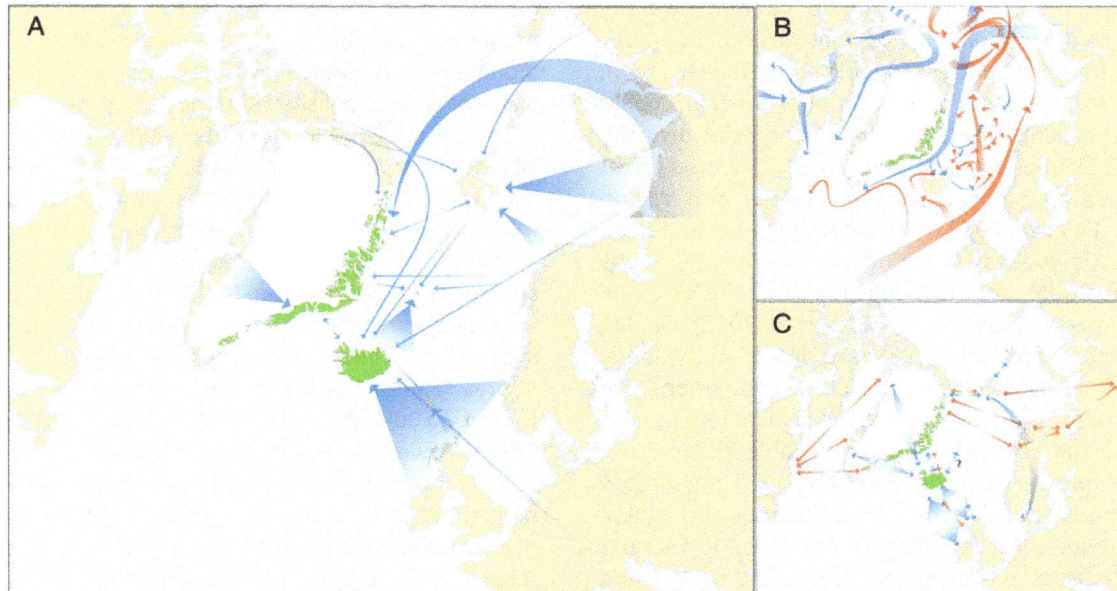

Figure 5. (A) Main (thick arrows) and additional (thin arrows) LDD routes of plants in the North Atlantic area inferred from genetic and floristic data (cf. Fig. 1). (B) Sea surface circulation patterns in the North Atlantic area (blue: cold water, red: warm water). (C) Main migration routes for geese species (thick blue arrows) and the supposedly efficient seed disperser *Plectrophenax nivalis* (snow bunting, thin red arrows) in the North Atlantic area (based on Madsen *et al.* 1999; Lyngs 2003).

factors also are important in determining source and direction of LDD events.

Dispersal vectors are an important deterministic factor affecting LDD patterns (Nathan *et al.* 2008). Although it has been claimed that North Atlantic floras are poorly adapted for LDD based on analyses of propagule morphology (Löve and Löve 1963), dispersal vectors such as wind, sea water or birds may still lead to dispersal over long distances, regardless of morphological dispersal adaptations (Higgins *et al.* 2003; Nathan 2006). The North Atlantic Current connects the Faroe Islands, Iceland and Jan Mayen eastwards to Northwest Europe, whereas the East Greenland Current connects East Greenland, Iceland and Jan Mayen to North Greenland, Svalbard and North Russia (Fig. 5). The similarity between the main current pattern and the inferred dispersal routes is intriguing (compare Fig. 5A and B), and observations of driftwood also suggest dispersal along these routes (Johansen and Hytteborn 2001). Analyses of global patterns of species diversity on islands also indicate that ocean currents can be important (Weigelt and Kreft 2013). The dominant wind directions in the Atlantic region largely follow the ocean currents (http://go.grolier.com/atlas?id=mtlr089), making it hard to distinguish dispersal by wind from dispersal by ocean currents. The main bird migration routes connect Iceland and the Faroe Islands to Great Britain (Löve and Löve 1963; Johansen and Hytteborn 2001). In contrast to the sea currents, the bird migration routes connect East

Greenland to Northwest Europe (Lyngs 2003). Many different bird species may carry seeds (Ridley 1930; Nogales *et al.* 2012). Especially Arctic geese migrate in huge numbers along these routes (Fig. 5C). Colonization of East Greenland from Svalbard and North Russia may have been caused by the only Arctic passerine bird, the snow bunting (Fig. 5C), as it is more likely to carry seeds than other birds in northern areas (Fridriksson 1970). Thus, we consider it likely that several dispersal vectors have contributed to colonization of the recipient islands analysed here. In addition, historical factors may have been important. More species might have been available to colonize from Europe during the last glaciation, where numerous fossils indicate presence of a widespread and well-developed Arctic flora during the LGM (Hultén and Fries 1986). In contrast, no LGM fossils have been found in South and East Greenland, where the possibility of glacial survival is still disputed (Cremer *et al.* 2008; Böcher 2012). However, as the patterns of colonization we inferred fit well with the main dispersal vectors, and as historical factors cannot explain the inferred colonization of East Greenland, our data support in the first place the importance of dispersal vectors.

Factors determining the founder effect

As expected, we found fairly strong correlations among the six measures of the founder effect (Fig. 3). Species that traced back to several source regions, or for which a high minimum number of propagules was inferred, experienced the least reduction in genetic diversity following colonization. A similar pattern has been observed for invasive species (Dlugosch and Parker 2008). We also found a stronger genetic founder effect in smaller islands, congruent with the patterns of species diversity (Figs 2 and 3), in agreement with the species-area effect as predicted by island theory (MacArthur and Wilson 1967; Triantis *et al.* 2012). This effect may be explained by stochastic processes acting on small populations (Frankham 2005) and/or a lower probability of small islands to receive diaspores (Patiño *et al.* 2013). On Jan Mayen, the active volcano may have amplified the initial founder effect by exterminating plant populations. For the small islands, the founder effect also increased with distance to source region (Fig. 4 and Table 2). Similarly, a stronger bottleneck has been observed on remote Pacific Islands than on the Canary Islands (Whittaker and Fernández-Palacios 2007). An increase in founder effect with distance is expected from island theory, and this effect is also expected to be stronger for small than for large islands (Jaenike 1973).

In our analyses, the founder effect was somewhat unexpectedly determined more by mode of pollination than by adaptation to seed dispersal, whereas dispersal distance was poorly related to adaptation to seed dispersal (Fig. 3) **[see Supporting Information—Table S2]**. In our previous study of Svalbard, we also found that the intensity of the founder effect was only weakly related to adaptation to dispersal (Alsos *et al.* 2007). Thus, at dispersal distances of more than 280 km, morphological adaptations to dispersal seem to be of minor importance although they are important for overall gene flow within species (Thiel-Egenter *et al.* 2009; Alsos *et al.* 2012). At larger distances, other factors such as stochasticity and dispersal vectors may be more important for long-distant colonization (Higgins *et al.* 2003; Nathan 2006; Vargas *et al.* 2012).

Long-distance dispersal of pollen in wind-pollinated species may have caused a less severe founder effect compared with insect-pollinated species (Figs 3 and 4). However, this appears unlikely since long-distance pollination typically has been documented only over a few hundred metres, rarely up to 160 km (Ashley 2010). The average dispersal distance of 370–856 km to our five recipient islands thus seems to make successful long-distance pollination unlikely. Rather, since most insect-pollinated species in our study are mixed maters, and as lack of pollinators can shift mating towards self-pollination (Kevan 1972; Tikhmenev 1985), we suggest that the more severe founder effect we found in insect-pollinated species may have been caused by increased inbreeding during the establishment phase. Whether the founder effect in general is stronger affected by pollination mode than by

dispersal distance could be investigated by, for example, comparing pollination ecology and inbreeding rates in pioneer populations on islands or glacier forelands with those of well-established sites at different distances.

The overall low founder effect and high species diversity we observed in East Greenland and Iceland support the hypothesis that LDD is frequent in Arctic plants (Alsos *et al.* 2007), contrary to the suggestion that most Arctic islands are unsaturated with species due to dispersal limitations (Hoffmann 2012). Also, rate of successful species colonization found for these islands (one per every 33–214 years) is high compared with, for example, Azores (1 per 40 000 years, Schaefer 2003) and Hawaii (1 per 20 000–250 000 years, Sohmer and Gustafson 1987). However, we also identified both island size and colonization distance as limiting factors for LDD. Independent of island size and distance (Figs 3 and 4), pollination mode was important for the extent of gene flow. A better knowledge of how these factors affect the founder effect can lead to more precise predictions about range shifts in species with different traits as well as to island (or fragmented habitats) of different sizes and distances to source regions.

Conclusions

Our analyses of floristic and genetic patterns in the North Atlantic area suggest that species diversity and genetic diversity may have been shaped to a large degree by similar processes. The large-scale patterns we inferred from both floristic and genetic data were congruent among many species and consistent with likely dispersal vectors, indicating that deterministic factors are important in determining LDD in addition to purely stochastic ones. This is supported by the clear effect of island size on the intensity of the genetic founder effect, mirrored by species diversity. As past colonization typically occurred from more than one source region, we may expect future colonization to be complex as well, but to be governed to some extent by deterministic processes. Assuming that dispersal vectors are constant, the same main dispersal routes may be expected in the future as in the past. However, the current reduction of the extent of sea ice may limit dispersal, whereas anthropogenic dispersal may increase it. By taking into account the main determinants of the genetic founder effect and the complexity of dispersal routes when modelling future distribution of species and genes, we may improve our ability to forecast effects of the ongoing climate change.

Sources of Funding

The work was supported by the Research Council of Norway (grant numbers 150322/720 and 170952/V40 to C.B. and 230617/E10 to I.G.A.).

Contributions by the Authors

I.G.A. conceived the idea and drafted the manuscript together with D.E.; I.G.A., D.E., P.B.E., K.B.W., P.S., A.T. and S.B. analysed the genetic data; C.B. lead the project compiling the genetic data; H.S. and R.E. compiled the floristic data; I.G.A. and D.E. did the statistical analyses and all co-authors commented on the manuscript.

Acknowledgements

We thank everyone contributing to generate the primary genetic datasets (see our earlier publications for names of collectors and laboratory assistants), Marie Kristine Føreid for assistance with Structure and AFLPOP analyses, Ernst Högtun for help with the graphics, and three anonymous referees for valuable comments on the manuscript.

Supporting Information

The following additional information is available in the online version of this article –

File S1. Supporting information containing details of genetic and statistical analyses, R script for estimating number of propagules.

Table S1. Data used for estimating founder effect and dispersal routes and traits of the 25 species analysed.

Table S2. Significant values of pair-wise association among the size of the island, distance to source region and four species traits.

Table S3. A compiled list of vascular plant taxa in recipient regions and occurrences of recipient region taxa in the potential source regions.

Table S4. Model selection using Akaike's information criterion.

Literature Cited

Alsos IG, Engelskjøn T, Brochmann C. 2002. Conservation genetics and population history of *Betula nana, Vaccinium uliginosum,* and *Campanula rotundifolia* in the arctic archipelago of Svalbard. *Arctic, Antarctic, and Alpine Research* 34:408–418.

Alsos IG, Eidesen PB, Ehrich D, Skrede I, Westergaard K, Jacobsen GH, Landvik JY, Taberlet P, Brochmann C. 2007. Frequent long-distance plant colonization in the changing Arctic. *Science* 316:1606–1609.

Alsos IG, Ehrich D, Thuiller W, Eidesen PB, Tribsch A, Schonswetter P, Lagaye C, Taberlet P, Brochmann C. 2012. Genetic consequences of climate change for northern plants. *Proceedings of the Royal Society B: Biological Sciences* 279:2042–2051.

Alsos IG, Elven R, Ware C. 2015. Past Arctic aliens have passed away, current ones may stay. *Biological Invasions*, in press.

Ashley MV. 2010. Plant parentage, pollination, and dispersal: how DNA microsatellites have altered the landscape. *Critical Reviews in Plant Sciences* 29:148–161.

Bates D, Maechler M, Bolker B, Walker S. 2014. *lme4: linear mixed-effects models using Eigen and S4.* R package version 1.1-7. http://CRAN.R-project.org/package=lme4.

Bennike O. 1999. Colonisation of Greenland by plants and animals after the last ice age: a review. *Polar Record* 35:323–336.

Bennike O, Bjorck S, Bocher J, Hansen L, Heinemeier J, Wohlfarth B. 1999. Early holocene plant and animal remains from North-east Greenland. *Journal of Biogeography* 26:667–677.

Bialozyt R, Ziegenhagen B, Petit RJ. 2006. Contrasting effects of long distance seed dispersal on genetic diversity during range expansion. *Journal of Evolutionary Biology* 19:12–20.

Birks HH. 1991. Holocene vegetational history and climatic change in west Spitsbergen—plant macrofossils from Skardtjørna, an Arctic lake. *The Holocene* 1:209–218.

Böcher J. 2012. Interglacial insects and their possible survival in Greenland during the last glacial stage. *Boreas* 41:644–659.

Brochmann C, Gabrielsen TM, Nordal I, Landvik JY, Elven R. 2003. Glacial survival or *tabula rasa*? The history of North Atlantic biota revisited. *Taxon* 52:417–450.

Brochmann C, Edwards M, Alsos IG. 2013. The dynamic past and future of arctic plants: climate change, spatial variation, and genetic diversity. In: Rhode K, ed. *The balance of nature and human impact.* Cambridge: Cambridge University Press, 133–152.

Burnham K, Anderson D. 2007. *Model selection and multimodel inference: a practical information-theoretical approach.* New York: Springer.

Caseldine C, Geirsdóttir À, Langdon P. 2003. Efstadalsvatn—a multiproxy study of a Holocene lacustrine sequence from NW Iceland. *Journal of Paleolimnology* 30:55–73.

Chambers JC, MacMahon JA. 1994. A day in the life of a seed: movements and fates of seeds and their implications for natural and managed systems. *Annual Review of Ecology and Systematics* 25: 263–292.

Cremer H, Bennike O, Wagner B. 2008. Lake sediment evidence for the last deglaciation of eastern Greenland. *Quaternary Science Reviews* 27:312–319.

de Queiroz A. 2005. The resurrection of oceanic dispersal in historical biogeography. *Trends in Ecology and Evolution* 20:68–73.

de Witte LC, Stöcklin J. 2010. Longevity of clonal plants: why it matters and how to measure it. *Annals of Botany* 106:859–870.

Dlugosch KM, Parker IM. 2008. Founding events in species invasions: genetic variation, adaptive evolution, and the role of multiple introductions. *Molecular Ecology* 17:431–449.

Duchesne P, Bernatchez L. 2002. AFLPOP: a computer program for simulated and real population allocation, based on AFLP data. *Molecular Ecology Notes* 2:380–383.

Ehlers J, Gibbard PL. 2004. *Quaternary glaciations-extent and chronology. Part I: Europe.* Amsterdam: Elsevier.

Eidesen PB, Ehrich D, Bakkestuen V, Alsos IG, Gilg O, Taberlet P, Brochmann C. 2013. Genetic roadmap of the Arctic: plant dispersal highways, traffic barriers and capitals of diversity. *New Phytologist* 200:898–910.

Elven R, Murray DF, Razzhivin VY, Yurtsev BA. 2011. Annotated checklist of the Panarctic Flora (PAF). Vascular plants. http://nhm2.uio.no/paf/.

Frankham R. 1997. Do island populations have less genetic variation than mainland populations? *Heredity* 78:311–327.

Frankham R. 2005. Genetics and extinction. *Biological Conservation* 126:131–140.

Fridriksson S. 1970. See dispersal by snow buntings in 1968. *Surtsey Research Progress Reports* **4**:43–49.

Gillespie RG, Baldwin BG, Waters JM, Fraser CI, Nikula R, Roderick GK. 2012. Long-distance dispersal: a framework for hypothesis testing. *Trends in Ecology and Evolution* 27:47–56.

Hamrick JL, Godt MJW. 1996. Effects of life history traits on genetic diversity in plant species. *Philosophical Transactions of the Royal Society B: Biological Sciences* 351:1291–1298.

Hannon GE, Rundgren M, Jessen CA. 2010. Dynamic early Holocene vegetation development on the Faroe Islands inferred from high-resolution plant macrofossil and pollen data. *Quaternary Research* 73:163–172.

Hegland SJ, Nielsen A, Lázaro A, Bjerknes A-L, Totland Ø. 2009. How does climate warming affect plant-pollinator interactions? *Ecology Letters* **12**:184–195.

Higgins SI, Nathan R, Cain ML. 2003. Are long-distance dispersal events in plants usually caused by nonstandard means of dispersal? *Ecology* 84:1945–1956.

Hoffmann MH. 2012. Not across the North Pole: plant migration in the Arctic. *New Phytologist* 193:474–480.

Hultén E, Fries M. 1986. *Atlas of North European vascular plants north of the Tropic of Cancer.* Königstein: Koeltz Scientific Books.

Jaenike JR. 1973. A steady state model of genetic polymorphism on islands. *The American Naturalist* **107**:793–795.

Johansen S, Hytteborn H. 2001. A contribution to the discussion of biota dispersal with drift ice and driftwood in the North Atlantic. *Journal of Biogeography* 28:105–115.

Kevan PG. 1972. Insect pollination of high arctic flowers. *The Journal of Ecology* 60:831–847.

Lenoir J, Virtanen R, Oksanen J, Oksanen L, Luoto M, Grytnes J-A, Svenning J-C. 2012. Dispersal ability links to cross-scale species diversity patterns across the Eurasian Arctic tundra. *Global Ecology and Biogeography* 21:851–860.

Löve D. 1963. Dispersal and survival of plants. In: Löve A, Löve D, eds. *North Atlantic biota and their History: a symposium held at the University of Iceland, Reykjavík, July 1962 under the auspices of the University of Iceland and the Museum of Natural History.* Oxford: Pergamon, 189–205.

Löve A, Löve D. 1963. *North Atlantic biota and their History: a symposium held at the University of Iceland, Reykjavík, July 1962 under the auspices of the University of Iceland and the Museum of Natural History.* Oxford: Pergamon.

Lyngs P. 2003. Migration and winter ranges of birds in Greenland. An analysis of ringing recoveries. *Dansk Ornitologisk Forenings Tidsskrift* 97:1–167.

MacArthur RH, Wilson EO. 1967. *The theory of island biogeography.* Princeton: Princeton University Press.

Madsen J, Crackness G, Fox T. 1999. *Goose populations of the Western Palearctic. A review of status and distribution.* Rönde, Denmark: National Environmental Research Institute; Wageningen, The Netherlands: Wetlands International.

Mazerolle MJ. 2011. *AICcmodavg: model selection and multi-model inference based on (Q)AIC(c).* 1.15 ed. R package version 1.17.

Nathan R. 2006. Long-distance dispersal of plants. *Science* 313: 786–788.

Nathan R, Schurr FM, Spiegel O, Steinitz O, Trakhtenbrot A, Tsoar A. 2008. Mechanisms of long-distance seed dispersal. *Trends in Ecology and Evolution* 23:638–647.

Nogales M, Heleno R, Traveset A, Vargas P. 2012. Evidence for overlooked mechanisms of long-distance seed dispersal to and between oceanic islands. *New Phytologist* **194**: 313–317.

Parducci L, Jørgensen T, Tollefsrud MM, Elverland E, Alm T, Fontana SL, Bennett KD, Haile J, Matetovici I, Suyama Y, Edwards ME, Andersen K, Rasmussen M, Boessenkool S, Coissac E, Brochmann C, Taberlet P, Houmark-Nielsen M, Larsen NK, Orlando L, Gilbert MTP, Kjaer KH, Alsos IG, Willerslev E. 2012. Glacial survival of boreal trees in northern Scandinavia. *Science* **335**:1083–1086.

Patiño J, Guilhaumon F, Whittaker RJ, Triantis KA, Gradstein SR, Hedenäs L, González-Mancebo JM, Vanderpoorten A. 2013. Accounting for data heterogeneity in patterns of biodiversity: an application of linear mixed effect models to the oceanic island biogeography of spore-producing plants. *Ecography* 36:904–913.

Pauls SU, Nowak C, Bálint M, Pfenninger M. 2013. The impact of global climate change on genetic diversity within populations and species. *Molecular Ecology* 22:925–946.

Pinheiro J, Bates D. 2000. *Mixed-effects models in S and S-plus.* New York: Springer.

R Core Team. 2013. *R: a language and environment for statistical computing.* Vienna, Austria: Foundation for Statistical Computing.

Ridley HN. 1930. *The dispersal of plants throughout the world.* Kent: L. Reeve & Co., Ltd.

Rundgren M. 1998. Early-Holocene vegetation of northern Iceland: pollen and plant macrofossil evidence from the Skagi peninsula. *The Holocene* 8:553–564.

Rundgren M, Ingólfsson O. 1999. Plant survival in Iceland during periods of glaciation? *Journal of Biogeography* 26:387–396.

Schaefer H. 2003. Chorology and diversity of the Azorean Flora. Dissertationes Botanicae 374. Stuttgart: J. Cramer, 130 pp. + CD rom (580 pp.).

Skrede I, Eidesen PB, Portela RP, Brochmann C. 2006. Refugia, differentiation and postglacial migration in arctic-alpine Eurasia, exemplified by the mountain avens (*Dryas octopetala* L.). *Molecular Ecology* **15**:1827–1840.

Sohmer SH, Gustafson R. 1987. *Plants and flowers of Hawaii.* Honolulu: Times Edition.

Taberlet P, Fumagalli L, Wust-Saucy AG, Cosson JF. 1998. Comparative phylogeography and postglacial colonization routes in Europe. *Molecular Ecology* 7:453–464.

Tamme R, Götzenberger L, Zobel M, Bullock JM, Hooftman DAP, Kaasik A, Pärtel M. 2014. Predicting species' maximum dispersal distances from simple plant traits. *Ecology* 95:505–513.

Thiel M, Gutow L. 2005. The ecology of rafting in the marine environment. I. The floating substrata. *Oceanography and Marine Biology: An Annual Review* 42:181–264.

Thiel-Egenter C, Gugerli F, Alvarez N, Brodbeck S, Cieślak E, Colli L, Englisch T, Gaudeul M, Gielly L, Korbecka G, Negrini R, Paun O, Pellecchia M, Rioux D, Ronikier M, Schönswetter P, Schüpfer F, Taberlet P, Tribsch A, van Loo M, Winkler M, Holderegger R. 2009. Effects of species traits on the genetic diversity of high-mountain plants: a multi-species study across the Alps and the Carpathians. *Global Ecology and Biogeography* **18**:78–87.

Thioulouse J, Dray S. 2007. Interactive multivariate data analysis in R with the ade4 and ade4TkGUI packages. *Journal of Statistical Software* 22:1–20.

Tikhmenev EA. 1985. Pollination and self-pollinating potential of entomophilic plants in arctic and mountain tundras of the northeastern USSR. *Soviet Journal of Ecology* **15**:166–172.

Triantis KA, Guilhaumon F, Whittaker RJ. 2012. The island species-area relationship: biology and statistics. *Journal of Biogeography* **39**:215–231.

Vargas P, Heleno R, Traveset A, Nogales M. 2012. Colonization of the Galápagos Islands by plants with no specific syndromes for long-distance dispersal: a new perspective. *Ecography* **35**:33–43.

Wasowicz P, Przedpelska-Wasowicz EM, Kristinsson H. 2013. Alien vascular plants in Iceland: diversity, spatial patterns, temporal trends, and the impact of climate change. *Flora—Morphology, Distribution, Functional Ecology of Plants* **208**:648–673.

Weigelt P, Kreft H. 2013. Quantifying island isolation—insights from global patterns of insular plant species richness. *Ecography* **36**: 417–429.

Westergaard KB, Jørgensen MH, Gabrielsen TM, Alsos IG, Brochmann C. 2010. The extreme Beringian/Atlantic disjunction in *Saxifraga rivularis* (Saxifragaceae) has formed at least twice. *Journal of Biogeography* **37**:1262–1276.

Westergaard KB, Alsos IG, Popp M, Engelskjøn T, Flatberg KI, Brochmann C. 2011. Glacial survival may matter after all: nunatak signatures in the rare European populations of two west-arctic species. *Molecular Ecology* **20**:376–393.

Whittaker RJ, Fernández-Palacios JM. 2007. *Island biogeography: ecology, evolution, and conservation.* Oxford: Oxford University Press.

3

Patterns of genetic diversity in three plant lineages endemic to the Cape Verde Islands

Maria M. Romeiras[1,2*], Filipa Monteiro[1], M. Cristina Duarte[2,3], Hanno Schaefer[4] and Mark Carine[5]

[1] Biosystems and Integrative Sciences Institute (BioISI), Faculty of Sciences, University of Lisbon, Campo Grande 1749-016, Lisbon, Portugal
[2] Tropical Research Institute (IICT/JBT), Trav. Conde da Ribeira 9, 1300-142 Lisbon, Portugal
[3] Centre in Biodiversity and Genetic Resources (CIBIO/InBIO), University of Porto, Campus Agrário de Vairão, 4485-661 Vairão, Portugal
[4] Technische Universitaet Muenchen, Biodiversitaet der Pflanzen, D-85354 Freising, Germany
[5] Plants Division, Department of Life Sciences, Natural History Museum, Cromwell Road, London SW7 5BD, UK

Guest Editor: Clifford Morden

Abstract. Conservation of plant diversity on islands relies on a good knowledge of the taxonomy, distribution and genetic diversity of species. In recent decades, a combination of morphology- and DNA-based approaches has become the standard for investigating island plant lineages and this has led, in some cases, to the discovery of previously over-looked diversity, including 'cryptic species'. The flora of the Cape Verde archipelago in the North Atlantic is currently thought to comprise ∼740 vascular plant species, 92 of them endemics. Despite the fact that it is considered relatively well known, there has been a 12 % increase in the number of endemics in the last two decades. Relatively few of the Cape Verde plant lineages have been included in genetic studies so far and little is known about the patterns of diversification in the archipelago. Here we present an updated list for the endemic Cape Verde flora and analyse diversity patterns for three endemic plant lineages (*Cynanchum*, *Globularia* and *Umbilicus*) based on one nuclear (ITS) and four plastid DNA regions. In all three lineages, we find genetic variation. In *Cynanchum*, we find two distinct haplotypes with no clear geographical pattern, possibly reflecting different ploidy levels. In *Globularia* and *Umbilicus*, differentiation is evident between populations from northern and southern islands. Isolation and drift resulting from the small and fragmented distributions, coupled with the significant distances separating the northern and southern islands, could explain this pattern. Overall, our study suggests that the diversity in the endemic vascular flora of Cape Verde is higher than previously thought and further work is necessary to characterize the flora.

Keywords: *Cynanchum*; DNA barcoding; *Globularia*; Macaronesian Islands; multi-island endemics (MIEs); *Umbilicus*.

* Corresponding author's e-mail address: mromeiras@yahoo.co.uk

Introduction

Efforts to conserve island floras and to understand their diversity are crucially dependent on baseline taxonomic knowledge: species are the unit of conservation actions, the focus of phylogenetic and phylogeographic work and the basic units for macro-ecological analyses. However, whilst island floras have been subject to study over a long period (e.g. Romeiras et al. 2014), and are often considered well explored (e.g. Joppa et al. 2011), recent discoveries of island taxa new to science have occurred even in groups of large organisms such as lizards and birds (Whittaker and Fernández-Palacios 2007).

Gray and Cavers (2013) showed that taxonomic effort expended contributes to the patterns investigated in theoretical biogeography and a recent survey of biologists working on oceanic islands suggested that 'Knowledge of the taxonomy, distribution and threat status of plants on oceanic islands is insufficient' (Caujapé-Castells et al. 2010). Clearly, the taxonomy of oceanic island plants is far from complete, and this is an important issue to address.

This paper considers recent developments in our understanding of the flora of the Cape Verde Islands and the potential of molecular data to provide further insights. The Cape Verdes are the most southerly archipelago of the Macaronesian Region that also comprises the archipelagos of the Azores, Canaries and Madeira in the North Atlantic. The Cape Verdes are a group of 10 volcanic islands located 1500 km southwest of the Canary Islands and ~570 km west of the African mainland. Lying at tropical latitudes and in close proximity to the coast of Senegal, their flora is of mainly tropical African origin and, accordingly, it was proposed to group the Cape Verde flora with the palaeotropical floras of the Saharan Tropical region (Rivas-Martínez 2009). Within the archipelago, the islands form three clusters: (i) northern group (Santo Antão, São Vicente, Santa Luzia and São Nicolau); (ii) southern group (Santiago, Fogo and Brava) and (iii) eastern group (Sal, Boavista and Maio) (see Fig. 1). The northern and the southern islands are characterized by high mountains [e.g. Monte Gordo (1304 m) in São Nicolau; Pico da Antónia (1392 m) in Santiago; Tope de Coroa (1979 m) in Santo Antão and Pico do Fogo (2829 m) in Fogo], offering a wide range of habitats over relatively short distances (Duarte and Romeiras 2009). The eastern islands are lower, drier and more homogeneous in their ecology. The islands' ages range from ~25.6 to ~21.1 Ma (for Sal and Maio, respectively) to <6 Ma (for Brava), with ages decreasing from east to west (Doucelance et al. 2003). Only Fogo Island currently has volcanic activity, with the most recent eruptions occurring in 1995 and 2014.

Extensive taxonomic and collecting activity during the last decade of the 20th century resulted in the publication of a monograph on the Cape Verde endemic flora by Brochmann et al. (1997). A total of 82 endemic taxa were recognized and 5 distributional elements were identified: (i) northern; (ii) southern, which together form the (iii) western element; (iv) eastern and (v) ubiquitous, which comprises taxa distributed on at least one island of western and eastern elements (see Fig. 1). The vast majority of taxa that occur in the northern, southern and western distributional elements are montane; all eastern taxa are coastal; most of the geographically ubiquitous taxa are also ecologically ubiquitous, and more than half are also altitudinally ubiquitous.

To date, molecular analyses of the Cape Verde flora have been limited. Analyses of the relationships of Macaronesian lineages have focussed largely on the floras of the Azores, Canaries and Madeira [e.g. for reviews see Carine and Schaefer (2010), Caujapé-Castells (2011), Pérez-de-Paz and Caujapé-Castells (2013)], and where Cape Verdean taxa have been included, the sampling has often been limited (e.g. Tornabenea, Spalik and Downie 2007; Lotus, Ojeda et al. 2014). However, molecular studies of Cape Verdean Campanula L. (Alarcón et al. 2013) and Echium L. (Romeiras et al. 2007, 2011) have sampled more extensively and revealed the geographic structure of genetic variation within those lineages. For Echium, a single colonization from the Canaries ~5 Ma was inferred, with subsequent diversification within the Cape Verdes during the Pleistocene (<1.8 Ma) and a split between the 'southern' (E. hypertropicum, E. vulcanorum) and 'northern' (E. stenosiphon s.l.) island species (García-Maroto et al. 2009; Romeiras et al. 2011). Within Campanula, Alarcón et al. (2013) inferred a recent divergence (1.0 Ma) of the Cape Verde endemic C. jacobaea from its sister species C. balfourii (endemic to Socotra). Within the archipelago, three groups were identified: two restricted to the northern islands (one each endemic to São Nicolau and Santo Antão) and a third restricted to the southern islands (Fogo, Santiago and Brava). Thus, both Romeiras et al. (2011) and Alarcón et al. (2013) found molecular patterns within lineages that are consistent with the distributional elements previously recognized by Brochmann et al. (1997) based on analyses of distributional data.

This paper has two goals. First, we provide a revised list of Cape Verde endemic taxa, discuss the new species discoveries and re-evaluate the distributional elements defined by Brochmann et al. (1997) in light of endemic taxa and distribution data published since 1997. Second, we investigate the patterns of genetic variation within three plant lineages endemic to Cape Verde to explore the geographical structuring of genetic variation in the highly fragmented insular landscape of this archipelago. We chose Globularia L., Cynanchum L. and Umbilicus DC. For each of these lineages, only one species is currently

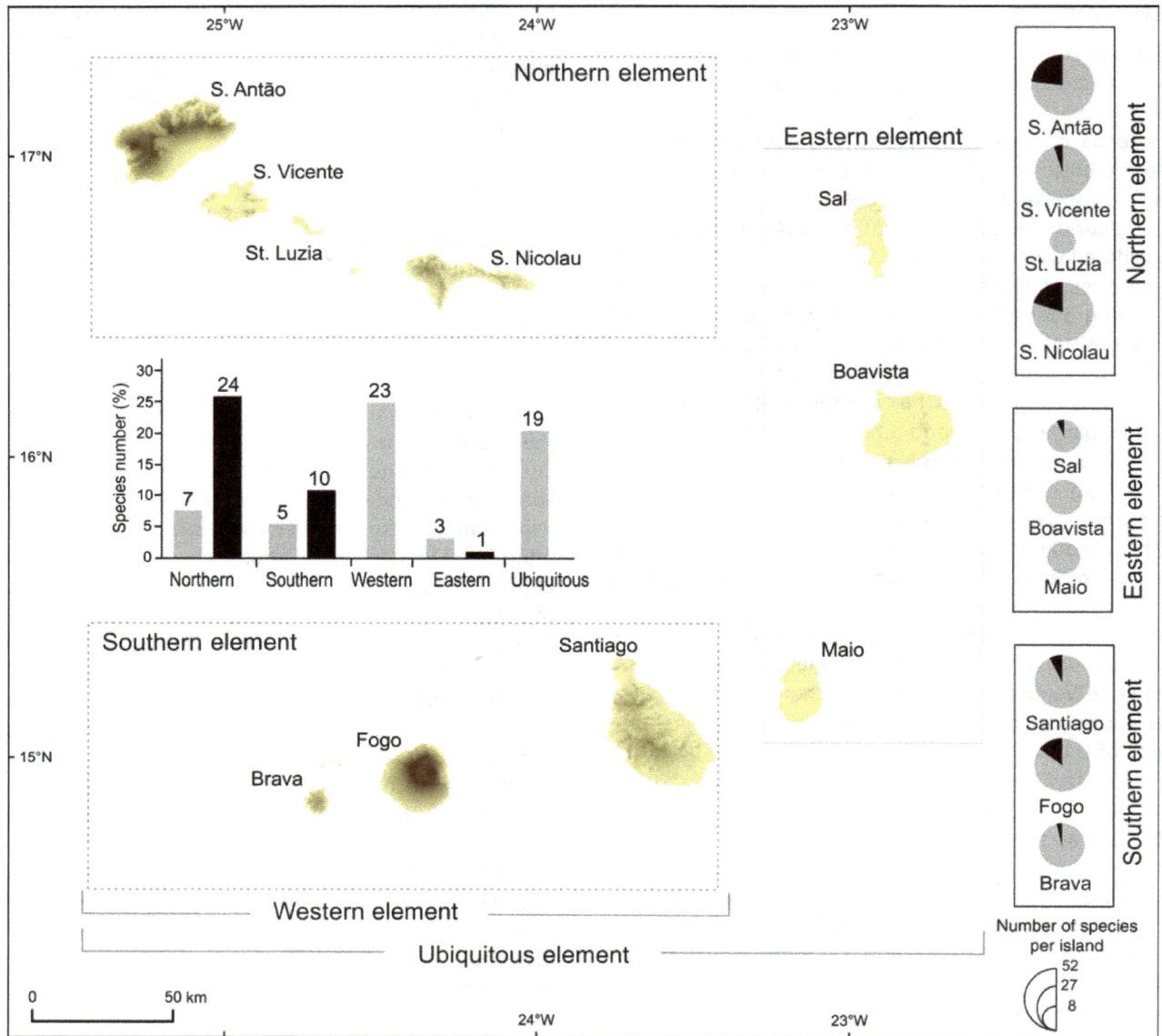

Figure 1. Relationship between the vascular endemic species of Cape Verde and their distribution within the five distributional elements: northern (i.e. Santo Antão, São Vicente, Santa Luzia and São Nicolau), southern (i.e. Santiago, Fogo and Brava), western (including species simultaneously present in northern and southern islands), eastern (i.e. Maio, Sal and Boavista) and ubiquitous (including species present in both western and eastern islands). The distribution of multi-island endemics—MIEs (grey) and single-island endemics—SIEs (black) in each island (right); and within the five distributional elements (bar graph in the centre; the number of taxa is placed above each bar).

recognized in the Cape Verdes: *Globularia amygdalifolia* Webb (Plantaginaceae), *Cynanchum daltonii* (Decne. ex Webb) Liede & Meve (Apocynaceae) and *Umbilicus schmidtii* Bolle (Crassulaceae). *Globularia amygdalifolia* and *U. schmidtii* constitute western elements *sensu* Brochmann *et al.* (1997) with distributions spanning five and four islands, respectively. *Cynanchum daltonii*, considered a ubiquitous element, occurs on seven islands in a variety of habitats from sea level to high mountain areas. The three taxa also differ in their phytogeographic

relationships: *G. amygdalifolia* has Canaro-Madeiran affinities (a Macaronesian element), *C. daltonii* Sudano–Zambesian–Sindian affinities (an African element) and *U. schmidtii* is a Mediterranean element (Brochmann *et al.* 1997). Sampling multiple accessions from across the distribution range of each species for both nuclear (ITS) and plastid DNA regions (*matK, psbA-trnH, rbcL, trnL-F*), we aim to determine whether a molecular variation in the three focal lineages correlates with geographic distance.

Methods

Characterization of the endemic flora

We compiled a list of the Cape Verde endemic vascular plants, including the recently described species, together with species distribution data, arranged according to the five distributional elements established by Brochmann *et al.* (1997): northern, southern, western (including species simultaneously present in northern and southern islands), eastern and ubiquitous (including species present in both western and eastern islands) (Fig. 1). Each species was also characterized as either hygrophytic, mesophytic or xerophytic following Brochmann *et al.* (1997). We determined the occurrence of single-island endemics (SIEs) and multi-island endemics (MIEs) for each distributional element and ecological class.

Eco-geographic data of studied plant lineages

Globularia amygdalifolia and *Umbilicus schmidtii* occur in small populations of few individuals in northeast-exposed areas over 400 m above sea level (a.s.l.). *Umbilicus schmidtii* populations are found in humid zones mainly associated with montane rupicolous vegetation and *Globularia amygdalifolia* is mainly found in montane scrub vegetation, remarking its presence in volcanic lapilli areas of Fogo Island. *Cynachum daltonii* mainly occurs in semi-arid rocky escarpments up to 900 m a.s.l., reaching more than 2000 m a.s.l. on Fogo. It is a characteristic component of the northeast-exposed coastal cliffs, where it forms large stands.

Molecular analyses

Sampling. Plant material for DNA extraction was collected and preserved in silica gel. Vouchers were deposited at LISC (Tropical Research Institute) and BM (Natural History Museum, London) herbaria [**see Supporting Information**].

Globularia amygdalifolia populations were sampled from four of the five islands on which it occurs: São Nicolau (one population, five individuals), Santo Antão (one population, two individuals), Fogo (two populations, five individuals) and Brava (two populations, five individuals) [**see Supporting Information**]. We did not sample plants from Santiago, since all individuals seen were cultivated.

For *Cynachum daltonii*, populations were sampled on all seven islands from which it has been recorded but DNA extraction from the samples from Santiago and São Vicente failed. The included samples are: São Nicolau (three individuals), Santo Antão (five individuals), Boavista (three individuals), Fogo (three individuals) and Brava (five individuals) with one population sampled on each island except Brava where we sampled two populations [**see Supporting Information**].

For *Umbilicus schmidtii*, individuals were sampled from all four islands on which it occurs: São Nicolau (one population, five individuals), Santo Antão (three populations, five individuals), Santiago (one population, five individuals) and Fogo (two populations, five individuals).

Molecular methods. DNA extraction from silica gel-dried leaf material followed a DNeasy Plant Mini Kit protocol (Qiagen, Crawley, UK), with a further purification using QIAquick PCR Purification Kit (Qiagen, Crawley, UK).

One individual per island per species was selected for a first molecular screening with five DNA regions [i.e. ITS and four cpDNA (*matK*, *psbA-trnH*, *rbcL*, *trnL-F*)] in order to ascertain which regions are variable within each species. Thereafter, only variable regions were amplified and sequenced for the remaining samples.

From the nuclear genome, the ITS1–5.8S–ITS2 region was amplified using ITS5 and ITS4 primers from White *et al.* (1990). From the chloroplast genome, four regions were sequenced: parts of the maturase K (*matK*) gene using the primers *matKF*_uni: 5′-AAT TTA CGA TCH ATT CAT TCM ATW TTT CC-3′ and *matKR*_uni: 5′-AGT TYT ARC ACA AGA AAG TCG AAR TAT ATA-3′ following Schaefer *et al.* (2011); the ribulose-1,5-bisphosphate carboxylase/oxygenase (*rbcL*) gene using the primers 1F and 724R of Olmstead *et al.* (1992); the *trnL-F* spacer using the primers 'e' and 'f' of Taberlet *et al.* (1991) and the *psbA-trnH* spacer using the primers of Sang *et al.* (1997). DNA amplification was performed in a 2720 Thermal Cycler (Applied Biosystems) in 25 μL-volume reactions. Standard polymerase chain reaction (PCR) procedures were applied to carry out amplifications. We used 1.25 units of DreamTaq™ DNA polymerase, and BSA (0.4 mg mL^{-1}) for all reactions. The PCR conditions were as follows: (i) 10 min pre-treatment at 94 °C, 28 cycles of 1 min at 95 °C, 1 min at 55 °C, 3 min at 72 °C and a final stage of 7 min at 72 °C, for ITS; (ii) 10 min pre-treatment at 94 °C, 30 cycles of 1 min at 96 °C, 3 min at 50 °C, 3 min at 72 °C and a final stage of 7 min at 72 °C, for the cpDNA region. Amplified products were purified with Sureclean Plus (Bioline, London, UK) and sent to STAB Vida—Investigação e Serviços em Ciências Biológicas, Lda (Monte da Caparica, Portugal) for Sanger sequencing. Sequences were deposited in GenBank under the accession numbers KP279325–KP279464 [**see Supporting Information**].

Raw sequences were edited with BioEdit v.7.0.9 (Hall 1999) and alignments were performed in ClustalX v.2.0.10 (Thompson *et al.* 1997) using default parameters. Chloroplast DNA regions were concatenated using Sequence Matrix 1.7.8 (Vaidya *et al.* 2011). Gaps and inversions were coded as single mutations and a statistical-parsimony network (Templeton *et al.* 1992) was constructed using

TCS vers. 1.21 (Clement *et al.* 2000) with a 95 % parsimony criterion.

Results

Characterization of the endemic flora

Ninety-two taxa (including several subspecies) are currently considered endemic to the Cape Verde Islands (Table 1). The majority of endemic taxa (~75 %) are distributed in the western islands, either in the (i) northern (34 %), (ii) southern (16 %) elements or in the (iii) western element (25 %); lower proportions are distributed in the eastern islands (4 %) or are ubiquitous elements (21 %) (Fig. 1 and Table 1). The northern islands are also the richest in SIEs (Fig. 1). Twenty-four of the 31 taxa that constitute the northern element are SIEs (77 %). The remaining seven taxa of the northern element occur on two to three islands with none found on Santa Luzia (Table 1). Ten SIEs occur in the southern element (six SIEs in Fogo, three in Santiago and one in Brava), constituting 66 % of the southern element. Of the four eastern element taxa, only one is a SIE (25 %).

Nineteen endemic taxa (SIEs + MIEs) are xerophytes (21 %), 46 are mesophytes (50 %) and 27 (29 %) are hygrophytes. Hygrophytes and mesophytic endemics are most prevalent in the northern, southern and western elements: of the 69 endemic taxa that constitute these three elements, only 2 (3 %) are xerophytes; 40 are mesophytes (58 %) and 27 (39 %) are hygrophytes. All four eastern island endemics are xerophytes. The ubiquitous endemics are mainly xerophytes (13 taxa, 68 %) with the remaining six taxa considered mesophytes. Among the 35 taxa that are SIEs in Cape Verde, 54 % are mesophytes and 40 % hygrophytes and 6 % xerophytes (Table 1).

Molecular analyses

Globularia amygdalifolia—genetic variation was detected only in the ITS region. Two substitutions in the 659 bp ITS fragment defined three ribotypes, one of which was found only in plants from Fogo, the second restricted to Brava and the third shared between plants from the northern islands of Santo Antão and São Nicolau (Fig. 2A). The network supported a split between plants from the northern and southern islands.

Umbilicus schmidtii—genetic variation was observed in the *matK* (844 bp), *psbA-trnH* (289 bp) and *trnL-F* spacer (286 bp) regions (the ITS region was not successfully sequenced for most samples). A total of 11 substitutions, 9 indels and 1 inversion (8 bp) defined five haplotypes for the combined 1420 bp of plastid DNA. Three haplotypes were private to the northern island populations; of which two were found in plants from Santo Antão (one comprising plants from Cova and the other comprising

plants from Pedra Rachada/Delgadinho da Corda) and one was found in São Nicolau. Three to four mutations separated the three northern island haplotypes. The two remaining haplotypes were found in plants from the southern islands. Both haplotypes occurred in the population on Santiago with one also found in Fogo. The two southern haplotypes are distinguished by 5 mutations with 13 mutations separating southern and northern haplotype groups (Fig. 2B).

Cynanchum daltonii—genetic variation was detected only in the ITS region: three substitutions were found in the 648 bp fragment and these defined two ribotypes (Fig. 2C). One was found in plants from Santo Antão, São Nicolau and Brava, and the second in plants from Fogo and Boavista.

Discussion

Spatial patterns of endemism

The endemic vascular plant list for Cape Verde has increased from 82 (Brochmann *et al.* 1997) to 92 taxa, which is an ~12 % increase in 18 years. Of the endemics more recently discovered or reclassified as endemics, one is a fern (*Dryopteris gorgonea*) and the remainder are angiosperms, of which six are recognized at species rank (*Fagonia mayana*, *Helichrysum nicolai*, *Lotus alianus*, *Tornabenea ribeirensis*, *Solanum rigidum* and *Withania chevalieri*) and three at subspecies rank (*Echium stenosiphon* subsp. *glabrescens*, *Teline stenopetala* subsp. *santoantonai* and *Dracaena draco* subsp. *caboverdeana*). Three of the 10 (*Echium stenosiphon* subsp. *glabrescens*, *Lotus alianus* and *Tornabenea ribeirensis*) have other conspecific Cape Verdean endemic taxa (Table 1) and their discovery reflects an enhanced understanding of lineages that have diversified within the Cape Verdes. The remaining taxa lack conspecific endemic taxa and may be considered anagenetic lineages *sensu* Stuessy *et al.* (2006). Their description has typically occurred within the context of wider revisionary or monographic studies. *Solanum rigidum*, for example, was previously considered introduced to the Cape Verdes from the Americas (Gonçalves 2002 sub *S. fuscatum* L.). However, monographic research on the genus has shown it to be a Cape Verde endemic that has rather been introduced into the Americas (Knapp and Vorontsova 2013). The discovery of the anagenetic element of the Cape Verde endemic flora is likely far from complete. Whilst many such taxa are widespread taxa of little conservation concern, this is not the case for all. For example, *Helichrysum nicolai*, described by Kilian *et al.* (2010), is restricted to the Alto das Cabaças range in the northeastern part of São Nicolau. Its conservation status has not been assessed, but population sizes (very few tens) and the extent and area of occurrence

Table 1. List of the endemic vascular plants of Cape Verde grouped by distributional elements (*sensu* Brochmann *et al.* 1997). The acronyms for each island are as follows: SA, Santo Antão; SV, São Vicente; SL, Santa Luzia; SN, São Nicolau; B, Boavista; M, Maio; ST, Santiago; F, Fogo and BR, Brava.

Distributional elements / Taxon	SA	SV	SL	SN	S	B	M	ST	F	Br	Ecological groups
Northern element											
Aeonium gorgoneum J. A. Schmidt	X	X		X							Mesophytic
Asteriscus smithii (Webb) Walp.				X							Hygrophytic
Campylanthus glaber Benth. subsp. *spathulatus* (A. Chev.) Brochmann, N. Kilian, Lobin & Rustan	X										Mesophytic
Carex antoniensis A. Chev.	X										Hygrophytic
Carex paniculata L. subsp. *hansenii* Lewej. & Lobin	X										Hygrophytic
Conyza schlechtendalii Bolle				X							Hygrophytic
Diplotaxis antoniensis Rustan	X										Xerophytic
Diplotaxis gorgadensis Rustan subsp. *brochmannii* Rustan	X										Hygrophytic
Diplotaxis gorgadensis Rustan subsp. *gorgadensis*	X										Mesophytic
Diplotaxis gracilis (Webb) O. E. Schulz				X							Mesophytic
Diplotaxis sundingii Rustan				X							Hygrophytic
Diplotaxis vogelli (Webb) Cout.		X									Mesophytic
Echium stenosiphon Webb subsp. *glabrescens* (Pett.) Romeiras & Maria C. Duarte				X							Mesophytic
Echium stenosiphon Webb subsp. *lindbergii* (Pett.) Bramwell	X										Mesophytic
Echium stenosiphon Webb subsp. *stenosiphon*		X									Mesophytic
Frankenia ericifolia Chr. Sm. ex DC. subsp. *caboverdeana* Brochmann, Lobin & Sunding	X	X		X							Mesophytic
Frankenia ericifolia Chr. Sm. ex DC. subsp. *montana* Brochmann, Lobin & Sunding				X							Hygrophytic
Helichrysum nicolai N. Kilian, Galbany & Oberpr.				X							Mesophytic
Kickxia elegans (G. Forst.) D. A. Sutton subsp. *webbiana* (Sunding) Rustan & Brochmann	X										Mesophytic
Launaea gorgadensis (Bolle) N. Kilian	X	X		X							Mesophytic
Launaea picridioides (Webb) Engler	X	X		X							Mesophytic
Limonium jovi-barba (Webb) Kuntze		X		X							Hygrophytic
Limonium sundingii Leyens, Lobin, N. Kilian & Erben				X							Hygrophytic
Lobularia canariensis (DC.) Borgen subsp. *spathulata* (J. A. Schmidt) Borgen		X		X							Mesophytic
Lotus alianus J.H. Kirkbr.	X	X									Xerophytic
Lotus arborescens Lowe ex Cout.				X							Hygrophytic
Lotus oliveirae A. Chev.	X										Mesophytic
Papaver gorgoneum Cout. subsp. *theresias* Kadereit & Lobin	X										Mesophytic
Teline stenopetala (Webb & Berthel.) Webb & Berthel. subsp. *santoantaoi* Marrero-Rodr.	X										Hygrophytic

Continued

Table 1. *Continued*

Distributional elements / Taxon	SA	SV	SL	SN	S	B	M	ST	F	Br	Ecological groups
Tornabenea bischoffii J. A. Schmidt	X										Hygrophytic
Tornabenea ribeirensis Schmidt & Lobin				X							Mesophytic
Southern element											
Asteriscus daltonii (Webb) Walp. subsp. *daltonii*								X			Mesophytic
Campanula bravensis (Bolle) A. Chev.								X	X	X	Hygrophytic
Centaurium tenuiflorum (Hoffmanns. & Link) Fritsch subsp. *viridense* (Bolle) A. Hansen & Sunding								X	X	X	Hygrophytic
Diplotaxis hirta (A. Chev.) Rustan & Borgen									X		Mesophytic
Diplotaxis varia Rustan								X		X	Mesophytic
Echium hypertropicum Webb								X		X	Mesophytic
Echium vulcanorum A. Chev.									X		Mesophytic
Erysimum caboverdeanum (A. Chev.) Sund.									X		Mesophytic
Launaea thalassica N. Kilian, Brochmann & Rustan										X	Mesophytic
Limonium lobinii N. Kilian & T. Leyens								X			Hygrophytic
Lotus jacobaeus L.								X	X		Mesophytic
Tornabenea annua Bég.								X			Hygrophytic
Tornabenea humilis Lobin & K. H. Schmidt									X		Mesophytic
Tornabenea tenuissima (A. Chev.) A. Hans. & Sunding									X		Hygrophytic
Verbascum cystolithicum (B. Petterson) Huber-Morath									X		Mesophytic
Western element (northern and southern elements)											
Artemisia gorgonum Webb	X							X	X		Mesophytic
Campanula jacobaea C. Sm. ex Webb	X	X		X				X			Mesophytic
Campylanthus glaber Benth. subsp. *glaber*	X	X		X				X	X	X	Mesophytic
Conyza feae (Bég.) Wild	X	X		X				X	X	X	Mesophytic
Conyza pannosa Webb	X	X		X				X		X	Hygrophytic
Conyza varia (Webb) Wild	X	X		X					X	X	Mesophytic
Dracaena draco (L.) L. subsp. *caboverdeana* Marrero-Rodr. & R. Almeida	X	X		X				X	X	X	Mesophytic
Dryopteris gorgonea J.P. Roux	X	X		X					X		Hygrophytic
Eragrostis conerti Lobin	X	X		X				X	X		Hygrophytic
Globularia amygdalifolia Webb	X			X				X	X	X	Mesophytic
Helianthemum gorgoneum Webb	X		X						X	X	Mesophytic
Kickxia elegans (G. Forst.) D. A. Sutton subsp. *dichondrifolia* (Benth.) Rustan & Brochmann	X	X		X				X			Hygrophytic
Lavandula rotundifolia Benth.	X	X		X				X	X		Mesophytic
Limonium braunii (Bolle) A. Chev.	X			X					X	X	Mesophytic
Lobularia canariensis (DC.) Borgen subsp. *fruticosa* (Webb) Borgen	X			X				X	X	X	Mesophytic
Micromeria forbesii Benth.	X			X				X	X	X	Mesophytic

Continued

Table 1. *Continued*

Distributional elements / Taxon	SA	SV	SL	SN	S	B	M	ST	F	Br	Ecological groups
Papaver gorgoneum Cout. subsp. *gorgoneum*				X					X		Hygrophytic
Periploca chevalieri Browicz	X			X				X	X	X	Mesophytic
Phagnalon melanoleucum Webb	X	X		X				X	X		Hygrophytic
Sonchus daltonii Webb	X	X		X				X	X		Hygrophytic
Tolpis farinulosa (Webb) Schmidt	X	X						X	X	X	Hygrophytic
Tornabenea insularis (Parl. ex Webb) Parl. ex Webb		X		X						X	Hygrophytic
Umbilicus schmidtii Bolle	X			X				X	X		Hygrophytic
Eastern element											
Diplotaxis glauca J. A. Schmidt					X	X					Xerophytic
Fagonia mayana Schlecht.						X	X				Xerophytic
Pulicaria burchardii Hutch. subsp. *longifolia* Gamal-Eldin					X						Xerophytic
Sporobolus minutus Link subsp. *confertus* (J. A. Schmidt) Lobin, N. Kilian & Leyens					X		X				Xerophytic
Ubiquitous element (northern, southern and eastern elements)											
Aristida cardosoi Cout.	X	X	X	X	X	X	X	X	X	X	Xerophytic
Asparagus squarrosus J. A. Schmidt	X	X	X	X	X	X	X				Xerophytic
Asteriscus daltonii subsp. *vogelii* (Webb) Greuter	X	X		X		X		X	X	X	Mesophytic
Brachiaria lata (Schumach.) C. E. Hubb. subsp. *caboverdeana* Conert & Ch. Köhler		X		X		X		X			Xerophytic
Cynanchum daltonii (Decne. ex Webb) Liede & Meve	X	X		X		X		X	X	X	Xerophytic
Euphorbia tuckeyana Steud. ex Webb	X	X	X	X	X	X		X	X	X	Mesophytic
Forsskaolea procridifolia Webb	X	X	X	X	X		X	X	X	X	Xerophytic
Kickxia elegans (G. Forst.) D. A. Sutton subsp. *elegans*	X	X		X	X	X	X	X	X	X	Xerophytic
Limonium brunneri (Webb) Kuntze		X	X		X						Xerophytic
Lotus brunneri Webb		X	X		X	X	X				Xerophytic
Lotus purpureus Webb	X	X		X		X		X	X	X	Xerophytic
Paronychia illecebroides Webb	X	X	X	X		X	X	X	X		Xerophytic
Phoenix atlantica A. Chev.					X	X	X	X			Xerophytic
Polycarpaea gayi Webb	X	X	X	X	X			X	X		Mesophytic
Pulicaria diffusa (Shuttlew. ex Brunn.) Pett.					X	X	X	X	X		Xerophytic
Sideroxylon marginata (Decne.) Cout.	X	X		X	X			X	X	X	Mesophytic
Solanum rigidum Lam.	X	X		X		X	X	X	X	X	Mesophytic
Verbascum capitis-viridis Hub.- Mor.	X	X		X		X	X	X			Mesophytic
Withania chevalieri A.E. Gonç.	X	X			X				X		Xerophytic

would all appear to be extremely limited, suggesting a high threat status.

The overall patterns in the endemic flora recognized by Brochmann *et al.* (1997) are still evident in the updated list of endemics presented in Table 1. Thus, the five distributional elements are clearly distinguished. Three quarters of endemic taxa are restricted to the western islands, nearly all of which are mesophytes or hygrophytes and a high proportion are SIEs. In general, populations of both SIEs and MIEs in the northern, southern and western elements are small, and restricted to specific habitats in N–NE facing moist cliffs of mountain areas

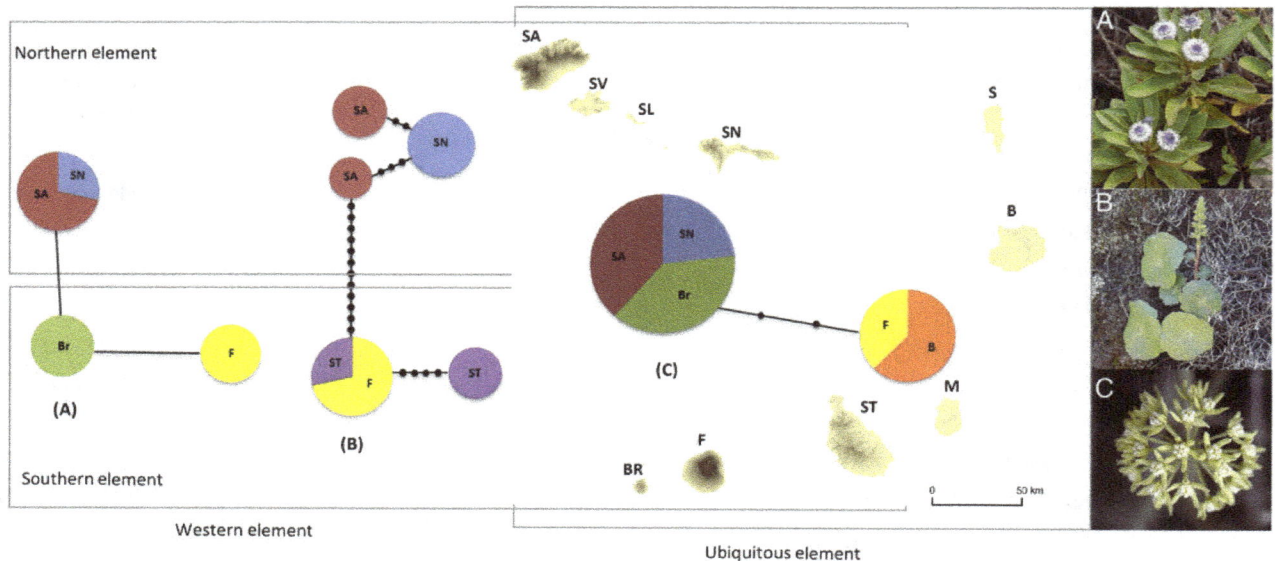

Figure 2. Distribution elements and haplotype network for three plant lineages produced with TCS 1.21: (A) *Globularia amygdalifolia* based on ITS; (B) *Umbilicus schmidtii* based on cpDNA (*matK, psbA-trnH* and *trnL-F*); (C) *Cynanchum daltonii* based on ITS. The size of circles representing each haplotype is proportional to the number of individuals possessing that haplotype. The acronyms for each island are as follows: SA, Santo Antão; SV, São Vicente; Sl, Santa Luzia; SN, São Nicolau; S, Sal; B, Boavista; M, Maio; ST, Santiago; F, Fogo; and BR, Brava. Plant species (flowering) on the right (photos M.R. and M.C.D.).

(Duarte *et al.* 2008). Their distributions are thus highly fragmented, and this is likely to have led to isolation and drift resulting in the high incidence of single-island endemism observed. Xerophytes dominate among eastern island endemics and the ubiquitous element wherein 68 % are xerophytes. Indeed, only one SIE is a xerophyte. The generally larger population sizes of lowland xerophytic taxa coupled with the relatively close proximity of islands (e.g. only 15 km separates the coastal areas of Santo Antão from São Vicente, in the northern islands) may facilitate gene flow between islands, maintaining the integrity of these taxa.

Genetic diversification within the plant lineages

The molecular analyses we present in this paper provide further insights into the patterns of endemic plant diversity in the Cape Verdes. *Globularia amygdalifolia* and *Umbilicus schmidtii* both constitute western elements of the flora (Table 1) and in both cases, differentiation is evident at the molecular level between plants from southern and northern islands (see Fig. 2A and B). Isolation and drift resulting from the small and fragmented distributions of these taxa coupled with the significant distances separating the northern and southern sub-archipelagos (230 km separates Santo Antão from Brava; and 140 km São Nicolau from Santiago, maximum and minimum distances, respectively) is likely to explain this pattern. A similar intraspecific pattern was documented in *Campanula jacobaea* (Alarcón *et al.* 2013), which is also a

western element. From a conservation perspective, it would be appropriate to treat northern and southern populations of western element taxa as distinct management units.

In the case of *U. schmidtii*, the differentiation of northern and southern populations appears to be consistent with differences observed in the inflorescence (e.g. morphological characters of the flowers, data not shown) suggesting the need of further research, in order to ascertain whether overlooked taxa occur in the archipelago. *Umbilicus* is a small succulent herb, which is perhaps most likely to harbour cryptic diversity because it is an undercollected species, and most of the herbarium specimens lack important characters for identification. Its tiny seeds are probably wind-dispersed, but gene flow might be reduced due to the significant distance between the islands.

Some variation in leaf size and shape was also evident within *G. amygdalifolia*, but no clear geographical pattern in the morphological variation could be identified. Nevertheless, a morphological re-examination in this species within the context of the molecular results presented here would be appropriate.

Intraspecific variation was also found in *C. daltonii* with two distinct haplotypes. However, the distribution of the two haplotypes is incongruent with Brochmann *et al.*'s (1997) elements since one is distributed in the eastern island of Boavista and the southern island of Fogo whereas the other is distributed in the northern islands of Santo

Antão and São Nicolau and the southern island of Brava. Further sampling including Santiago and São Vicente is needed to better understand the patterns of genetic diversity in this species. An analysis of ploidy level of plants might also be informative since two ploidy levels have been provisionally reported for *C. daltonii* (Brochmann et al. 1997; Meve and Liede-Schumann 2012) and this may explain the occurrence of the two distinct haplotypes.

Our study has surveyed three plant lineages and it would be premature to make generalizations regarding the entire flora. Nevertheless, it does appear that levels of intraspecific genetic diversity could be similar to that reported from the Azores Islands (Schaefer et al. 2011; Rumsey et al. 2014). Although there are more SIEs in the Cape Verdes than in the Azores, a high proportion of MIE relative to SIE is reported for both archipelagos in comparison with the Canaries and Madeira (Carine and Schaefer 2010). In both cases, plant lineages that are widespread across the archipelago and that exhibit little morphological diversification have nevertheless been found to exhibit geographically structured molecular patterns.

Oceanic islands such as the Cape Verdes have been the focus of taxonomic research during the last 200 years (Romeiras et al. 2014), but our baseline taxonomic knowledge still needs to be refined to effectively support conservation and provide a context for understanding the evolution and biogeography of island plants. Knowledge of the endemic flora has developed significantly in the last two decades but our molecular results suggest that further taxonomic revisions, integrating genetic, morphological and ecological data, among others, are still necessary to better understand the patterns of diversity in the endemic flora. Widespread species, in particular, should be the focus of taxonomic work since they may contain overlooked taxa, and species as they are currently circumscribed, may not be the most appropriate management unit from a conservation perspective.

Sources of Funding

This work was supported by the Portuguese Foundation for Science and Technology (FCT) and European Social Funds through project PTDC/BIA-BIC/4113/2012.

Contributions by the Authors

M.R. and M.C. conceived the study. M.R., M.C.D. and M.C. conducted the sampling in Cape Verde. M.R. and F.M. carried out molecular work and M.C.D. carried out morphological observations. All authors analyzed data. M.R. wrote the first draft: M.C. and H.S. improved upon versions. All authors read and approved the final manuscript.

Acknowledgements

We thank the editors and two anonymous reviewers for the valuable comments and suggestions that improved the manuscript. We are grateful to numerous Cape Verde colleagues for their collaboration in the field, in particular C. Fernandes; I. Gomes (Santiago); H. Diniz (Fogo); E. Ramos, A. Fortes; G. Monteiro (S. Antão); C. Monteiro (S. Vicente); A. Fernandes, H. Santos (Sal) and I. Duarte (Boavista; S. Nicolau). And we also thank J.C. Costa (ISA-UL), E. Correia, D. Batista (IICT) and F. Pina-Martins (FCUL). We are grateful to the Cape Verde authorities for research and collecting permits no. 06/2013 (Direcção Geral do Ambiente/MAHOT).

Literature Cited

Alarcón M, Roquet C, García-Fernández A, Vargas P, Aldasoro JJ. 2013. Phylogenetic and phylogeographic evidence for a Pleistocene disjunction between *Campanula jacobaea* (Cape Verde Islands) and *C. balfourii* (Socotra). *Molecular Phylogenetics and Evolution* **69**:828–836.

Brochmann C, Rustan ØH, Lobin W, Kilian N. 1997. The endemic vascular plants of the Cape Verde Islands, W Africa. *Sommerfeltia* **24**: 1–356.

Carine MA, Schaefer H. 2010. The Azores diversity enigma: why are there so few Azorean endemic flowering plants and why are they so widespread? *Journal of Biogeography* **37**:77–89.

Caujapé-Castells J. 2011. Jesters, red queens, boomerangs and surfers: a molecular outlook on the Canarian endemic flora. In: Bramwell D, Caujapé-Castells J, eds. *The biology of island floras*. London: Cambridge University Press, 284–324.

Caujapé-Castells J, Tye A, Crawford DJ, Santos-Guerra A, Sakai A, Beaver K, Lobin W, Florens FBV, Moura M, Jardim R, Gómes I, Kueffer C. 2010. Conservation of oceanic island floras: present and future global challenges. *Perspectives in Plant Ecology, Evolution and Systematics* **12**:107–129.

Clement M, Posada D, Crandall KA. 2000. TCS: a computer program to estimate gene genealogies. *Molecular Ecology* **9**:1657–1659.

Doucelance R, Escrig S, Moreira M, Gariépy C, Kurz MD. 2003. Pb-Sr-He isotope and trace element geochemistry of the Cape Verde archipelago. *Geochimica et Cosmochimica Acta* **67**:3717–3733.

Duarte MC, Romeiras MR. 2009. Cape Verde Islands. In: Gillespie R, Clague D, eds. *Encyclopedia of Islands*. California: University of California Press, 143–148.

Duarte MC, Rego F, Romeiras MM, Moreira I. 2008. Plant species richness in the Cape Verde Islands—eco-geographical determinants. *Biodiversity and Conservation* **17**:453–466.

García-Maroto F, Mañas-Fernández A, Garrido-Cárdenas JA, Alonso DL, Guil-Guerrero JL, Guzmán B, Vargas P. 2009. Δ6-desaturase sequence evidence for explosive Pliocene radiations within the adaptive radiation of Macaronesian *Echium* (Boraginaceae). *Molecular Phylogenetics and Evolution* **52**:563–574.

Gonçalves AE. 2002. Solanaceae 71. In: Paiva J, Martins ES, Diniz MA, Moreira I, Gomes I, Gomes S, eds. *Flora de Cabo Verde*. Lisboa/Praia: Instituto de Investigação Científica Tropical/Instituto Nacional de Investigação e Desenvolvimento Agrário.

Gray A, Cavers S. 2013. Island Biogeography, the effects of taxonomic effort and the importance of island niche diversity to single-island endemic species. *Systematic Biology* 63:55–65.

Hall TA. 1999. BioEdit: a user-friendly biological sequence alignment editor and analysis program for Windows 95/98/NT. *Nucleic Acids Symposium Series* 41:95–98.

Joppa LN, Roberts DL, Pimm SL. 2011. How many species of flowering plants are there? *Proceedings of the Royal Society B: Biological Sciences* 278:554–559.

Kilian N, Galbany-Casals M, Oberprieler C. 2010. *Helichrysum nicolai* (Compositae, Gnaphalieae), systematics of a new dwarf local endemic of the Cape Verde Islands, W Africa. *Folia Geobotanica* 45:183–199.

Knapp S, Vorontsova M. 2013. From introduced American weed to Cape Verde Islands endemic: the case of *Solanum rigidum* Lam. (Solanaceae, *Solanum* subgenus *Leptostemonum*). *PhytoKeys* 25: 35–46.

Meve U, Liede-Schumann S. 2012. Taxonomic dissolution of *Sarcostemma* (Apocynaceae: Asclepiadoideae). *Kew Bulletin* 67: 751–758.

Ojeda DI, Santos-Guerra A, Oliva-Tejera F, Jaen-Molina R, Caujapé-Castells J, Marrero-Rodríguez A, Cronk Q. 2014. DNA barcodes successfully identified Macaronesian *Lotus* (Leguminosae) species within early diverged lineages of Cape Verde and mainland Africa. *AoB PLANTS* 6: plu050; doi:10.1093/aobpla/plu050

Olmstead RG, Michaels HJ, Scott KM, Palmer JD. 1992. Monophyly of the Asteridae and identification of their major lineages inferred from DNA sequences of *rbcL*. *Annals of the Missouri Botanical Garden* 79:249–265.

Pérez-de-Paz J, Caujapé-Castells J. 2013. A review of the allozyme data set for the Canarian endemic flora: causes of the high genetic diversity levels and implications for conservation. *Annals of Botany* 111:1059–1073.

Rivas-Martínez S. 2009. Ensayo geobotánico global sobre la Macaronesia. In: Beltrán-Tejera E, Afonso-Carrillo J, García-Gallo A, Rodríguez-Delgado O, eds. *Homenaje al Prof. Wolfredo Wildpret de la Torre*. La Laguna (Santa Cruz de Tenerife): Instituto de Estudios Canarios. Serie Monografía LXXVIII, 255–296.

Romeiras MM, Cotrim HC, Duarte MC, Pais MS. 2007. Genetic diversity of three endangered species of *Echium* L. (Boraginaceae) endemic to Cape Verde Islands. *Biodiversity and Conservation* 16:547–566.

Romeiras MM, Paulo OS, Duarte MC, Pina-Martins F, Cotrim MH, Carine MA, Pais MS. 2011. Origin and diversification of genus *Echium* (Boraginaceae) in Cape Verde Islands: a phylogenetic study based on ITS (rDNA) and cpDNA sequences. *Taxon* 60: 1375–1385.

Romeiras MM, Duarte MC, Santos-Guerra A, Carine MA, Francisco-Ortega J. 2014. Botanical exploration of the Cape Verde Islands: from the pre-Linnaean records and collections to the late 18th century accounts and expeditions. *Taxon* 63:625–640.

Rumsey FJ, Schaefer H, Carine M. 2014. *Asplenium auritum* Sw. sensu lato (Aspleniaceae: Pteridophyta)—an overlooked Neotropical fern native to the Azores. *Fern Gazette* 19:259–271.

Sang T, Crawford DJ, Stuessy TF. 1997. Chloroplast DNA phylogeny, reticulate evolution, and biogeography of *Paeonia* (Paeoniaceae). *American Journal of Botany* 84:1120–1136.

Schaefer H, Moura M, Maciel MGB, Silva L, Rumsey FJ, Carine MA. 2011. The Linnean shortfall in oceanic island biogeography: a case study in the Azores. *Journal of Biogeography* 38: 1345–1355.

Spalik K, Downie SR. 2007. Intercontinental disjunctions in *Cryptotaenia* (Apiaceae, Oenantheae): an appraisal using molecular data. *Journal of Biogeography* 34:2039–2054.

Stuessy TF, Jakubowsky G, Gómez RS, Pfosser M, Schlüter PM, Fer T, Sun B-Y, Kato H. 2006. Anagenetic evolution in island plants. *Journal of Biogeography* 33:1259–1265.

Taberlet P, Gielly L, Pautou G, Bouvet J. 1991. Universal primers for amplification of three non-coding regions of chloroplast DNA. *Plant Molecular Biology* 17:1105–1109.

Templeton AR, Crandall KA, Sing CF. 1992. A cladistic analysis of phenotypic associations with haplotypes inferred from restriction endonuclease mapping and DNA sequence data. III. Cladogram estimation. *Genetics* 132:619–633.

Thompson JD, Gibson TJ, Plewniak F, Jeanmougin F, Higgins DG. 1997. The CLUSTAL_X Windows interface: flexible strategies for multiple sequence alignment aided by quality analysis tools. *Nucleic Acids Research* 25:4876–4882.

Vaidya G, Lohman DJ, Meier R. 2011. SequenceMatrix: concatenation software for the fast assembly of multi-gene datasets with character set and codon information. *Cladistics* 27:171–180.

White TJ, Bruns T, Lee S, Taylor J. 1990. Amplification and direct sequencing of fungal ribosomal RNA genes for phylogenetics. In: Innis MA, Gelfand DH, Sninsky JJ, White TJ, eds. *PCR Protocols: a guide to methods and applications*. New York: Academic Press, 315–322.

Whittaker R, Fernández-Palacios JM, eds. 2007. *Island biogeography*, 2nd edn. Oxford: Oxford University Press.

4

Barriers to seed and seedling survival of once-common Hawaiian palms: the role of invasive rats and ungulates

Aaron B. Shiels[1]* and Donald R. Drake[2]

[1] USDA, National Wildlife Research Center, Hawai'i Field Station, Hilo, HI 96721, USA
[2] Department of Botany, University of Hawai'i, 3190 Maile Way, Honolulu, HI 96822, USA

Guest Editor: Peter Bellingham

Abstract. Mammalian herbivores can limit plant recruitment and affect forest composition. Loulu palms (*Pritchardia* spp.) once dominated many lowland ecosystems in Hawai'i, and non-native rats (*Rattus* spp.), ungulates (e.g. pigs *Sus scrofa*, goats *Capra hircus*) and humans have been proposed as major causes of their decline. In lowland wet forest, we experimentally determined the vulnerability of seeds and seedlings of two species of *Pritchardia*, *P. maideniana* and *P. hillebrandii*, by measuring their removal by introduced vertebrates; we also used motion-sensing cameras to identify the animals responsible for *Pritchardia* removal. We assessed potential seed dispersal of *P. maideniana* by spool-and-line tracking, and conducted captive-feeding trials with *R. rattus* and seeds and seedlings of both *Pritchardia* species. Seed removal from the forest floor occurred rapidly for both species: >50 % of *Pritchardia* seeds were removed from the vertebrate-accessible stations within 6 days and >80 % were removed within 22 days. Although rats and pigs were both common to the study area, motion-sensing cameras detected only rats (probably *R. rattus*) removing *Pritchardia* seeds from the forest floor. Captive-feeding trials and spool-and-line tracking revealed that vertebrate seed dispersal is rare; rats moved seeds up to 8 m upon collection and subsequently destroyed them (100 % mortality in 24–48 h in captivity). Surprisingly, seedlings did not suffer vertebrate damage in field trials, and although rats damaged seedlings in captivity, they rarely consumed them. Our findings are consistent with the hypothesis generated from palaeoecological studies, indicating that introduced rats may have assisted in the demise of native insular palm forests. These findings also imply that the seed stage of species in this Pacific genus is particularly vulnerable to rats; therefore, future conservation efforts involving *Pritchardia* should prioritize the reduction of rat predation on the plant recruitment stages preceding seedling establishment.

Keywords: Island biology; *Rattus exulans*; *Rattus rattus*; seed dispersal; seed predation; seedling herbivory; *Sus scrofa*; tropical palm forest.

* Corresponding author's e-mail address: ashiels@hawaii.edu

Introduction

Alterations to island ecosystems resulting from introduced mammals are well documented across the world (Vitousek *et al*. 1997; Courchamp *et al*. 2003; Blackburn *et al*. 2004; Traveset and Richardson 2006). Aside from humans, perhaps the most ubiquitous introduced (non-native) mammals negatively affecting island flora and fauna are ungulates and rats; these non-native vertebrates are invasive because they spread rapidly and cause ecological or economic harm (Lockwood *et al*. 2007). Indeed ungulates and rats are commonly implicated in local extinctions and species reductions at multiple trophic levels, and they disrupt ecological processes (Singer 1981; Hone 1995; Vitousek *et al*. 1997; Courchamp *et al*. 2003; Towns *et al*. 2006). A well-accepted reason for these invasive vertebrates causing such ecosystem-changing effects in insular environments stems from the absence of similar native vertebrate species, and therefore includes the novel behaviours that are characteristic of such invasive herbivores (e.g. chewing, rooting, trampling).

The seed and seedling stages of the plant life cycle are typically more vulnerable than adults to the negative effects of rats and many ungulates. Seed predation by native and non-native mammals can limit plant recruitment and ultimately affect forest composition and structure (DeSteven and Putz 1984; Cabin *et al*. 2000; Campbell and Atkinson 2002). Alternatively, consumption of seeds may sometimes result in dispersal of native and non-native plant species (Abe 2007; Shiels and Drake 2011; O'Connor and Kelly 2012). Until relatively recently, the role of non-native rats (*Rattus* spp.) as a primary cause of insular forest and ecosystem change received little attention, perhaps because changes in tree communities typically occur over greater time scales than human life spans, written documentation was often uncommon during the years following rat introductions to islands, palaeo-ecological baselines were not widely documented and because rats are generally nocturnal and thus their links to causing change are not always obvious (Shiels *et al*. 2014).

Palaeoecological studies have documented the decline of many Pacific Island palm (Arecaceae) species following the arrival of humans (Prebble and Dowe 2008). At least two studies provide evidence supporting non-native rats as partially responsible for island-wide vegetation changes (Athens *et al*. 2002; Hunt 2007). In both cases, the plant life-form suffering decline was a palm—the Jubaea palm (now extinct, but related to the extant *Jubaea chilensis* of South America) in Rapa Nui (Easter Island; Hunt 2007) and the Loulu palm (*Pritchardia* spp.) in Hawai'i (Athens *et al*. 2002; Athens 2009). Of the 25 species of Hawaiian

Pritchardia currently recognized by Wagner *et al*. (1999), most survive only in small numbers and few locations, and at least eight are endangered (Chapin *et al*. 2004). Palms, which typically have large seeds relative to other co-occurring species, are well-known food items in rat diets (Pérez *et al*. 2008, Auld *et al*. 2010). Humans and rats colonized many Pacific islands simultaneously, because *R. exulans* was either a stowaway or was intentionally carried on early Polynesian watercraft. Most Pacific islands, like Hawai'i and Rapa Nui, lacked native ground mammals. Most *Jubaea* fruit endocarps recovered from archaeological sites on Rapa Nui bear rat incisor marks (Hunt 2007), and radio-carbon dates on these specimens match the time period of island-wide *Jubaea* deforestation. A few hundred years after humans and rats colonized Hawai'i, there was a major shift in vegetation from *Pritchardia*-dominated lowland forests to a grass- and shrub-dominated ecosystem (Athens *et al*. 2002). By examining palaeoecological evidence (e.g. pollen records, rat bones, radio-carbon dating charcoal) in the most arid region of Oahu, Athens *et al*. (2002) described a chronology of *Pritchardia* forest decline that was coincidental with little human impact or settlement but an abundant local rat population. Their findings support the hypothesis that rats were partially responsible for the large-scale deforestation of *Pritchardia*-dominated dry landscapes in Hawai'i. The most likely mechanism for large-scale rat-induced deforestation of native palms in both Rapa Nui and Hawai'i is through rat predation of seeds and perhaps seedlings (Hunt 2007; Athens 2009).

Invasive pigs (*Sus scrofa*) and goats (*Capra hircus*) have also been linked to native plant damage and mortality in insular ecosystems (Coblentz 1978; Barrios-Garcia and Ballari 2012). Whereas goats were introduced to Hawai'i following European contact in 1778, Polynesians introduced pigs (probably descendants from the Asiatic form of *S. scrofa*) upon their arrival ~700 years ago. Through rooting, trampling and herbivory, pigs reduce native plant abundance and cover in many wet forests in Hawai'i (Drake and Pratt 2001; Cole *et al*. 2012; Murphy *et al*. 2014). In drier habitats in Hawai'i, goats threaten native plants, particularly seedlings, by browsing and trampling (Scowcroft and Hobdy 1987). Although no formal studies have investigated ungulate herbivory on native *Pritchardia*, a combination of fencing to exclude ungulates (particularly goats) and rodenticide bait application to control rats resulted in elevated seedling recruitment and a 9-fold increase in juvenile abundance in the understory after 7 years in one of the largest remaining stands of the endangered *Pritchardia kaalae* on O'ahu (Mosher *et al*. 2007). Therefore, goats, pigs and rats may play an important role in suppression of *Pritchardia* regeneration and ultimately the survival of these now uncommon palm stands (Chapin *et al*. 2004).

In this study, we addressed the following three questions. (i) Are *Pritchardia* seeds and seedlings vulnerable to predation by vertebrates in a wet forest in Hawai'i? (ii) Are there particular animals responsible for the negative impacts on these once-common palms? (iii) Are *Pritchardia* seeds likely to be dispersed by non-native rats? In contemporary conditions, this study tests the hypothesis that invasive rats can limit *Pritchardia* plant recruitment; study outcomes will also provide corroborative evidence for the palaeoecological inferences implicating rats in the demise of the Hawaiian *Pritchardia* forests.

Methods

Study site

This study took place in a tropical wet forest on the eastern border of Lyon Arboretum on the island of O'ahu, Hawai'i (21°17'N 157°50'W). The Arboretum is a 50 ha reserve at the back of Mānoa Valley, and it is bounded on three sides by steep, forested slopes. During the study period (2005–06), the average temperature at the Arboretum was ~21–24 °C and rainfall was 3694 mm year^{-1} (R. Baker, unpubl. data). Elevation ranged from ~210 to 230 m above sea level (a.s.l.). Most of the Arboretum is forested with a high dominance of non-native plants; many of the plant species were planted and later spread and established to other parts of the valley and island. Vegetation ground cover to 50 cm height across the study area averages 34 % (range: 0–84 %; based on measurements in 24 1 m^2 plots). The canopy is continuous, reaching heights of >25 m, with albizia (*Falcataria moluccana*; Fabaceae), blue marble (*Elaeocarpus angustifolius*; Elaeocarpaceae) and figs (*Ficus* spp.; Moraceae) common. Non-native palms (e.g. *Livistona* spp., *Veitchia* spp.; Arecaceae) are abundant in both the canopy and the understory, but there are few (<10 indiv.) *Pritchardia* spp. in the Arboretum.

As elsewhere in Hawai'i, non-native vertebrates are common in the Arboretum's forests, and these vertebrates include at least three species of non-native rodents (*R. rattus*, *R. exulans*, *Mus musculus*; M. Wong, unpubl. data), mongoose (*Herpestes auropunctatus*), pig (*S. scrofa*) and many birds (e.g. passerines, doves, parrots). Feral cats and dogs have also been observed in the area, and pig hunting is common in the uninhabited parts of Mānoa Valley, including occasionally within the Arboretum. The only native vertebrates in and around the Arboretum include O'ahu 'amakihi (*Hemignathus flavus*) and 'apapane (*Himatione sanguinea*), which are both small forest birds (10–20 g) that are unlikely to be capable of dispersing native *Pritchardia* seeds. Carlquist (1974) noted that the seeds and fruits of the Hawaiian *Pritchardia* are much larger than those of the southwest

Pacific. Furthermore, there are no obvious contemporary native animal dispersers of Hawaiian *Pritchardia* seeds, yet Culliney *et al.* (2012) demonstrated that species such as the endangered crow (*Corvus hawaiiensis*), or perhaps its extinct congeners, could have been important dispersers prehistorically.

Within the 50 ha site, the focal area of the study included a 6-ha section of closed-canopy forest bordering State forestry land and located on the east side of the stream flowing from Aihualama Falls. This forest section was chosen because of the infrequency of visitation by people (mainly visitors to the Arboretum); however, the Arboretum staff visits the area ~3–4 times per year to remove weeds around some of the planted ornamentals.

Post-dispersal seed removal

Two *Pritchardia* species (*P. hillebrandii* and *P. maideniana*) were used in field experiments to assess fruit and seed attractiveness and removal from the forest floor by vertebrates. *Pritchardia maideniana* (endangered; syn. *P. affinis*; endemic to Hawai'i Island) has ripe fruits that are approximately twice the size (fresh mass: 6.15 ± 0.32 g; length × width: 2.48 ± 0.04 × 2.15 ± 0.04 cm; $N = 15$) of those of *P. hillebrandii* (species of conservation concern; fresh mass: 2.50 ± 0.09 g; length × width: 1.72 ± 0.02 × 1.69 ± 0.03 cm; $N = 20$; endemic to Moloka'i Island). For each species, 24 stations were established along three separate transects (48 stations and 6 transects in total). Owing to the availability of ripe fruit and treatment material (e.g. cameras and vertebrate exclusion material; see below), the two experiments occurred separately. Trials for *P. hillebrandii* began on 1 November 2005, and trials for *P. maideniana* began on 5 April 2006; each trial lasted 42 days. For each species, stations were at least 25 m apart on the forest floor, and one of the three treatment levels was randomly assigned ($n = 8$ for each treatment level): (i) no-vertebrate-access (NVA), which consisted of a wire metal-mesh (0.5 cm aperture) rectangular box (20 × 15 × 5 cm; length × width × height) that excluded all vertebrates (e.g. rodents, pigs, birds) and served as the control for subsequent treatments, (ii) small-vertebrate-access (SVA), which was composed of metal mesh (1 cm aperture) that enclosed a 20 × 20 × 20 cm (length × width × height) area but had an 8 × 8 cm opening on each side that allowed small vertebrates (e.g. rodents, possibly mongoose) to access the interior but excluded larger vertebrates and (iii) open forest floor (OPEN), where all animals were able to freely access the station. Each exclosure was held in place using 8-cm-long turf staples, and the open sites were marked with the same turf staples so that fruits could be easily relocated by the investigators. Although ground cover vegetation was variable, each microsite where fruit was placed was similar in soil type, soil depth and percentage rockiness.

Approximately 100 ripe fruits of *P. maideniana* and 300 ripe fruits of *P. hillebrandii* were collected from trees at the University of Hawai'i at Mānoa (UH) campus (~4 km from the study site). Two conspecific fruits were placed at each station (i.e. 48 fruits in total for each species). Fruit (and hereafter seed) removal was monitored periodically (1, 2, 6, 8, 13, 20, 22, 34, 42 days) over the course of the 6-week study for each species, and removed seeds were not replaced with fresh seeds. Additionally, two motion-sensing cameras (Bushnell brand) were installed on a randomly chosen OPEN and SVA station for each trial.

Potential seed dispersal

In an attempt to determine the fate of the *Pritchardia* seeds, 10 *P. maideniana* fruits from the same batch originally collected for the trials described above were set out singly on the forest floor between 20 November and 5 December 2005. Each fruit was attached to a spool of coloured thread by passing the thread through the pericarp and knotting the end before suspending the spool on a turf staple such that it would spin freely when pulled. After 48 h, the fruit was revisited, the thread was followed from its origin and the distance to the end of the thread was measured to estimate the distance that the fruit and seed was moved by the animal. *Pritchardia hillebrandii* was not assessed for potential seed dispersal.

Seedling predation in the field

In order to test the vulnerability of *Pritchardia* seedlings to vertebrates, we grew seedlings of *P. maideniana* and *P. hillebrandii* from fresh fruits/seeds collected from the same trees used in the seed removal trials. Ripe fruits were buried just below the soil surface in a potting soil: sand mixture (3 : 1 ratio) in pots placed outdoors in partial shade and watered approximately every second day. For *P. maideniana*, 84.6 % of seeds germinated ($N = 22$), and for *P. hillebrandii* 66.7 % of seeds germinated ($N = 24$); the average time to germination (number of days to first emergence of the shoot) was 94.4 ± 5.2 days for *P. maideniana* (*P. hillebrandii* not measured).

On 2 March 2006, 24 seedlings of *P. maideniana* (mean $\pm$ SE height: 17.9 ± 0.6 cm), and on 18 September 2006, 24 seedlings of *P. hillebrandii* (mean $\pm$ SE height: 16.7 ± 0.4 cm), were planted near to, and in the same three treatment levels as, the seed removal trials. The dimensions of the NVA were $20 \times 20 \times 20$ cm (length $\times$ width $\times$ height) and the metal mesh (1 cm aperture) that surrounded the seedling lacked a floor. Seedlings were randomly assigned treatment levels (NVA, SVA, OPEN; $n = 8$ for each) along transects and each station was at least 25 m from an adjacent seedling station.

Seedlings were ~4–5 months old and had at least two leaves when outplanted. Seedlings were monitored at least weekly for 3 months, then monthly for an additional 3 months, to assess seedling damage and mortality. Three motion-sensing cameras were randomly positioned at OPEN and SVA stations to monitor vertebrate visitation and seedling consumption during the first 3 months of the trials. All seedlings of both species were measured for height (base of the shoot to the furthest green structure, which was usually the tip of the longest leaf) at the time of outplanting, and 6 months after outplanting.

Seed and seedling predation trials with captive wild rats

We conducted a series of captive-feeding trials by offering either fresh fruits (and seeds) or seedlings of *P. maideniana* and *P. hillebrandii* to *R. rattus* individuals, using the methods described in Shiels and Drake (2011). For each trial, at least seven adult rats (*R. rattus*) were captured from wild populations in mesic forest sites within the Wai'anae Mountains, O'ahu, transported to a rodent housing facility at Lyon Arboretum, and held in $38 \times 22 \times 18$ cm metal-mesh cages (one rat per cage). Rats were allowed to acclimate for at least 1 week before beginning feeding trials and for at least 5 days between trials, during which time the rats were fed a diet of mixed seeds (e.g. corn, sunflower, wheat, barley, oats, sorghum) and fruit wedges (tangerine). Rats were checked daily in order to ensure ample food and fresh water.

Two feeding trials took place to assess seed predation, one for *P. hillebrandii* and one for *P. maideniana*. Each of the seven rats was offered three ripe fruits of *P. hillebrandii* (13 July 2007) and 18 days later two ripe fruits of *P. maideniana*. A dish of fresh water was always present in each rat's cage. Visual inspection occurred after 24 and 48 h to estimate the percentage of fruit and seed mass remaining. Seeds were classified as destroyed if the embryo was eaten or >50 % of the seed mass had been eaten (Shiels and Drake 2011).

Seedling damage by rats was assessed in two trials where 10 rats were offered a single seedling, each ~10–25 cm tall with one to two leaves, of either *P. hillebrandii* or *P. maideniana*. On each trial date (21 June 2008 and 11 July 2008), five rats received *P. maideniana* seedlings and five rats received *P. hillebrandii* seedlings such that no rat was exposed to the same species more than once. Each seedling, offered in a 10-cm-diameter and 10-cm-tall pot of soil, was placed in each rat's cage with a dish of water but no other food. After 24 h, each seedling was inspected and quantified (by visual estimate) for the amount of aboveground seedling mass consumed, and damaged, by each rat.

Statistical analysis

At the end of the 42-day post-dispersal seed removal trials, and at the end of the 6-month seedling predation trials in the field, Kruskall–Wallis tests (assumptions of analysis of variance were not met) were administered for each species to compare the proportion of seed removal (or seedling mortality) as a function of our vertebrate treatment (all vertebrates excluded, access to small vertebrates, and access to all animals). When the vertebrate treatment was significant, we performed multiple comparison tests to determine differences among means. For captive-feeding trials, the damage to seedlings was compared between plant species using Student's t-test upon meeting assumptions of parametric testing. Statistical analyses were performed using R version 2.12.0.

Results

Post-dispersal seed removal

Vertebrates readily removed *Pritchardia* seeds positioned on the forest floor. For *P. maideniana* and *P. hillebrandii*, all seeds remained in the NVA (control) (Fig. 1), which resulted in significant differences when the three treatment levels were compared for *P. hillebrandii* (d.f. = 2; $\chi^2 = 16.1$; $P < 0.001$; Fig. 1A) and *P. maideniana* (d.f. = 2; $\chi^2 = 16.9$; $P < 0.001$; Fig. 1B). Identical proportions of available seed (SVA and OPEN) were removed (87.5 %) at the end of the 6-week (42-day) study for *P. hillebrandii* (d.f. = 1; $W = 32$; $P > 0.999$; Fig. 1A), and the proportion of seed remaining after the same duration for *P. maideniana* was not significantly different between SVA and OPEN (d.f. = 1; $W = 20$; $P = 0.076$; Fig. 1B). The only animals photographed by the motion-sensing cameras were rats (probably *R. rattus*; Fig. 2), and the images of rats visiting seed of both species coincided with the removal of seed from the forest floor. Seed removal by vertebrates occurred rapidly, as indicated by at least 50 % seed removal within 6 days in stations accessible to vertebrates (SVA and OPEN) (Fig. 3). The rapid removal during the first week was followed by a slower, more gradual, removal over the next 2 weeks, whereas the final 3 weeks had a total of only two seeds removed for each species (Fig. 3).

Potential seed dispersal

Five of the 10 *P. maideniana* fruits that were attached to spools of threads were moved a distance of >10 cm, two fruits were moved only 15–20 cm and three fruits were moved >3 m. Fruits attached to the spools of thread were never recovered, and both moved and unmoved fruits appeared to have the thread chewed off where it

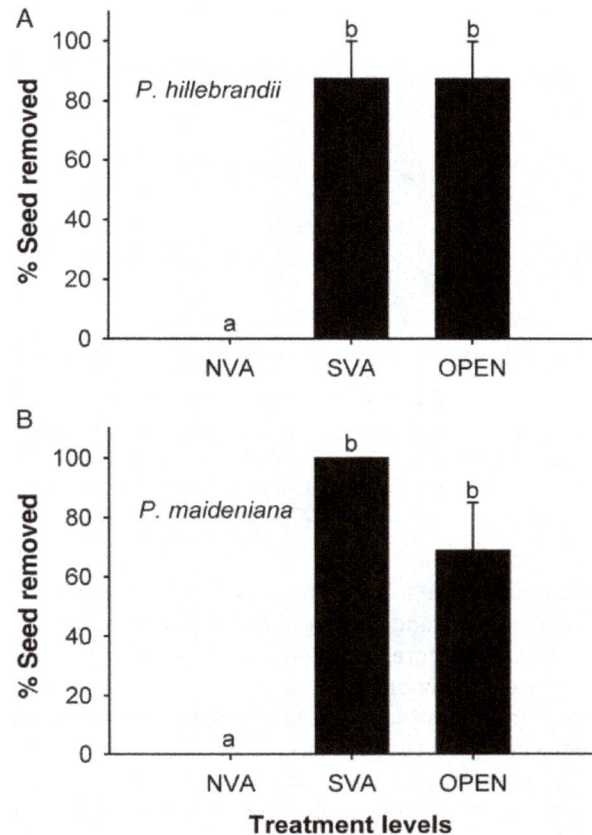

Figure 1. Mean ($\pm$ SE) seed removal of (A) *P. hillebrandii* and (B) *P. maideniana* from the forest floor after 42 days of study in Hawai'i. Different lowercase letters represent significant ($P < 0.05$) differences among treatment levels ($n = 8$ for each treatment level) within each species. NVA, no-vertebrate-access; SVA, small-vertebrate-access and OPEN, access for all animals.

Figure 2. Photograph taken by a motion-sensing camera depicting *R. rattus* removing fruit (and seed) of *P. hillebrandii* from an OPEN station, in Hawai'i wet forest (Lyon Arboretum). One characteristic that identifies the photographed rat as *R. rattus* is the very long tail (longer than the body; see Shiels *et al.* 2014).

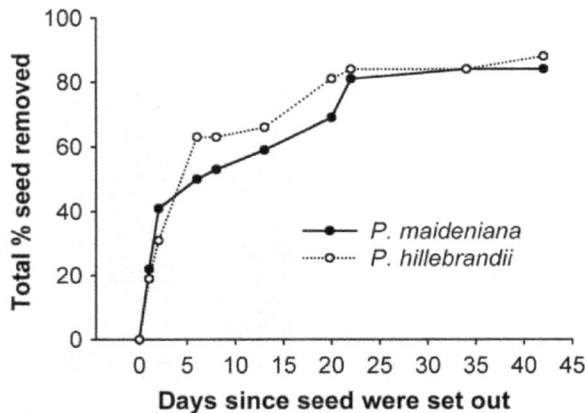

Figure 3. Percent *P. maideniana* and *P. hillebrandii* seed removal from the forest floor over 42 days of study in Hawai'i ($N = 32$ seeds/species). Seeds included here were those available to vertebrates (i.e. those in the SVA and OPEN and not in the NVA).

is attached to the fruit, or more commonly the fruit was pulled off the thread, leaving only the knot remaining on the string. Therefore, the measured fruit movement distances were likely underestimates of the true distances moved. The fruits that were not moved >10 cm had the threads caught on vegetation, or the spool was not functioning properly (not freely spinning when pulled) when revisited. The maximum distance that fruit was moved in a single trial was 8.05 m, whereas the average distance moved for those moved >10 cm was 3.28 ± 1.48 m. In all cases where fruit was moved, the thread revealed the movement was either lateral or uphill in reference to the sloping topography, and never up into trees. In one case the fruit was moved 3.4 m up a 30° slope. In all cases where fruit was moved >10 cm, the thread was taut, low to the ground and often under vegetation that was <30 cm high; these findings are consistent with the removal by a small (<30 cm tall) animal.

Seedling predation in the field

In the field, there was no evidence of seedling predation or herbivory damage to either *Pritchardia* species during the 6-month study. Seedling survival was high for *P. maideniana* (95.8 %) and *P. hillebrandii* (87.5 %). The only *P. maideniana* seedling that died was in the NVA, and there was no significant difference among treatment levels (d.f. = 2; $\chi^2 = 2.0$; $P = 0.368$). Similarly for *P. hillebrandii*, one of the three seedlings that died was in NVA while the other two were in the OPEN; there was no significant difference among treatment levels (d.f. = 2; $\chi^2 = 2.2$; $P = 0.335$). All dead *Pritchardia* seedlings were upright in the spot where they had been planted, fully intact (no evidence of herbivory) and dessicated. Several times during the experiments pig disturbance

was evident within the study site, as indicated by expanses of overturned soil, overturned exclosures and pig tracks found in close proximity (sometimes <30 cm) to the seedlings. Motion-sensing cameras also photographed pigs and rats in the vicinity and passing by the seedlings, but never closer than 30 cm from the seedlings. Both species of *Pritchardia* exhibited growth in the field; *P. maideniana* grew 8.4 ± 1.2 cm and *P. hillebrandii* grew 5.1 ± 0.6 cm in a 6-month period. Similar to field growth, *P. maideniana* grew 9.3 ± 0.8 cm and *P. hillebrandii* grew 4.2 ± 0.6 cm in pots on outdoor benchtops at UH during the equivalent 6-month period.

Seed and seedling predation trials with captive wild rats

Captive wild *R. rattus* readily consumed most seed tissue of both *Pritchardia* species, and such active feeding on the seed resulted in 100 % mortality (Table 1). In fact, when rats were offered *P. hillebrandii* fruit and seed, they had consumed ~97 % of the seed tissue within 24 h, at which point the trial was discontinued. The distinction between the 24 h *P. hillebrandii* trial and the 48 h *P. maideniana* trial may provide an explanation for the appearance of a greater amount of fruit mass consumed for *P. maideniana* (Table 1).

Very little (<5 % on average) of the aboveground seedling tissue was consumed by *R. rattus* in the captive-feeding trials, and there was no significant difference between plant species for average percentage of seedling damage (d.f. = 18; $t = 0.2$; $P = 0.838$; Table 2). However, it was common for rats to tip over the pots and uproot seedlings and displace soil; they would then occupy the inside of the empty pot. It was also common for the rats to clip leaves and stack them on the cage bottom and apparently use them for bedding.

Discussion

We examined contemporary impacts of vertebrates on *Pritchardia* palms to determine seed and seedling vulnerability to predation. Both *P. maideniana* and *P. hillebrandii* experience rapid removal of fruits and seeds from the floor of a wet forest in Hawai'i, and the vertebrate responsible for such removal and likely seed destruction (based on motion-sensing camera photos and captive-feeding trials) is the rat (*Rattus* spp., probably *R. rattus*). Therefore, these findings support the hypothesis put forward by palaeoecological evidence implicating rats as a significant factor responsible for the demise of *Pritchardia* forests in Hawai'i (Athens *et al.* 2002; Hunt 2007). A further finding of our study that is perhaps less amenable to testing by palaeoecological evidence is that *Pritchardia*

Table 1. Summary of *Rattus rattus* feeding trials involving fruit and seed of *Pritchardia*. Fruits of each of the two species of *Pritchardia* were individually offered to seven rats (three fruits [~7.5 g] per individual for *P. hillebrandii* and two fruits [~12 g] per individual for *P. maideniana*) for 48 h. The percentage of fruit mass and seed mass remaining was estimated visually; seed survival was based on the presence of an intact embryo or >50 % seed mass remaining. *All seeds were destroyed (i.e. zero survivors) after 24 h for *P. hillebrandii*.

Species	Range of fruit remaining (%)	Mean ± SE fruit remaining (%)	Range of seed remaining (%)	Mean ± SE seed remaining (%)	Mean ± SE seed survival (%)
*P. hillebrandii**	88–100	96.1 ± 2.0	0–15	3.0 ± 2.1	0.0 ± 0.0
P. maideniana	0–45	24.3 ± 7.5	0–4	0.7 ± 0.6	0.0 ± 0.0

Table 2. Summary of *Rattus rattus* feeding trials involving seedlings of *Pritchardia*. A seedling (~1.9 g above-ground tissue) of each of the two species of *Pritchardia* were individually offered to each of the 10 rats for 24 h. The percentage of seedling mass remaining and damaged was estimated visually.

Species	Range of mass remaining (%)	Mean ± SE mass remaining (%)	Range of damage (%)	Mean ± SE of damage (%)
P. hillebrandii	65–100	95.5 ± 3.4	0–95	58.0 ± 13.7
P. maideniana	65–100	96.5 ± 3.5	0–90	62.0 ± 13.6

is most vulnerable to rats at the seed stage, because seedlings did not suffer damage or mortality from rats or other common vertebrates in the field. Such findings also have direct implications for the conservation and restoration of *Pritchardia* in Hawai'i, by suggesting that restoration will best be achieved by preventing seed predation or by outplanting seedlings.

Post-dispersal fruit removal (and seed predation) of *P. maideniana* and *P. hillebrandii* occurs rapidly, and few (12.5 %) available fruits/seeds escape removal by animals before germination. The high fruit and seed removal rate (>50 % after 6 days; >80 % after 42 days) in this study is similar to past rodent studies in the tropics. In a seed removal experiment conducted in Los Tuxtlas, Mexico, 60–68 % of seeds of four common species were removed after 5 days, and further study revealed that small rodents were the main post-dispersal removal agents (Sánchez-Cordero and Martínez-Gallardo 1998). In Barro Colorado Island, Forget (1992) found that 85.5 % of *Gustavia superba* seeds were removed by animals in 28 days, which included 3.8 % gnawed by rodents and 47.5 % buried by agoutis. Studying a common tropical tree, Wenny (2000) found that 50 % of the total seeds destroyed (99.7 %) were attributable to rodents. In a study excluding ungulates in a dry forest in Hawai'i, Cabin et al. (2000) observed the seed crop and recruitment of many native species suffered from rodent predation. On Lord Howe Island in the South Pacific, fruit and seed removal from two native palms by *R. rattus* ranged from 54 % (*Hedyscepe canterburyana*) to 94 % (*Lepidorrhachis mooreana*) (Auld et al. 2010). Following captive-feeding trials in New Zealand, Daniel (1973) reported that *R. rattus* consumed fruit and destroyed the two offered seeds of the

native New Zealand palm (*Rhopalostylis sapida*). Seed removal rates by rodents also depend on the plant species examined, and Forget (1996) found species removal of seeds ranged from 25 to 95 %, whereas Hulme (1997) found 5–87 % removal among species. This pattern of a wide range of seed removal rates of native species was also documented with invasive rodents on Maui (0–100 % seed removal for four species; Chimera and Drake 2011) and O'ahu (15–85 % seed removal for eight species; Shiels and Drake 2011). Documentation of the negative effects of invasive *Rattus* spp. on native seeds has also occurred on many other islands (Meyer and Butaud 2009; Traveset et al. 2009; Wegmann 2009; Grant-Hoffman et al. 2010). Not only is a large range of fruits and seeds removed from the forest floor by animals, but also rodents are pervasive and are commonly responsible for such rapid removal rates.

Pritchardia seed removal by animals is patchy in space and time; seed in some locations was removed in hours or days, whereas in others it persisted for 42 days or more. The cause of the patchy seed removal remains unexplained, yet many factors can influence vertebrate foraging behaviour, including the density of vegetation cover, densities of conspecifics or other predators and available food supply. Several studies show seed predators prefer specific habitats. For example, in Los Tuxtlas, Mexico, rodent seed predators were less likely to visit and remove seeds from gaps than from primary and secondary forest (Sánchez-Cordero and Martínez-Gallardo 1998). In temperate regions, rodents completely avoided seeds in the open and foraged only under trees and shrubs (Hulme 1997). In our study, understory ground cover was variable (0–84 %), and this may have partially contributed to microhabitats differing in likelihood of seed remaining at the end of the study.

A critical component for predicting the positive and negative effects of seed removal is the ability to determine whether removed seeds are deposited in a condition that enables them to germinate (Forget *et al.* 2004). In this study, evidence from field and captive-feeding trials suggest that rats are destroying the seeds of *Pritchardia*. On rare occasions in which seeds are not destroyed, then the distance that seed was moved in this study (>8 m), as well as the microhabitat that the seed was deposited in, can have direct effects on the spatial distribution of plant regeneration. Vander Wall *et al.* (2005) studied seed movements by rodents in the high desert of Nevada and found that rodents carried seeds 2.5 ± 3.2 m (maximum 12 m) before caching them about 1 cm beneath the soil surface. Rats in Hawai'i are not known to cache seeds (Shiels and Drake 2011), but rat husking stations (i.e. food processing stations) are commonly observed in Hawai'i (Shiels and Drake 2011) and other Pacific islands (McConkey *et al.* 2003; Wegmann 2009). Therefore, although invasive *Rattus* commonly move seeds, seed storage is unlikely. The potential for animal-mediated seed dispersal of *Pritchardia* remains unlikely given evidence from this study and that of Pérez *et al.* (2008), which found that *R. rattus* rarely leaves viable seeds of *P. hillebrandii* and *P. kaalae* during captive-feeding trials. Although the fate of the diaspores could not be determined in our field study, evidence from captive-feeding trials suggests that the rats removing the seeds are most likely consuming and destroying them (e.g. 100 % mortality for both species).

Despite non-native vertebrates being present and foraging on the ground at the study site, it does not appear that vertebrates affect the seedling stage of these two species of *Pritchardia*. Motion-sensing cameras photographed pigs and rats in the vicinity of, and passing by, the seedlings. Perhaps in more arid environments, additional non-native vertebrates such as goats may threaten *Pritchardia* seedlings (Chapin *et al.* 2004; Mosher *et al.* 2007). It is more common for invasive *Rattus* to consume seeds rather than seedlings (Grant-Hoffman and Barboza 2010; Shiels *et al.* 2013, 2014). However, on subantarctic Macquarie Island, Shaw *et al.* (2005) demonstrated that *R. rattus* reduced initial seedling establishment and seedling survival of the megaherb *Pleurophyllum hookeri*. On Palmyra Atoll, tropical Pacific, Wegmann (2009) used motion-cameras to determine that *R. rattus* caused 57–100 % of seedling mortality for three native plant species (*Terminalia catappa*, *Tournefortia argentea* and *Cordia subcordata*) during a 27-day period. One explanation for the relatively high seedling herbivory by *R. rattus* on Macquarie Island and Palmyra Atoll may be a result of few alternative food resources because of seasonality (Macquarie; Shaw *et al.* 2005) or high competition with land crabs (Palmyra; Wegmann 2009). Protecting fruit and seed from invasive rats, as well as collecting, growing and outplanting *Pritchardia* seedlings may therefore be necessary conservation measures to sustain and expand remnant *Pritchardia* stands in the Hawaiian Islands.

Conclusions

Rat foraging on the forest floor of Hawai'i wet forest (Lyon Arboretum) is pervasive. The rapid removal of *P. maideniana* and *P. hillebrandii* fruit and seed is evidence that rats forage in a range of forest-floor habitats at high frequencies, and are capable of moving relatively large fruits at least 8 m from where they were encountered. Despite fruit freshness declining from environmental exposure, most fruits that were more than a week old were still found and moved by animals. Because *Pritchardia* seeds take several months to germinate (average of 3 months for *P. maideniana* enclosed in the fruit), the fruits/seeds on the forest floor are much more vulnerable to rat predation than species that germinate quickly. The evidence from this study reveals that rats can remove, and probably destroy, high abundances of *Pritchardia* seeds rapidly from the forest floor. These findings support the notion that invasive rats could have facilitated widespread transformation of native plant communities within a few hundred years. The findings of this study suggest that better understanding of rat foraging behaviour, fruit and seed handling, as well as seed fate of both common and rare species is needed to better assess the impacts of these seed foragers on insular plant community structure.

Sources of Funding

This research was supported by two Achievement Rewards for College Scientists (ARCS) scholarships to A.B.S., the Maybelle Roth Award for Conservation Biology and the Sarah Martin Award in Botany.

Contributions by the Authors

Both authors contributed to the project conception, experimental design and writing of the manuscript. A.B.S. also collected and analysed field and lab data.

Acknowledgements

We dedicate this paper to the late Ray Baker who dedicated much of his life to improving Lyon Arboretum, and who, as a palm enthusiast, provided frequent logistical support throughout this study. We also thank Clifford Morden and Mashuri Waite for providing additional logistical support, Michelle Akamine and Hector Peréz for materials to construct the exclosures and Alex Wegmann for loaning his motion-sensing cameras. Two anonymous reviewers and Peter Bellingham provided helpful comments on an earlier version of the manuscript.

Literature Cited

Abe T. 2007. Predator or disperser? A test of indigenous fruit preference of alien rats (*Rattus rattus*) on Nishi-jima (Ogasawara Islands). *Pacific Conservation Biology* 13:213–218.

Athens JS. 2009. *Rattus exulans* and the catastrophic disappearance of Hawai'i's native lowland forest. *Biological Invasions* 11: 1489–1501.

Athens JS, Toggle HD, Ward JV, Welch DJ. 2002. Avifaunal extinctions, vegetation change, and Polynesian impacts in prehistoric Hawai'i. *Archaeology in Oceania* 37:57–78.

Auld TD, Hutton I, Ooi MKJ, Denham AJ. 2010. Disruption of recruitment in two endemic palms on Lord Howe Island by invasive rats. *Biological Invasions* 12:3351–3361.

Barrios-Garcia MN, Ballari SA. 2012. Impact of wild boar (*Sus scrofa*) in its introduced and native range: a review. *Biological Invasions* 14:2283–2300.

Blackburn TM, Cassey P, Duncan RP, Evans KL, Gaston KJ. 2004. Avian extinction and mammalian introductions on oceanic islands. *Science* 305:1955–1958.

Cabin RJ, Weller SG, Lorence DH, Flynn TW, Sakai AK, Sandquist D, Hadway LJ. 2000. Effects of long-term ungulate exclusion and recent alien species control on the preservation and restoration of a Hawaiian tropical dry forest. *Conservation Biology* 14:439–453.

Campbell DJ, Atkinson IAE. 2002. Depression of tree recruitment by the Pacific rat (*Rattus exulans* Peale) on New Zealand's northern offshore islands. *Biological Conservation* 107:19–35.

Carlquist S. 1974. *Island biology.* New York: Columbia University Press.

Chapin MH, Wood KR, Perlman SP, Maunder M. 2004. A review of the conservation status of the endemic *Pritchardia* palms of Hawaii. *Oryx* 38:273–281.

Chimera CG, Drake DR. 2011. Could poor seed dispersal contribute to predation by introduced rodents in a Hawaiian dry forest? *Biological Invasions* 13:1029–1042.

Coblentz BE. 1978. The effects of feral goats (*Capra hircus*) on island ecosystems. *Biological Conservation* 13:279–286.

Cole RJ, Litton CM, Koontz MJ, Loh RK. 2012. Vegetation recovery 16 years after feral pig removal from a wet Hawaiian forest. *Biotropica* 44:463–471.

Courchamp F, Chapuis J-L, Pascal M. 2003. Mammal invaders on islands: impact, control and control impact. *Biological Reviews* 78: 347–383.

Culliney S, Pejchar L, Switzer R, Ruiz-Gutierrez V. 2012. Seed dispersal by a captive corvid: the role of the 'Alalā (*Corvus hawaiiensis*) in shaping Hawai'i's plant communities. *Ecological Applications* 22: 1718–1732.

Daniel MJ. 1973. Seasonal diet of the ship rat (*Rattus r. rattus*) in lowland forest in New Zealand. *Proceedings of the New Zealand Ecological Society* 20:21–30.

DeSteven D, Putz FE. 1984. Impact of mammals on early recruitment of a tropical canopy tree, *Dipteryx panamensis*, in Panama. *Oikos* 43:207–216.

Drake DR, Pratt LW. 2001. Seedling mortality in Hawaiian rain forest: the role of small-scale physical disturbance. *Biotropica* 33:319–323.

Forget P-M. 1992. Seed removal and seed fate in *Gustavia superba* (Lecythidaceae). *Biotropica* 24:408–414.

Forget P-M. 1996. Removal of seeds of *Carapa procera* (Meliaceae) by rodents and their fate in rainforest in French Guiana. *Journal of Tropical Ecology* 12:751–761.

Forget P-M, Lambert JE, Hulme PE, VanderWall SB. 2004. *Seed fate: predation, dispersal, seedling establishment.* Oxfordshire: CABI Publishing.

Grant-Hoffman MN, Barboza PS. 2010. Herbivory in invasive rats: criteria for food selection. *Biological Invasions* 12:805–825.

Grant-Hoffman MN, Mulder CP, Bellingham PJ. 2010. Invasive rats alter woody seedling composition on seabird-dominated islands in New Zealand. *Oecologia* 163:449–460.

Hone J. 1995. Spatial and temporal aspects of vertebrate pest damage with emphasis on feral pigs. *Journal of Applied Ecology* 32: 311–319.

Hulme PE. 1997. Post-dispersal seed predation and the establishment of vertebrate dispersed plants in Mediterranean scrublands. *Oecologia* 111:91–98.

Hunt TL. 2007. Rethinking Easter Island's ecological catastrophe. *Journal of Archaeological Science* 34:485–502.

Lockwood JL, Hoopes MF, Marchetti MP. 2007. *Invasion ecology.* Malden: Blackwell Publishing.

McConkey KR, Drake DR, Meehan HJ, Parsons N. 2003. Husking stations provide evidence of seed predation by introduced rodents in Tongan rain forests. *Biological Conservation* 109:221–225.

Meyer J-Y, Butaud J-F. 2009. The impacts of rats on the endangered native flora of French Polynesia (Pacific islands): drivers of plant extinction or *coup de grâce* species? *Biological Invasions* 11:1569–1585.

Mosher SM, Kawelo HK, Rohrer JL, Costello V, Burt MD, Keir M, Beachy J, Mansker M. 2007. Rat control for the protection of endangered species in the Waianae and Koolau Mountains on Oahu, Hawaii. *Rats, humans, and their impacts on islands: Integrating historical and contemporary ecology (conference).* University of Hawaii at Manoa, Honolulu, HI.

Murphy MJ, Inman-Narahari F, Ostertag R, Litton CM. 2014. Invasive feral pigs impact native tree ferns and woody seedlings in Hawaiian forest. *Biological Invasions* 16:63–71.

O'Connor SJ, Kelly D. 2012. Seed dispersal of matai (*Prumnopitys taxifolia*) by feral pigs, Sus scrofa. *New Zealand Journal of Ecology* 36: 228–231.

Pérez HE, Shiels AB, Zaleski HM, Drake DR. 2008. Germination after simulated rat damage in seeds of two endemic Hawaiian palm species. *Journal of Tropical Ecology* 24:555–558.

Prebble M, Dowe JL. 2008. The late Quaternary decline and extinction of palms on oceanic Pacific islands. *Quaternary Science Reviews* 27:2546–2567.

Sánchez-Cordero V, Martínez-Gallardo R. 1998. Postdispersal fruit and seed removal by forest-dwelling rodents in a lowland rainforest in Mexico. *Journal of Tropical Ecology* 14:139–151.

Scowcroft PG, Hobdy R. 1987. Recovery of goat-damaged vegetation in an insular tropical montane forest. *Biotropica* 19:208–215.

Shaw JD, Hovenden MJ, Bergstrom DM. 2005. The impact of introduced ship rats (*Rattus rattus*) on seedling recruitment and distribution of a subantarctic megaherb (*Pleurophyllum hookeri*). *Austral Ecology* **30**:118–125.

Shiels AB, Drake DR. 2011. Are introduced rats (*Rattus rattus*) both seed predators and dispersers in Hawaii? *Biological Invasions* **13**:883–894.

Shiels AB, Flores CA, Khamsing A, Krushelnycky PD, Mosher SM, Drake DR. 2013. Dietary niche differentiation among three species of invasive rodents (*Rattus rattus, R. exulans, Mus musculus*). *Biological Invasions* **15**:1037–1048.

Shiels AB, Pitt WC, Sugihara RT, Witmer GW. 2014. Biology and impacts of Pacific island invasive species. 11. *Rattus rattus*, the black rat (Rodentia: Muridae). *Pacific Science* **68**:145–184.

Singer FJ. 1981. Wild pig populations in the national parks. *Environmental Management* **5**:263–270.

Towns DR, Atkinson IAE, Daugherty CH. 2006. Have the harmful effects of introduced rats on islands been exaggerated? *Biological Invasions* **8**:863–891.

Traveset A, Richardson DM. 2006. Biological invasions as disruptors of plant reproductive mutualisms. *Trends in Ecology and Evolution* **21**:208–216.

Traveset A, Nogales M, Alcover JA, Delgado JD, López-Darias M, Godoy D, Igual JM, Bover P. 2009. A review on the effects of alien rodents in the Balearic (Western Mediterranean Sea) and Canary Islands (Eastern Atlantic Ocean). *Biological Invasions* **11**:1653–1670.

Vander Wall SB, Kuhn KM, Gworek JR. 2005. Two-phase seed dispersal: linking the effects of frugivorous birds and seed-caching rodents. *Oecologia* **145**:282–287.

Vitousek PM, D'Antonio CM, Loope LL, Rejmanek M, Westbrooks R. 1997. Introduced species: a significant component of human-caused global change. *New Zealand Journal of Ecology* **21**: 1–16.

Wagner WL, Herbst DR, Sohmer SH. 1999. *Manual of the flowering plants of Hawaii*, revised edn, Vol. 1 and 2. Honolulu: University of Hawaii Press.

Wegmann AS. 2009. *Limitations to tree seedling recruitment at Palmyra Atoll*. PhD Thesis, University of Hawai'i, Mānoa.

Wenny DG. 2000. Seed dispersal, seed predation, and seedling recruitment of a neotropical montane tree. *Ecological Monographs* **70**:331–351.

5

Towards a more holistic research approach to plant conservation: the case of rare plants on oceanic islands

Luís Silva[1]*, Elisabete Furtado Dias[1], Julie Sardos[2], Eduardo Brito Azevedo[3], Hanno Schaefer[4] and Mónica Moura[1]

[1] InBIO, Rede de Investigação em Biodiversidade, Laboratório Associado, CIBO, Centro de Investigação em Biodiversidade e Recursos Genéticos, Polo-Açores, Departamento de Biologia, Universidade dos Açores, 9501-801 Ponta Delgada, Açores, Portugal
[2] Bioversity-France, Parc Scientifique Agropolis II, 34397 Montpellier Cedex 5, France
[3] Research Center for Climate, Meteorology and Global Change (CMMG - CITA-A), Departamento de Ciências Agrárias, Universidade dos Açores, Angra do Heroísmo, Açores, Portugal
[4] Plant Biodiversity Research, Technische Universität München, D-85354 Freising, Germany

Guest Editor: Christoph Kueffer

Abstract. Research dedicated to rare endemic plants is usually focused on one given aspect. However, holistic studies, addressing several key issues, might be more useful, supporting management programmes while unravelling basic knowledge about ecological and population-level processes. A more comprehensive approach to research is proposed, encompassing: phylogenetics/systematics, pollination biology and seed dispersal, propagation, population genetics, species distribution models (SDMs), threats and monitoring. We present a holistic study dedicated to *Veronica dabneyi* Hochst. ex Seub., an endangered chamaephyte endemic to the Azores. *Veronica dabneyi* was mainly found associated with other endemic taxa; however, invasive plants were also present and together with introduced cattle, goats and rabbits are a major threat. Most populations grow at somewhat rocky and steep locations that appeared to work as refuges. Seed set in the wild was generally high and recruitment of young plants from seed seemed to be frequent. In the laboratory, it was possible to germinate and fully develop *V. dabneyi* seedlings, which were planted at their site of origin. No dormancy was detected and time for 50 % germination was affected by incubation temperature. Eight new microsatellite markers were applied to 72 individuals from 7 sites. A considerable degree of admixture was found between samples from the two islands Flores and Corvo, with 98 % of the genetic variability allocated within populations. Levels of heterozygosity were high and no evidence of inbreeding was found. Species distribution models based on climatic and topographic variables allowed the estimation of the potential distribution of *V. dabneyi* on Flores and Corvo using ecological niche factor analysis and Maxent. The inclusion of land-use variables only slightly increased the information explained by the models. Projection of the expected habitat in Faial largely coincided with the only historic record of *V. dabneyi* on that island. This research could be the basis for the design of a recovery plan, showing the pertinence of more holistic research approaches to plant conservation.

* Corresponding author's e-mail address: lsilva@uac.pt

Keywords: Azores; conservation; germination; population genetics; species distribution models; threats; *Veronica dabneyi*.

Introduction

Island endemic plants are among the most threatened group of organisms worldwide (Caujapé-Castells *et al.* 2010). Research dedicated to rare endemic plants is usually focused on one or few aspects, such as conservation genetics, propagation or distribution (e.g. Dubuis *et al.* 2013; Evans *et al.* 2014; Mir *et al.* 2014). Multidisciplinary studies, addressing several key issues, are much more useful but still the exception (e.g. Menges 1990; Halbur *et al.* 2014). They have the potential to provide science-based evidence to management or recovery programmes while at the same time unravelling basic knowledge about the population processes and the ecology of endangered species. We propose a general framework towards more holistic conservation research, particularly when devoted to rare plants on oceanic islands. We suggest that such approach should include the following critical areas: (i) phylogenetics/systematics (i.e. DNA sequences and morphology) to determine origin and close relatives as well as the existence of unaccounted taxa (Bateman *et al.* 2013); (ii) population genetics to estimate genetic structure and diversity (Dias *et al.* 2014), identify possible cases of inbreeding depression (Li *et al.* 2012) and ensure adequate provenance of propagation material (Silva *et al.* 2011; Hancock and Hughes 2014); (iii) germination biology and propagation methods to identify possible biological constraints (Beaune *et al.* 2013) and support the species recovery (Pence 2013); (iv) pollination biology and research on dispersal mechanisms to identify possible biological/ecological constraints (Ollerton *et al.* 2011; Rodríguez *et al.* 2015); (v) the identification of threats, including invasive plants (Foxcroft *et al.* 2013) and animals, particularly herbivores (Donlan *et al.* 2003; Garzón-Machado *et al.* 2010, Barrios-Garcia *et al.* 2014); (vi) species distribution models (SDMs) to determine environmental constraints and potentially favourable areas, possible impacts of climate change and of other anthropogenic alterations (Costa *et al.* 2012, 2013a; Marcer *et al.* 2013) and (vii) long-term monitoring (minimum 10 years, depending on the duration of the species' generation time) to evaluate population fluctuations and the effect of management actions (Godefroid *et al.* 2011). Besides addressing those biological issues, conservation research should also integrate possible societal aspects (i.e. the different stakeholders that will have a direct or indirect role in the conservation process).

Here we present an example of a multidisciplinary study, using *Veronica dabneyi* Hochst. ex Seub. (Plantaginaceae Juss.), known as 'veronica' or 'Azorean speedwell', a rare chamaephyte (subshrub) endemic to the Azores islands.

The genus *Veronica* L. is the largest of Plantaginaceae with ~450 species. It is distributed worldwide, with a large range of life forms from diverse habitats (Albach *et al.* 2005). Based on morphological affinities and preliminary molecular data, *V. dabneyi* is a close relative of *V. officinalis* L., belonging to the subgenus *Veronica* L., a clade also including *V. alpina* L. and *V. montana* L. (Albach and Meudt 2010).

Veronica dabneyi was first described by Karl C.F. Hochstetter in 1838, after a visit to Faial Island. Since the plant was cultivated as ornamental in the garden of the American consul, Charles William Dabney in Horta (Faial), Hochstetter attributed the specific name presently used. Seubert (1844) published the description with drawings in his *Flora Azorica*. Later, Watson (1870) reported *V. dabneyi* for São Miguel, Faial and Corvo Islands, without specifying the exact places where the species was found. Cunha and Sobrinho (1939) collected *V. dabneyi* in Faial, at the inner side of the Caldeira summit, the only record published during the 20th century (specimen in LISU, seen by HS). The species was later cited as extinct by Catarino *et al.* (2001). However, it turned out that the species was still extant in inaccessible parts of the western islands (Pereira *et al.* 2002). Those authors observed a number of mature plants on Flores and Corvo, but noticed that young plants were rare and seed production was low, probably due to predation by goats and rabbits, and to the occurrence of environmental disturbance (e.g. landslides, trampling by cattle). The species was found to be associated with a vegetation type described by Sjögren (1973) as *Festucetum jubatae*, dominated by *Festuca francoi* Fern.Prieto, C.Aguiar, E.Dias & M.I.Gut (Poaceae Barnhart). The conservation status was reevaluated according to the International Union for Conservation of Nature criteria and it was classified as extinct in Faial and as critically endangered in Flores and Corvo (Pereira *et al.* 2002), based on an area of occurrence of 63 km², with 16 subpopulations in Flores, and one in Corvo. In general, the populations only included a few individuals, while no individuals were found in the wild at Faial. *Veronica dabneyi* was thus classified as extinct in Faial and as critically endangered in Flores and Corvo. In a global analysis of the conservation status of Azorean indigenous species, Silva *et al.* (2009) suggested possible natural threats (storms, strong wind, landslides), biological limitations (isolation of populations), as well as threats of human origin (expansion of invasive plants and introduced herbivores, changes in land use) as causes of species decline.

Two aspects that should be addressed as important contributions to the species long-term survival are estimates of genetic diversity and population genetic structure. These are crucial for any sort of recovery plan for endangered plants, particularly in islands, where genetic diversity has been often expected to be lower than in mainland populations (e.g. Caujapé-Castells *et al.* 2008; Silva *et al.* 2011; Martins *et al.* 2013; Moreira *et al.* 2013; Moura *et al.* 2013; Dias *et al.* 2014).

Propagation measures, particularly those allowing the maintenance of the genetic variability, such as seed germination, are also critical in the recovery of endangered plants (e.g. Moura and Silva 2010; Martins *et al.* 2012; Moreira *et al.* 2012).

Modelling is nowadays a common approach for predicting species distributions. This is based on statistically or theoretically derived response surfaces that link the known distribution of a species to the pertinent environmental descriptors, allowing to estimate its potential distribution, as well as to determine the environmental factors limiting its range (Guisan and Zimmermann 2000). Species distribution modelling has been used in a wide range of applications (Elith and Leathwick 2009), including the evaluation and management of endangered species (e.g. Engler *et al.* 2004; Marcer *et al.* 2013).

Integrating the above aspects in the design of a multidisciplinary research programme is an example of a more holistic approach to be applied in plant conservation, as a basis for recovery plans. Here, we reanalyse the available data for *V. dabneyi*, and provide new information on germination rate and population genetic structure and diversity. We also analyse species distribution in order to understand what ecological factors might be constraining it. Due to the small size of the populations, and to the degree of isolation and fragmentation, we expect to find reduced levels of seed set, comparably low levels of genetic diversity and some degree of differentiation between the studied populations. Because it is rare, we expect the species to have a relatively restricted ecological niche and high levels of ecological specialization.

Methods

Sampling

Veronica dabneyi populations. *Veronica dabneyi* was searched in areas with potentially suitable habitat on the islands of Corvo, Flores and Faial (Fig. 1), between the years 2000 and 2014. The search was not successful on Faial but on Flores and Corvo a total of seven

(sub)populations were found. All cited locations here are recorded in the Atlantis database of the 'Azorean Biodiversity Portal' (Borges *et al.* 2010), including data from Pereira *et al.* (2002) and Schaefer (2003).

Demographic data. Three populations were studied in detail: Miradouro Craveiro Lopes and Tapada da Forcada (Flores) and Madeira Seca (Corvo). Variables measured included: individual number and size, stem length, number of inflorescences and inflorescence height (measured in loco), number of fruits per inflorescence, number of seeds per fruit, seed diameter and weight, using a digital calliper and an electronic scale (measured/counted at the laboratory). Fruits were sampled in July 2008 from Miradouro Craveiro Lopes, and in June 2010 form Madeira Seca and Tapada da Forcada.

Associated flora and invasive species. Characterization of the associated flora was based on the work by Pereira *et al.* (2002) and on data collected by the authors in 2010, using a 1 m² plot centred on *V. dabneyi* individuals or groups of individuals. The presence of plant invaders (see Silva *et al.* 2008) was recorded from the immediate surroundings of *V. dabneyi* populations. The presence of cattle, goats and rabbits was confirmed through direct visual observation of the animals or by the presence of characteristic faeces.

Plant material for genetic analysis. In 2008 and 2010, a collection of leaf material was carried out to complement the samples already available at the DNA bank collection of the AZB herbarium (Biology Department, Azores University). Depending on leaf size, one or two leaves per individual were collected and immediately stored in a plastic bag with silica gel. After drying, the leaves were vacuum sealed and stored in folders. The number of individuals sampled per site varied from 10 to 30 individuals depending on the population size. In total, 72 individuals from 7 different sites were sampled (Fig. 1, Table 1). The plant material obtained from the F1 of Miradouro Craveiro Lopes (i.e. obtained from the germinated seedlings) was also used for comparison with the mother population.

Propagation

Seed germination. Fresh seeds collected in 2008 and 2010, as well as 2-year-old seeds stored at room temperature, were used for germination tests. Germination tests were done in Petri dishes using growth chambers with automatic temperature control (error margin of ~1 °C) and a light period of 12 h per day, provided by six fluorescent lamps with a photosynthetic photon flux density (PPFD) of 19–22 μmol m^{-2} s^{-1}. The chambers were set to the incubation temperatures of 25/20, 20/15,

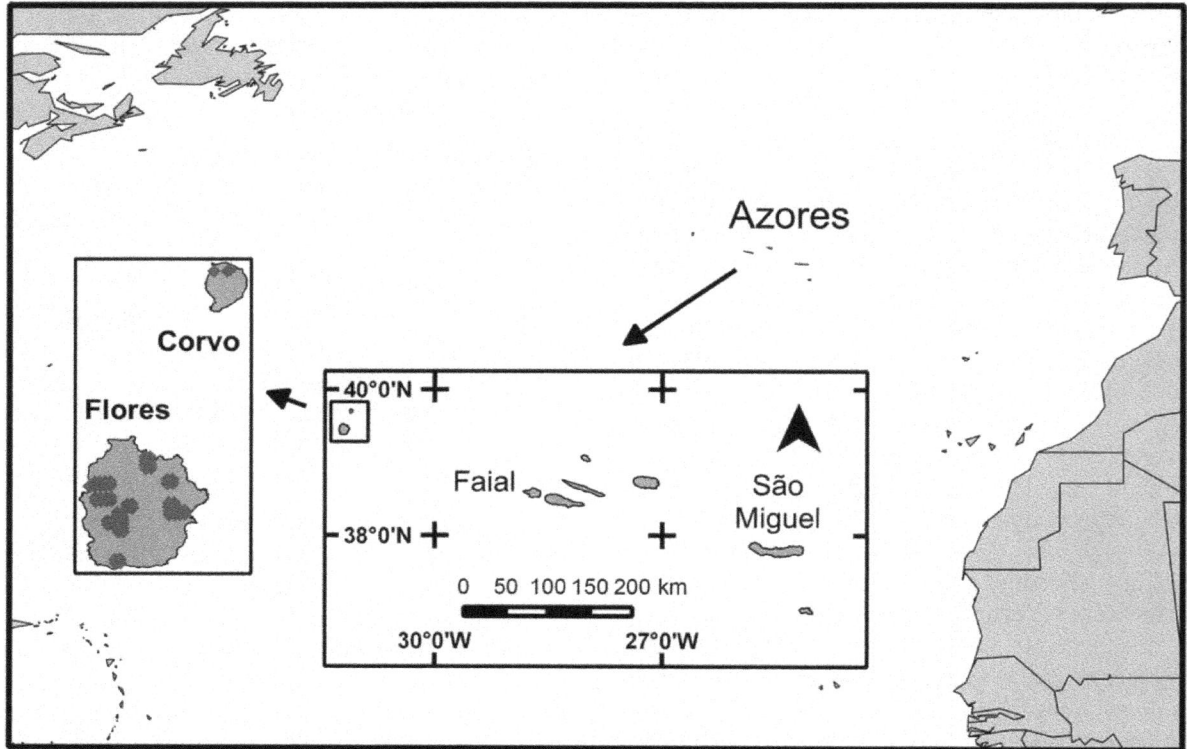

Figure 1. The location of the Azores Archipelago in the North Atlantic Ocean, and recorded presence of *Veronica dabneyi* in Flores and Corvo islands.

Table 1. Location and number of samples of *V. dabneyi* for population genetic analysis in the Islands of Flores and Corvo (Azores). Voucher code for the material deposited at AZB (Herbarium Ruy Telles Palhinha), population code used in the study, local designation for each site, elevation (m above sea level), universal transverse mercator coordinates (*X, Y*; WGS84 25S) and number of samples per site (*N*). Three individuals resulting from seed germination were also included (i.e. F1 Miradouro Craveiro Lopes).

Island	Voucher	Code	Site	Elevation	X	Y	N
Flores	VD-FLML-001	VDFLML	Miradouro Craveiro Lopes	482	651 717	4 365 683	26
Flores	VD-FLTP-001	VDFLTP	Tapada da Forcada	490	654 964	4 373 203	12
Corvo	VD-COMS-005	VDCOMS	Madeira Seca	442	662 989	4 398 153	18
Corvo	VD-COMC-002	VDCOMC	Caldeirão I	600	662 653	4 397 918	4
Corvo	VDCO1A	VDCO1	Caldeirão II	569	662 640	4 397 900	3
Corvo	VDCO3A	VDCO3	Arribas do Caldeirão	255	661 310	4 397 900	3
Corvo	VDCO2A	VDCO2	Arribas	255	661 330	4 397 900	3

15/10 and 10/5 °C, and the highest temperature coincided with the 12 h of the photophase. Germination proceeded in light or in darkness (Petri dish covered with aluminium foil) with three replicates per treatment and 17, 18 or 56 seeds per replicate, depending on the total number of seeds sampled at each site. Seeds were monitored daily and considered to be germinated when the radicle extruded. Seeds from the dark treatments were observed under a green light. Accumulated germination curves were adjusted to a Gompertz model to allow the calculation of the time, in days, necessary for 50 % seed germination (T50). The latter model has been successfully used to describe accumulated germination while allowing a biological interpretation (see Latera and Bazzalo 1999; Moura and Silva 2010).

Seedling growth and establishment. Germinated seedlings were planted in Jiffy® peat pellets in a growth chamber with regulated temperature (20 °C) and photoperiod (12 h of photophase) for 2 months, and regularly watered to avoid substrate desiccation. The seedlings were transported to Flores Island and planted near the mother plants at Miradouro Craveiro Lopes, and were followed up every 6 months during 2 years.

Genetic analysis

General DNA extraction. Deoxyribonucleic acid was extracted from dry leaves using a modified hexadecyl-trimethyl-ammonium bromide extraction method (Doyle and Dickson 1987) without the final ethanol wash. Deoxyribonucleic acid was then precipitated by adding 450 μL of isopropanol and re-suspended in 50 μL of pure water. The DNA quality and quantity were measured using a Nanodrop 2000 (Thermo Fisher Scientific) spectrophotometer. Samples were conserved at −20 °C until use.

Microsatellite development. Total DNA from fresh leaves of one individual of *V. dabneyi* was sent to the Savannah River Ecology Laboratory (University of Georgia, USA), where the enrichment procedure described in Glenn and Schable (2005), with the exceptions described in Lance *et al.* (2010), was followed for microsatellite isolation. CAP3 (Huang and Madan 1999) was used to assemble sequences at 98 % sequence identity using a minimal overlap of 75 bp. Search for microsatellite DNA loci was conducted using the programme MSATCOMMANDER version 0.8.1 (Faircloth 2008) and primers designed with Primer3 (Rozen and Skaletsky 2000). One primer from each pair was extended on the 5′-end with an engineered sequence (M13R tag 5′-GGAAACAGCTATGACCAT-3′) to enable the use of a third primer identical to the M13R (Schuelke 2000), and a GTTT 'pigtail' was added to the 5′-end of the untagged primer to facilitate accurate genotyping (Brownstein *et al.* 1996). Out of the 202 sequences of primer pairs provided by the Savannah River Ecology Lab, we selected 24 primer pairs, 12 with expected polymerase chain reaction (PCR) products ranging between 100 and 200 bp (A series) and 12 exhibiting expected PCR products ranging between 200 and 300 bp (B series) to allow later the multiloading of PCR products. All the primer pairs (with the tag sequence included) were selected on criteria of non-complementarities within and between primers, low secondary structures and 3′-end instability (Rychlik 1995).

Microsatellite selection and full-scale genotyping. All 24 primer pairs were tested on eight samples of *V. dabneyi* using a unlabelled tag primer (M13R) in a final volume of 25 μL consisting of 25 ng of DNA, 75 μg mL^{-1} BSA, 1× NH$_4$ buffer, 2 mM MgCl$_2$, 0.4 μM of untagged primer, 0.08 μM of tagged primer, 0.36 μM of Universal dyed M13R, 200 μM of dNTPs, 1 U of Immolase (Bioline) and using a Biometra TGradient thermocycler. Touchdown thermal cycling programmes (Don *et al.* 1991) encompassing a 10 °C span of annealing temperatures ranging between 63 and 53 °C were used for all loci. The PCR programme included the following steps: 95 °C for 7 min (hot start); 96 °C for 3 min; 20 cycles of 95 °C for 30 s, the highest annealing temperature of 63 °C (decreased by 0.5 °C per cycle) for 30 s, and 72 °C for 30 s; 20 cycles of 95 °C for 30 s, 53 °C for 30 s, and 72 °C for 30 s and finally 72 °C for 10 min for the final extension of the PCR products. Five microlitres of PCR products were then run on a 3.5 % agarose gel, stained with SafeView™ Classic Nucleic Acid Stain (ABM, Inc.) and visualized under UV to check for amplification, polymorphism and scorability of the bands. Ten primer pairs exhibited scorable amplified products of the expected length range and with at least two alleles.

After analysis of the quality of the PCR products obtained with the universal primer M13R, 10 primers with acceptable to high scorability were selected to run the complete study (Table 2). After optimization, the amplifications for the whole sample were performed using the protocol indicated above with the alterations presented in Table 3, and the M13R labelled either with PET, FAM, NED or VIC. Amplification products were diluted, multiloaded, run on an ABI-3130xl Genetic Analyzer and sized with LIZ500 size standard. The genotypes obtained were scored using the software GeneMarker V.1.97 Demo version (Softgenetics).

Analysis of genetic data. Population structure was analysed with GenAlEx 6.5 (Peakall and Smouse 2012), to obtain mean values per population of the total number of alleles, the number of alleles with a minimum allele frequency of 5 %, the number of effective alleles, the Shannon's Information Index, the number of private alleles, the expected heterozygosity, R_{st} and the estimation of gene flow. A principal coordinates analysis (PCoA) and an analysis of molecular variance (AMOVA) were also performed. Furthermore, we used a Bayesian approach to estimate the number of genetic clusters present in the whole sample. This model-based analysis was run with the software STRUCTURE version 2.3.3 (Pritchard *et al.* 2000), using a batch-oriented web programme package for construction of super matrices ready for phylogenomic analyses (Kumar *et al.* 2009). We ran 10 replicates for each *K* value ranging from 1 to 10 with a burn-in length of 50 000 followed by 500 000 iterations of each chain using the admixture model along with the assumption of correlated allele frequencies between groups (Falush *et al.* 2003). STRUCTURE then partitioned individuals of the sample according to the

Table 2. Set of simple sequence repeats primers that showed scorable and polymorphic products for the analysis of *Veronica dabneyi* population genetic structure in the Azores.

Primer	Motif	Forward	Reverse
Vd2B04	AC(9)	GTTTAGTGACGAGGACATTGATTG	GGAAACAGCTATGACCATCCTTCTAACATCGCAAACTG
Vd2B09	AC(10)	GTTTGCACACTGAAGGGTATCAAC	GGAAACAGCTATGACCATAAATCGGTGAATGTTTGATC
Vd3A01	AAG(9)	GTTTGTGTTCAGCTTGGAATTGAG	GGAAACAGCTATGACCATCTCTTCGACCAAATTCTTG
Vd3A03	ATC(18)	GGAAACAGCTATGACCATAAGTTCTTGCTCTGCTTGTC	GTTTCTTGTAGCCCAGATTGAAAC
Vd3A05	ATC(14)	GGAAACAGCTATGACCATCTAAACTCCCTTTCACTGG	GTTTGCGTCGAAGTACAAGAACAG
Vd3B07	AAG(15)	GTTTAGCTCGGAAACTTTGTAATG	GGAAACAGCTATGACCATGCAATAAAGTGATTAAGTGG
Vd4A01	AATG(6)	GTTTCCCACTATCCAACCATAATC	GGAAACAGCTATGACCATAACTCAGCTCAGCGTGAC
Vd4B01	AAAC(11)	GGAAACAGCTATGACCATAACCACATCACTCCAAACAG	GTTTGACTGGGCTAGAGTTGTC
Vd4B04	ACAT(20)	GTTTAATCCATTGTGTGCAGTCTC	GGAAACAGCTATGACCATCACCTCCCACACTTAATC
Vd4A03	AAAC(6)	GGAAACAGCTATGACCATGCTTTAATTTGTGCGTATC	GTTTCCTATCCCTTAACCTTTCTTC

Table 3. Optimization conditions used in the full-scale genotyping of *Veronica dabneyi*.

Primer	Optimization
Vd2B04	0.36 μM VIC, 0.5 U immolase
Vd2B09	0.36 μM NED, 0.5 U immolase
Vd3A01	0.36 μM FAM, 0.5 U immolase
Vd3A03	0.04 μM tagged primer, 0.36 μM FAM, 0.5 U immolase
Vd3A05	0.04 μM tagged primer, 0.36 μM VIC, 0.5 U immolase
Vd3B07	0.04 μM tagged primer, 0.36 μM NED, 0.5 U immolase, 50 ng DNA
Vd4A01	0.04 μM tagged primer, 0.36 μM NED, 0.5 U immolase
Vd4B01	0.36 μM VIC, 0.5 U immolase
Vd4B04	0.2 μM tagged primer, 0.2 μM FAM, 0.75 U immolase, 50 ng DNA
Vd4A03	0.36 μM PET, 0.5 U Immolase

membership coefficient Q, that ranges from 0 (lowest affinity to the group) to 1 (highest affinity to a group), across K groups. Estimation of the best K value was conducted with STRUCTURE Harvester (Earl and von Holdt 2012) following the Evanno *et al.* (2005) method. The optimal K repetitions were permuted in Clumpp version 1.1.2 (Jakobsson and Rosenberg 2007), using the Greedy algorithm, with results graphically represented using Distruct version 1.1 (Rosenberg 2004). The population matrix is available at DEMIURGE (http://www.demiurge-project.org/) with digest code D-NMICR-98.

Species distribution modelling

Modelling approaches. Since it is likely that the current distribution range on the studied islands is much reduced as a consequence of human activities (e.g. changes in land use and biological invasions), true absences are not available in this case. We thus opted to use modelling methods based on presences only, namely ecological niche factor analysis (ENFA) and maximum entropy modelling. Such an approach allows us not only to estimate the potential distribution and the habitat suitability for the species but also to identify the macroecological factors that might affect species distribution (e.g. altitude, climate, land use). The ENFA (Hirzel *et al.* 2002, 2006, 2007; Martinez *et al.* 2006; Hirzel and Le Lay 2008) provides smooth responses to environmental factors (Václavík and Meentemeyer 2012). This is desirable for modelling potential distributions, as models fitting complex responses may not accurately predict the distribution of species that are not at equilibrium. This approach was used successfully to model Azorean plant species, both invasive and native (Costa *et al.* 2012, 2013a; Moreira *et al.* 2014; Martins *et al.* 2015). Due to its wide application, Maxent was used for comparison (Phillips *et al.* 2004, 2006; Phillips and Dudík 2008).

Distribution data. The species presences (Fig. 1) recorded at the Atlantis data base (shape file format) were transformed into raster format at the same resolution as the ecogeographical variables (EGVs) for input in Biomapper (Idrisi raster format) and in Maxent (ASCI format).

Ecogeographical variables. As the number of presences was relatively low (<100), we followed Lomba *et al.* (2010) regarding the number of EGVs that should be used. We used three EGVs categories: climate, topography and land cover. Climatic variables were selected from the CIELO model (Azevedo 1996; Azevedo et al. 1999), a raster GIS environment with 100 m spatial resolution which models local scale climate variables relying on limited available data from synoptic coastal meteorological stations, and based on physical models that simulate the movement of air masses and their interaction with island topography [for more information, see http://www.climaat.angra.uac.pt or Azevedo (2003)]. We used the annual average of minimum, maximum, mean and range values of temperature (TMIN, TMAX, TM, TRAG), relative humidity (RHMIN, RHMAX, RHM, RHRAG) and precipitation (PMIN, PMAX, PM, PRAG). In addition, these climatic variables were submitted to a principal component analysis (PCA) as most of them were highly correlated. The principal components explaining more than 90 % of variance in the original variables were held and used alternatively. Those components, used as five alternative EGVs, corresponded to the first two components extracted from temperature (TPC1-2) and relative humidity variables (RHPC1-2), and to the first component extracted from precipitation variables (PCP). The topographic and land cover EGVs were acquired from the supporting data available in the CIELO model database, which matches the same spatial resolution of 100 m. To characterize the topography, we used the elevation (ELE) and the slope (SLP). Land cover was defined in six classes: (i) forest, (ii) natural vegetation, (iii) pasture, (iv) agriculture, (v) barren/bare areas and (vi) urban/industrial areas. We tested two different approaches for land cover data. The land cover classes were sorted in the foregoing order to define an ordinal land cover (OLC) variable, from 'like forest' (forest) to 'unlike forest' (urban/industrial areas). Moreover, distance variables were calculated for each land cover class (DLC1-6). Distance variables express the distance between the focal cell and the closest cell belonging to a given land cover class. In total, 26 EGVs were tested. A global model was calculated in Maxent including all available variables. Only variables providing more information to the model were kept for each of the following descriptors: elevation, temperature, relative humidity and rainfall. These four variable groups were differently combined with other physiographic descriptors: aspect, slope, flow accumulation and hill shade.

Modelling. Models were run with Maxent, and the best model was selected, based on the analysis of jackknife permutations and AUC (area under the curve). The same data were used to run the model in Biomapper. An R script was used to calculate the Boyce index and the Boyce curve, as well as to compare the habitat suitability maps obtained in the two methods, using the McNemar test (for further details on model validation and comparison, see Costa et al. 2012, 2013a, and references within). The best model was then changed to include one or two variables describing the distance to different types of land use. The best model was selected and compared as previously. The projection of the potential distribution of *V. dabneyi* in Faial Island was obtained using Maxent.

Results

Demographic analysis

Veronica dabneyi individuals usually grow linearly and horizontally, according to the procumbent nature of the stems. At Miradouro Craveiro Lopes we found 15 plants, with a stem length of 16-70 cm. Only 4 (27 %) showed inflorescences, ranging from 1 to 2, with 7.5-18 cm in height and 6-23 fruits per inflorescence (mean = 14.6, sd = 6.9). At Tapada da Forcada, we found 12 plants, with a stem length of 8-65 cm, and 5 seedlings. Only six (50 %) showed inflorescences, ranging from 1 to 15, with 8-21 cm in height and 8-31 fruits per inflorescence (mean = 19.1, sd = 5.8). At Madeira Seca (Corvo) we found 10 plants, with a stem length of 6-35 cm, and 1 seedling. Of those, 9 (90 %) showed inflorescences, ranging from 1 to 24, with 3-13.5 cm in height and 4-20 fruits per inflorescence (mean = 8.8, sd = 3.6).

In Flores the number of seeds per fruit ranged from 1 to 24 (mean = 10.4, sd = 6.3) while in Corvo, a larger variation was found (1-31 seeds, mean = 15.4, sd = 7.2). Seed diameter was similar on both islands (Flores: mean = 1.20 mm, sd = 0.13 mm; Corvo: mean = 1.09 mm, sd = 0.17 mm). However, the weight of 100 seeds ranged from 0.00049 g in Flores, to up to 0.00074 g in Corvo. Based on those values, we estimated seed production as 5870 ± 309.9 (mean ± se) seeds at Madeira Seca, 4969 ± 500.4 seeds at Tapada da Forcada and only 760 ± 76.6 at Miradouro Craveiro Lopes.

Associated flora and invasive species

The flora closely associated with the presence of *V. dabneyi* included mostly other endemic taxa (Table 4). However, surrounding *V. dabneyi* populations, *Hydrangea macrophylla* (Thunb.) Ser. (Hydrangeaceae Dumort.), one of the most problematic invasive species in Flores and Corvo Islands (Silva et al. 2008), was found. In Flores, the largest populations were located along steep road sides. In Corvo, the population of Madeira Seca is established on vertical walls of a volcanic chimney. At this location, the main threats identified were the presence of feral goats (Fig. 2A) and the proximity of *H. macrophylla* clumps (Fig. 2B).

Table 4. Vascular plants associated with *Veronica dabneyi*, observed by Pereira *et al.* (2002) and by the authors.

Taxa	Family	Life form	Origin	Pereira et al.	Authors
Blechnum spicant	Blechnaceae	Hemicryptophyte	Native	X	X
Calluna vulgaris	Ericaceae	Chamaephyte	Native		X
Centaurium scilloides	Gentianaceae	Chamaephyte	Endemic		X
Deschampsia foliosa	Poaceae	Hemicryptophyte	Endemic	X	X
Euphrasia azorica	Orobanchaceae	Chamaephyte	Endemic		X
Festuca francoi	Poaceae	Hemicryptophyte	Endemic	X	X
Frangula azorica	Rhamnaceae	Phanerophyte	Endemic		X
Hedera azorica	Araliaceae	Phanerophyte (scandent)	Endemic		X
Holcus rigidus	Poaceae	Hemicryptophyte	Endemic	X	X
Hypericum foliosum	Hypericaceae	Phanerophyte	Endemic		X
Ilex azorica	Aquifoliaceae	Phanerophyte	Endemic		X
Juniperus brevifolia	Cupressaceae	Phanerophyte	Endemic		X
Lotus pedunculatus	Fabaceae	Hemicryptophyte	Introduced		X
Luzula purpureosplendens	Juncaceae	Hemicryptophyte	Endemic		X
Lysimachia azorica	Primulaceae	Chamaephyte	Endemic	X	X
Myosotis azorica	Boraginaceae	Chamaephyte	Endemic		X
Picconia azorica	Oleaceae	Phanerophyte	Endemic		X
Rubia agostinhoi	Rubiaceae	Chamaephyte	Endemic		X
Rubus hochstetterorum	Rosaceae	Phanerophyte (scandent)	Endemic		X
Scabiosa nitens	Caprifoliaceae	Hemicryptophyte	Endemic	X	X
Selaginella kraussiana	Selaginellaceae	Hemicryptophyte	Native	X	X
Viburnum treleasei	Adoxaceae	Phanerophyte	Endemic		X
Woodwardia radicans	Blechnaceae	Hemicryptophyte	Native		X

Seed germination and seedling establishment

The germination percentage of the seeds collected at Craveiro Lopes in July 2008 was high for all temperature regimes (90–100 %), with faster germination occurring at the higher ones (Fig. 3). However, seed batches lost viability after 2 years of storage at room temperature, with no germination in 2010. Seeds collected in June 2010 at Tapada da Forcada and at Madeira Seca did not germinate; however, they were most likely not mature at the time of collection.

Population genetics

A total of 72 samples from 7 populations were analysed at 10 microsatellite loci. The results of the AMOVA indicated that the majority of the genetic variation was found within the populations (98 %), and only a small portion among populations (2 %). The permutation test showed that the R_{st} value (0.033) was not significant ($P = 0.160$) with an estimated gene flow of 7.249. Also,

the R_{it} value (-0.451) was significant and negative (Prand $\geq$ data, 1), revealing a lack of genetic population structure. The inbreeding coefficient R_{is} (-0.501) was significant and negative (Prand $\geq$ data, 1). Diversity patterns were similar across the sampled populations (Fig. 4). The average number of alleles ranged from 2.5 to 5, with the average number of private alleles per population below 1 (Fig. 4). Expected heterozygosity was somewhat homogeneous across the sampled populations, ranging from 0.46 to 0.59 (Fig. 4). The results obtained with the PCoA and STRUCTURE (Figs 5 and 6) showed a considerable degree of admixture, with three genetic clusters identified and represented at various degrees in all the sampled populations. Miradouro Craveiro Lopes individuals seem to encompass all the genetic variability found in the other populations of both Flores and Corvo (Figs 5 and 6). The F1 generation (i.e. resulting from seed germination) obtained from Miradouro Craveiro Lopes population showed genetic patterns compatible with those obtained in the source population (Fig. 6). The Corvo populations

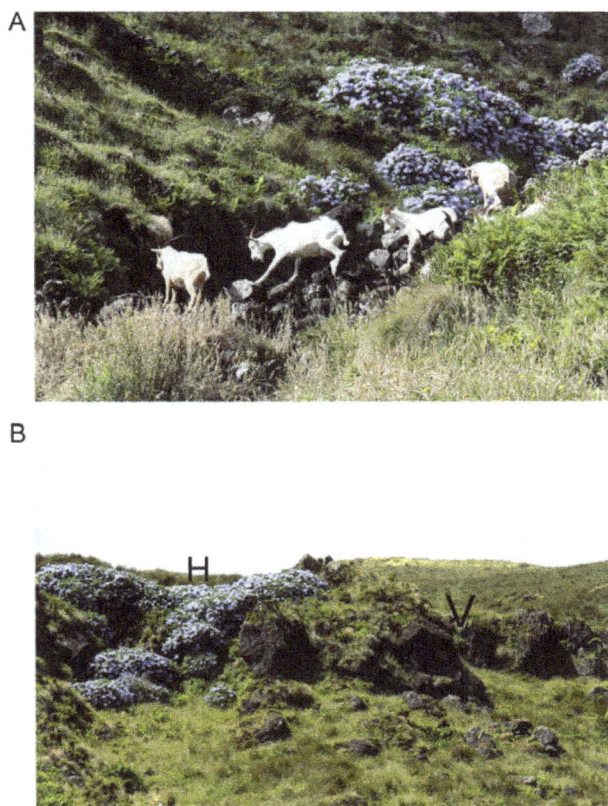

Figure 2. Two of the most relevant threats affecting *V. dabneyi* populations in the Azores: (A) feral goats free roaming and grazing very close to Madeira Seca in Corvo Island; (B) *Hydrangea macrophylla* clumps very close to a rock outcrop in Madeira Seca.

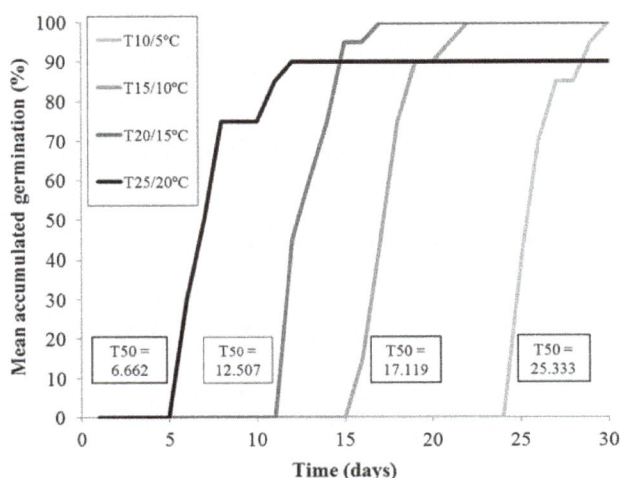

Figure 3. *Veronica dabneyi* accumulated seed germination curves obtained under 12 h of light and four alternating temperature regimes in Petri dishes in acclimatized chambers, after 30 days. Time for 50 % germination, in days, is shown for each treatment, based on the Gompertz model that was adjusted to each curve (all models with an R^2 of >0.9).

showed slightly different genetic patterns, at the exception of the Arribas population that is similar to those found in Flores (Fig. 6).

Species distribution modelling

The best distribution model found in Maxent (AUC = 0.840) also corresponded to a good model in Biomapper (total information explained = 0.852, for the first three factors; Boyce index = 0.990) (Fig. 7A). Although there was a slight increase in the AUC (0.865) and in the total information explained (0.871 for the first three factors; Boyce Index = 0.953), according to the shape of the Boyce curve, the best model including land-use information was not as good as the model based on physiographic and climatic data alone (Table 5, Fig. 7B). When considering only the cells with habitat suitability above the third quartile, the McNemar test did not show significant differences between the results obtained with Maxent and Biomapper (model without land-use data, $\chi^2 = 0.264$, $P = 0.607$; model with land-use data, $\chi^2 = 0.250$, $P = 0.617$; Fig. 8). Likewise, marginality and specialization gave similar results for both models (marginality 0.426 and 0.349; specialization 1.108 and 1.144). The low marginality indicated that the habitat actually occupied by the species is similar to the average conditions of the available habitat. The relatively low specialization suggested that the conditions for species occurrence were not narrow. The analysis of the score matrix for the best models showed a positive link with elevation and slope, which affect the niche of the species, a negative link with the temperature, a positive link with rainfall and a negative association with high relative humidity ranges. It also demonstrates a negative association with the distance to uncultivated areas and a positive association with the distance to cultivated areas (Table 5).

Discussion

Demographic analysis

The heterogeneity of plant sizes in the three study populations suggests that there is frequent recruitment of new individuals from seed. This is more evident at Tapada da Forcada (Flores), a population with an almost uniform distribution of life stages, including 30 % seedlings and 35 % seed-producing individuals. At Madeira Seca (Corvo), more than 80 % of the individuals were reproducing, whereas at Miradouro Craveiro Lopes individuals were usually large but produced fewer inflorescences, resulting in a much lower total seed set. The existence of larger individuals with low number of inflorescences might be associated with regular brush cutting (see below). The shortest plant with an inflorescence measured 6 cm, although fruit production was more common in

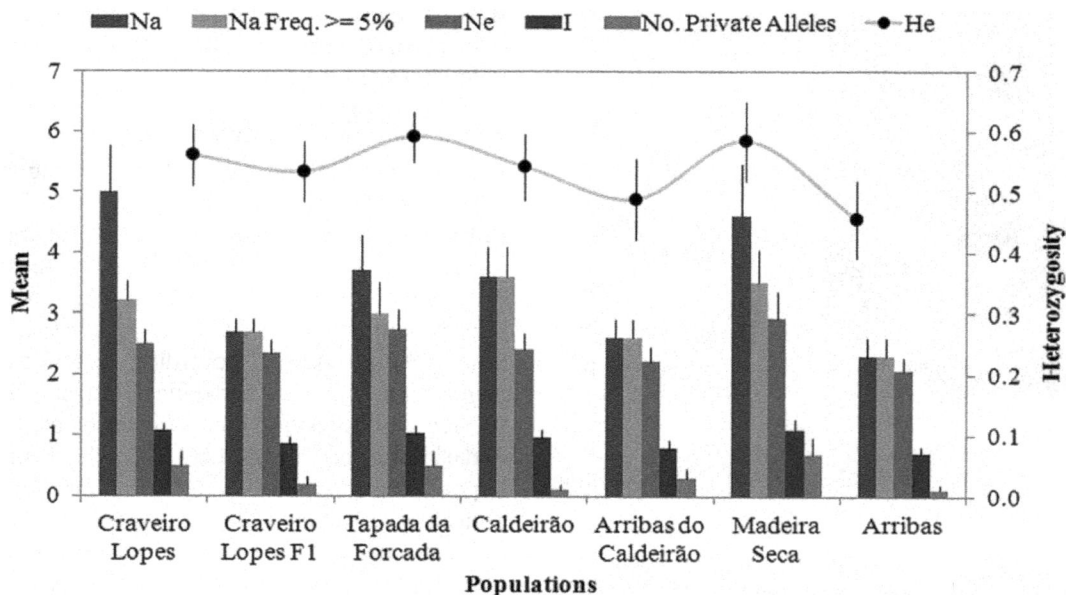

Figure 4. The mean allelic patterns across populations. Results of an analysis using GenAlEx on a total of 72 samples from 7 populations of *V. dabneyi* from Flores and Corvo islands (Azores) analysed with 10 microsatellite loci. Na, total number of alleles; Na (Freq ≥5 %), number of alleles with a frequency ≥5 %; Ne, number of effective alleles; I, Shannon's Information Index; No. Private Alleles, number of alleles unique to a single population. The line represents expected heterozygosity.

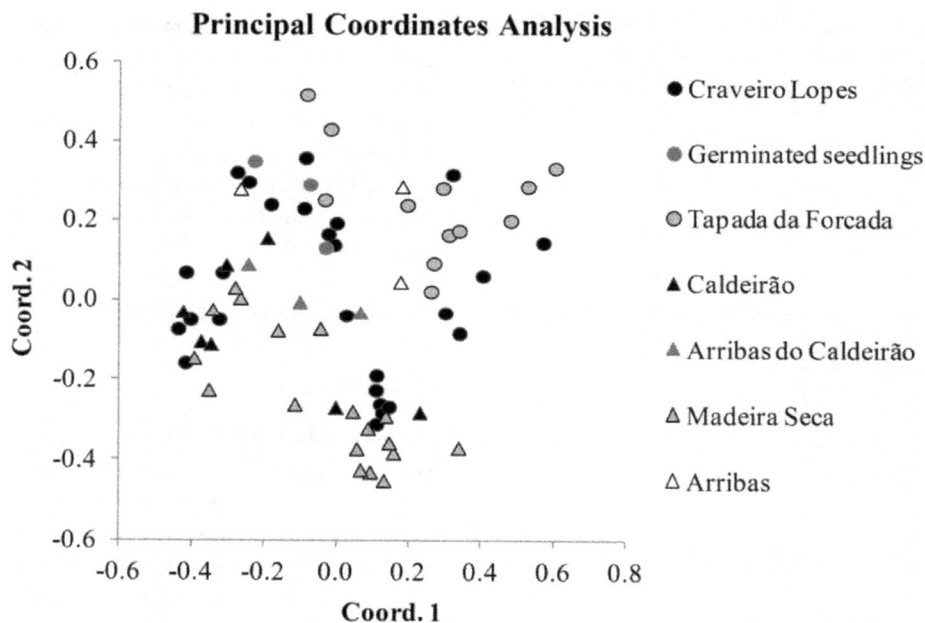

Figure 5. Graphic representation of the two first PCoA axes which explained 63.8 % of the detected genetic variation. Results obtained with GenAlEx for 10 microsatellite loci applied to 72 samples of *V. dabneyi* collected at six sites in Flores and Corvo islands, Azores. Germinated seedlings were derived from seeds obtained from Miradouro de Craveiro Lopes.

individuals with a length of ≥12 cm. This might imply that seed-producing plants can be relatively young. Despite the considerable variations in plant size and number of inflorescences, the estimated seed set for two of the studied populations was relatively high. The very small size/weight of the seeds, rapid germination, absence of dormancy and relatively fast decline on viability suggested that persistence as seed bank should be low

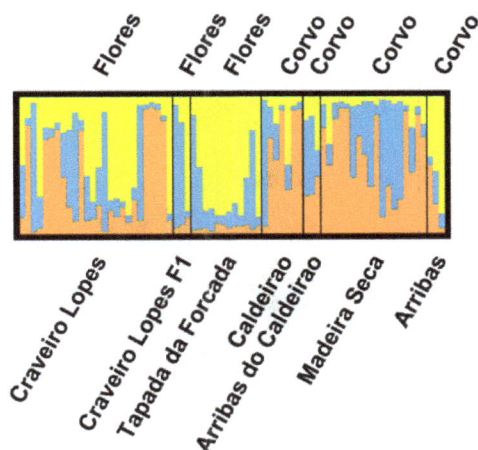

Figure 6. Results of the analysis of genetic clusters using STRUC-TURE and DISTRUCT on a total of 72 samples from 7 populations of *V. dabneyi* from Flores and Corvo islands (Azores) analysed with 10 microsatellite loci. The number of genetic clusters was estimated at three, using a model including admixture.

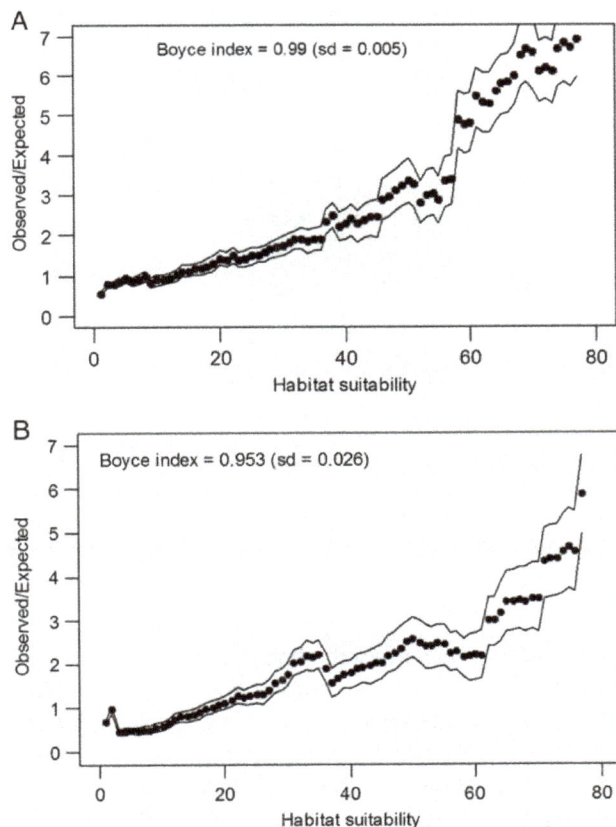

Figure 7. Boyce curve and Boyce index for the best distribution models for *V. dabneyi*, calculated using six EGVs in Biomapper: (A) the model with climatic and physiographic variables only (elevation, hillshade summer, annual rain fall, relative humidity annual range, annual mean temperature, slope); (B) the model including land-use information (elevation, annual rain fall, annual mean temperature, slope, distance to the nearest cell with agriculture, distance to the nearest cell with abandoned/unused land).

Table 5. Results of modelling *Veronica dabneyi* distribution in Flores and Corvo islands using ENFA (Biomapper). Score matrix of each variable for the first three extracted factors for best models including or not, information regarding land use.

EGV	Factors		
	1	2	3
Model without land use			
Elevation	0.50	0.58	−0.04
Hillshade Summer	−0.32	0.18	−0.57
Slope	0.28	0.25	−0.23
Mean annual temperature	−0.47	−0.17	−0.46
Annual rainfall	0.28	0.00	−0.57
Relative humidity annual range	−0.52	0.74	0.30
Model with land use			
Elevation	0.61	0.30	0.64
Mean annual temperature	−0.57	0.38	0.50
Annual rainfall	0.34	−0.01	0.08
Distance to agricultural land	0.21	0.45	−0.52
Distance to uncultivated land	−0.13	0.74	0.24
Slope	0.34	0.12	0.02

(Yu *et al.* 2007), and this was confirmed in our study by the absence of germination following 2 years of storage at room temperature.

Habitat and threats

Veronica dabneyi was found to be mainly associated with other endemic species. The conservation of native vegetation cover is thus of the utmost importance for the preservation of this species. It grows at sites with low vegetation cover mostly dominated by *Festuca francoi* and *Deschampsia foliosa* Hack. (Poaceae), generally found in forest openings and at steep locations such as waterfalls or road side slopes (Sjögren 1973). Therefore, the conservation of this type of vegetation, by avoiding changes in steep areas, will also be necessary. While populations of *V. dabneyi* were generally recorded on steep locations, the largest ones were found along steep road sides. On Flores, the population found at Miradouro Craveiro Lopes is located in front of a viewpoint, where brush cutting is regular. Changes in the road or at the top of the road side slope might affect this population. A similar situation was found at Tapada da Forcada, where the existing population occurs in a very steep slope, along the road. At some sites, the presence of *H. macrophylla* is a threat that should not be ignored, since this plant invader develops pure stands, and can outcompete all other plant species (Silva *et al.*

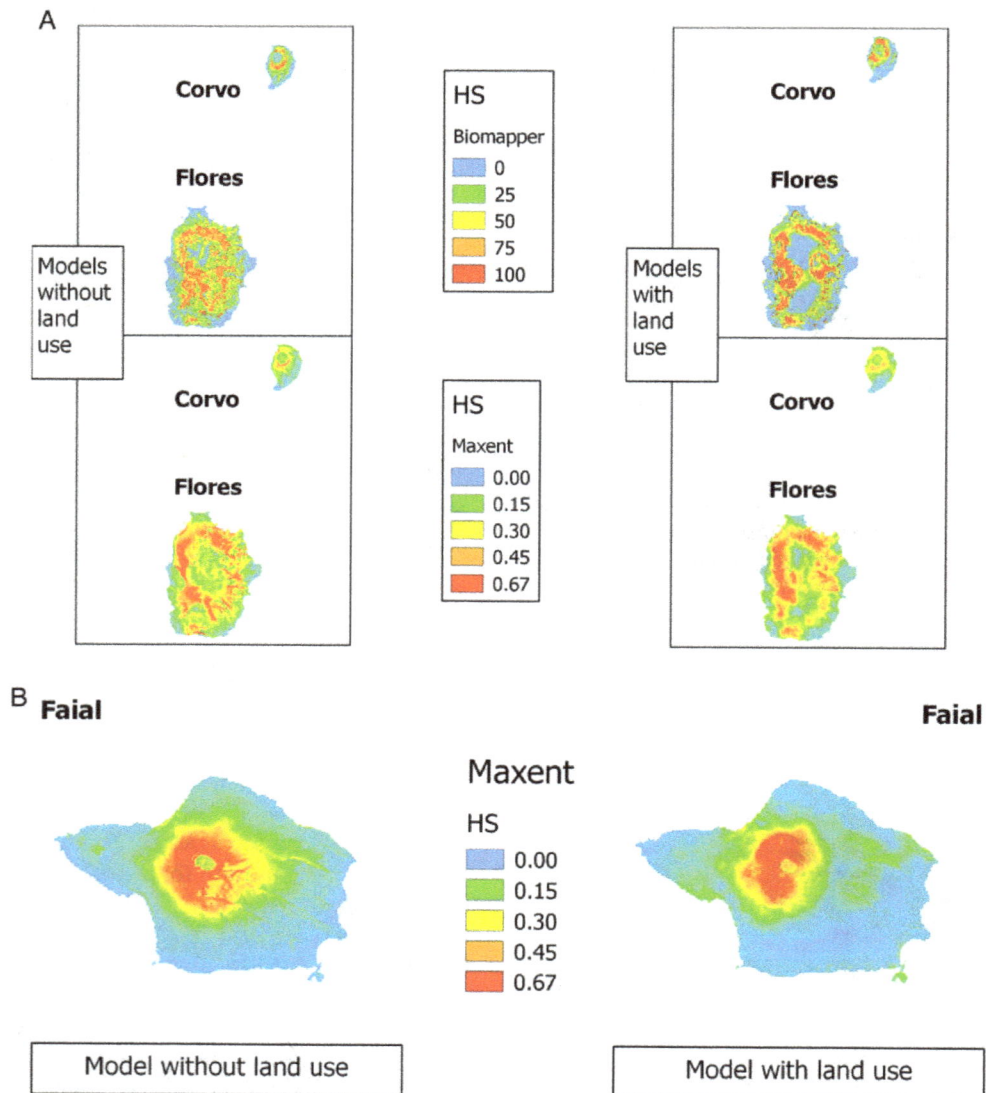

Figure 8. Habitat suitability maps for *V. dabneyi* in Corvo, Flores and Faial islands, calculated using six EGVs. (A) Models derived using Biomapper and Maxent for Flores and Corvo islands, either using physiographic and climatic variables only, or including land-use information also. (B) Projection of the potential habitat on Faial Island calculated using Maxent. Habitat suitability values were approximately divided according to the respective quartiles.

2008). In the Azores (Costa *et al.* 2013*b*), as in similar island groups and in many regions worldwide, plant invaders are commonly found within protected areas (see Foxcroft *et al.* 2013) and hydrangeas are even actively planted by Azorean farmers and the forest department as 'green fences'. The hydrangea shrub can form impenetrable stands and stop cattle from falling down cliffs. These cliff-top hydrangea plantations spread throughout the cliff sides, outcompeting endemic species that survive in these inaccessible refugia. Hydrangea clumps are very close to several *V. dabneyi* populations both in Flores and Corvo and might overgrow those populations in the near future. Thus, *H. macrophylla* clumps that are detected

near *V. dabneyi* occurrences should be considered as priority targets for removal/control measures, and its plantation on cliff tops should be banned. Instead, barbed wire fences should be preferred to keep cattle away from dangerous areas with no significant impact on natural vegetation. Other important herbaceous invasive species like *Hedychium gardnerianum* Ker Gawl. (Zingiberaceae Martinov) (Silva *et al.* 2008) should also be monitored at *V. dabneyi* sites.

Although direct evidence of *V. dabneyi* consumption by goats, cattle and rabbits is rare, its distribution at Caldeirão in Corvo Island, mostly on steep, rocky outcrops, suggests a possible retreat from areas more fully accessible to herbivores (Milchunas and Noy-Meir 2002). During our visits we found that cattle were free to roam at Caldeirão, even on sensitive vegetation like peat bogs that are important systems for water absorption and retention, and that goats were generally feral or unherded. This poses a big threat not only to V. dabneyi but also to other herbaceous endemic species (Houston et al. 1994; Silva et al. 2008) like Tolpis azorica (Nutt.) P.Silva (Asteraceae Bercht. & J.Presl) or Euphrasia azorica H.C.Watson (Scrophulariaceae Juss.), which are eaten in all accessible places. In the Azores it was not yet possible to use herbivore-exclusion plots, but this approach was used in the Canary Islands, showing that herbivores exert a strong negative effect on plant establishment, demanding the implementation of conservation measures, such as large fenced areas, control activities and eradication (Garzón-Machado et al. 2010). Meanwhile, the necessary changes in herbivore management clearly include a societal component, involving stakeholders that are not directly linked to conservation (e.g. hunters, farmers, the agriculture services).

Seed germination and seedling establishment

The fastest germination treatment occurred under the 25/20 °C temperature regime (T50 = 6.7 days), germination reaching 90 %. The rate of 100 % germination was obtained with a temperature regime of 20/15 °C, given the second best T50 (12.6 days). Similarly, high germination percentages were also obtained for the related Veronica arvensis L. (King 1975). For conservation purposes, seeds should be collected after full development of the fruits which generally occurs by mid-July, and promptly sown. It was possible to grow seedlings obtained from germination in the laboratory and to plant them at the site of origin. For the first 2 years, the plants were monitored every 6 months and it was found that the seedlings planted next to the mother population survived at a rate of ~50 %. Some early mortality due to transportation/establishment, and later mortality associated with roadside maintenance activities, was found. Furthermore, the genetic analysis of the germinated seedlings showed no considerable reduction in the genetic variability. Thus, the use of seeds might be a valuable option in a future recovery plan, in cases where population reinforcement might be preferred.

Population genetics

Contrary to our initial expectations that were based on the existence of small and possibly isolated populations, the levels of genetic diversity found in the populations were relatively high, as was the level of genetic admixture, with no evidence of inbreeding. Expected heterozygosity was similar to that found for other Azorean herbaceous endemic taxa (Dias et al. 2014), which are much more common than V. dabneyi. The concentration of most of the genetic variation within populations was similar to patterns found for some of the endemic trees of the archipelago (Martins et al. 2013; Moreira et al. 2013). In a study using amplified fragment length polymorphism (AFLP) markers with Veronica hederifolia L., a European species invasive in China, high levels of genetic diversity were also found, and most of the total variance was attributed to that within (76 %) rather than between the populations (24 %) (Wu et al. 2010). Regarding Hebe speciosa (R.Cunn. ex A.Cunn.) Andersen (Plantaginaceae), a threatened endemic New Zealand shrub, using AFLP markers it was found that there is negligible contemporary gene flow, and that some of the populations exhibited extremely low genetic diversity (Armstrong and De Lange 2005). In V. dabneyi, high levels of gene flow and genetic admixture among the sampled populations presently impede clear population differentiation, even between Flores and Corvo. It should be noted that flower morphology of V. dabneyi suggests an entomophily syndrome (Garnock-Jones 1976): many-flowered inflorescence well above the level of the leaves; the background colour of the corolla is generally lavender with darker guide marks radiating from this ring to the surrounding corolla lobes, especially the posterior lobe. The related Veronica chamaedrys L. is mainly pollinated by hoverflies and short-tongued bees. Other possible pollinators include the Ichneumonidae (Garnock-Jones 1976). We frequently observed a range of Diptera species on V. dabneyi flowers, especially syrphid flies, dung flies and small, unidentified dipterans. Even though they probably do not regularly cross the channel between Flores and Corvo, they could easily be blown between islands in stormy weather, thus transferring pollen from island to island. Alternatively, the present population genetic structure might be the result of a bottleneck effect, consequence of the fragmentation of a previously wider species distribution range that was reduced due to land-use changes (Arenas et al. 2012). In fact, our modelling results do suggest that the potential distribution of V. dabneyi could have been wider in the past. Still another possibility is that the population at the Miradouro Craveiro Lopes, which shares most of the sampled genetic diversity, could be the result of a human translocation of plant material, although written records are not available. In this regard, although seed germination might be a good option to reinforce the most depauperate populations, we do not support translocation of plants between different populations and, particularly, between different islands. The species seems to comprise one global meta-population with gene flow among populations, but there is no reason for translocation, which would be an artificial intervention in the natural gene flow patterns. Meanwhile, the occurrence of natural gene flow might be viewed as a positive factor for the conservation of *V. dabneyi*, ensuring the transfer of genetic information among populations, and avoiding extreme cases of inbreeding depression (Li *et al.* 2012).

Species distribution modelling

The macroecological factors modelled showed that, at this scale, *V. dabneyi* is neither a marginal nor a highly specialized species. Both modelling approaches used, ENFA and Maxent, provided similar results, showing that its potential distribution largely coincides with the inter-mediate elevation zone in Flores and with the Caldeirão zone in Corvo. While ENFA had already been used in the Azores to model invasive and native trees (Costa *et al.* 2012, 2013a; Moreira *et al.* 2014; Martins *et al.* 2015), this is a first result devoted to modelling the distribution of a rare plant in the Azores, further supported by the agreement obtained with Maxent. *Veronica dabneyi* was shown to prefer intermediate elevations, high slopes, relatively low temperature, high rainfall and small varia-tions in relative humidity. Land-use data did not increase the model quality in a sensible way but suggested the existence of a negative correlation with agricultural land. This largely coincides with the type of habitat known for *V. dabneyi*, suggesting that microenvironmen-tal factors like vegetation cover at a specific location or the presence of a rock outcrop might also be relevant for its establishment (Batik *et al.* 1992; Svenning 2001; Crain *et al.* 2014). It should also be stressed that, as stated above, *V. dabneyi* is mostly found associated with other native and endemic plants at sites with relatively low stature vegetation, which is frequently found at steep locations (e.g. volcanic craters, road side slopes). However, even those sites are not completely protected from herbi-vores (e.g. goats, rabbits) or human disturbance (e.g. road-side maintenance). Interestingly, the model based on Flores and Corvo occurrence data correctly predicted that *V. dabneyi* would have adequate habitat at the Faial Caldeira, in agreement with the previous record (Cunha and Sobrinho 1939).

Conservation measures

Since the species shows a considerable degree of genetic diversity, high seed production and high germination rate, conservation measures should be devoted to: (i) monitoring of natural populations to detect possible changes associated with human impacts; (ii) effective management of herbivores, especially feral goats, in Corvo and Flores Island Natural Parks, delimiting areas from which cattle should be prohibited and feral goats removed and (iii) the use of population circumscribed seed germination and seedling growth to recover the

most depauperate populations. Eventually, more field work should be directed to Faial Island, particularly in the whole Caldeira area, in order to detect a potentially still existing *V. dabneyi* population on that island. We hope that this study will stimulate the development of a scientifically based recovery plan for this species, while serving as a model to similar studies devoted to other rare or endangered endemic plant species worldwide.

Towards a more holistic approach to research in plant conservation

Although it is a common approach to dedicate attention to specific areas involved in plant conservation, we con-sider that a more holistic approach, devoted to multidis-ciplinary studies of endangered plants, should serve as basis for designing management or recovery plans. The latter are an important gap in the Azores where conserva-tion efforts dedicated to endemic plants do not follow integrated recovery plans, contrary to the situation, for example, in the Canary Islands (BOC 2012). In fact, more than 500 000 native plants are produced annually in the Azorean nurseries but their use does not follow approved recovery plans or strategies.

Why should conservation research follow a more holis-tic approach?

First, the absence of phylogenetic/systematic revues dedicated to endemic species might lead to erroneous conservation decisions or to the lack of action where it is needed. It was recently found that the most endan-gered plant in the Azores, known only from one location, is in fact an introduced species (Schaefer *et al.* 2011). In contrast, for several native genera, the number of taxa present in the islands is still unclear and often underesti-mated as shown by two recent studies that discovered overlooked endemic taxa with specific distribution pat-terns (Bateman *et al.* 2013; Moura *et al.* 2015). At another level, accumulated evidence for the population genetics of endemic trees in the Azores shows that levels of gen-etic diversity, and patterns of population structure, vary considerably among the evaluated taxa (Silva *et al.* 2011; Martins *et al.* 2013; Moreira *et al.* 2013; Moura *et al.* 2013), demanding a detailed study per taxon. More-over, the absence of long-term monitoring data will not only preclude the evaluation of recovery programmes (Godefroid *et al.* 2011), but also impede a sound evalu-ation of conservation status, which has then to be deter-mined based on distribution areas and not on observed population trends (Moreira *et al.* 2014; Martins *et al.* 2015). Monitoring is also linked to other relevant factors such as the detection of high mortality rates associated with herbivore pressure, making the propagation of high numbers of individuals an almost complete loss if no measures are taken to control predation (Donlan *et al.*

2003; Garzón-Machado *et al.* 2010). In the future, other aspects such as climate change will have to be integrated in long-term management or recovery programmes, making modelling approaches fundamental tools to support decision making (Fordham *et al.* 2012).

This more holistic approach can be accomplished by evaluating conservation status and possible management actions, based on a series of previous publications devoted to the target species (e.g. Moreira *et al.* 2014; Martins *et al.* 2015), or by developing multidisciplinary projects from the onset of the research programme, like it was done in the present paper.

Undoubtedly the need to address different aspects involved in the assessment and restoration of endangered plants arises directly from the Global Strategy for Plant Conservation, namely from its Objective 1 (*Plant diversity is well understood, documented and recognized*, CBD 2012). In our view, the different targets of this strategy will only be implemented if holistic approaches to research in plant conservation are effectively implemented in the near future.

Multidisciplinary studies like the one presented here, although sometimes longer, would increase the success of recovery and long-term maintenance of rare species, therefore improving the outcome of conservation investment, besides being a more powerful tool to halt plant extinctions.

Sources of Funding

This research was supported by project VERONICA, funded by the Azorean Government, and by project DEMIURGO, funded by MAC-TCP.

Contributions by the Authors

L.S. participated in sampling, performed species distribution modelling, conceived the idea for the paper and wrote the first draft; M.M., E.F.D. and J.S. were involved in the genetic analysis; M.M. performed the germination tests; E.B.A. provided climatic data; H.S. participated in field work. All authors helped revise the manuscript.

Acknowledgements

We are grateful to Hugo Costa for help in managing eco-geographical variables, to Selícia Cordeiro for help in the lab and to Raquel Alcaria, Graciete Maciel, José Martins, Orlanda Moreira, Mark Carine and Fred Rumsey for help in field work. We thank two anonymous reviewers who helped to improve the manuscript quality.

Literature Cited

Albach DC, Meudt HM. 2010. Phylogeny of *Veronica* in the Southern and Northern Hemispheres based on plastid, nuclear ribosomal and nuclear low-copy DNA. *Molecular Phylogenetics and Evolution* **54**:457–471.

Albach DC, Meudt HM, Oxelman B. 2005. Piecing together the "new" Plantaginaceae. *American Journal of Botany* **92**:297–315.

Arenas M, Ray N, Currat M, Excoffier L. 2012. Consequences of range contractions and range shifts on molecular diversity. *Molecular Biology and Evolution* **29**:207–218.

Armstrong TTJ, De Lange PJ. 2005. Conservation genetics of *Hebe speciosa* (Plantaginaceae) an endangered New Zealand shrub. *Botanical Journal of the Linnean Society* **149**:229–239.

Azevedo EB. 1996. Modelação do Clima Insular à Escala Local—Modelo CIELO Aplicado à Ilha Terceira. PhD Thesis, Universidade dos Açores, Angra do Heroísmo, Portugal.

Azevedo EB. 2003. *Projecto CLIMAAT – Clima e Meteorologia dos Arquipélagos Atlânticos*. PIC Interreg_IIIB-Mac2, 3/A3.

Azevedo EB, Pereira LS, Itier B. 1999. Modelling the local climate in island environments: water balance applications. *Agricultural Water Management* **40**:393–403.

Barrios-Garcia MN, Classen AT, Simberloff D. 2014. Disparate responses of above- and belowground properties to soil disturbance by an invasive mammal. *Ecosphere* **5**:art44.

Bateman RM, Rudall PJ, Moura M. 2013. Systematic revision of *Platanthera* in the Azorean archipelago: not one but three species, including arguably Europe's rarest orchid. *PeerJ* **1**:e218. doi: 10.7717/peerj.218.

Batik SK, Pandey HN, Tripathi RS, Rao P. 1992. Microenvironmental variability and species diversity in treefall gaps in a sub-tropical broadleaved forest. *Vegetatio* **103**:31–40.

Beaune D, Bollache L, Muganza MD, Bretagnolle F, Hohmann G, Fruth B. 2013. Artificial germination activation of *Dialium corbisieri* by imitation of ecological process. *Journal of Sustainable Forestry* **32**:565–575.

BOC. 2012. Decreto 329/2011, de 22 de diciembre, por el que se aprueba el Plan de Recuperación de la especie vegetal "flor de mayo leñosa" (*Pericallis hadrosoma*). *Boletín Oficial de Canarias núm. 2, Martes 3 de enero de 2012*:71–84.

Borges PAV, Gabriel R, Arroz AM, Costa A, Cunha RT, Silva L, Mendonça E, Martins AMF, Reis F, Cardoso P. 2010. The Azorean Biodiversity Portal: an internet database for regional biodiversity outreach. *Systematics and Biodiversity* **8**:423–434.

Brownstein MJ, Carpten JD, Smith JR. 1996. Modulation of non-templated nucleotide addition by Taq DNA polymerase: primer modifications that facilitate genotyping. *Biotechniques* **20**: 1004–1006, 1008–1010.

Catarino FM, Carvalho JA, Dias E, Fernandes F, Fontinha S, Jardim R, Roselló-Graell A. 2001. Acções de conservação da flora em Portugal. In: *Fundación Ramón Areces, ed. Conservación de Especies Vegetales Amenazadas En la Región Mediterránea Occidental*. Madrid: Editorial Centro de Estudios Ramón Areces, S.A., 63–92.

Caujapé-Castells J, Marrero-Rodríguez A, Baccarani-Rosas M, Cabrera-García N, Vilches-Navarrete B. 2008. Population genetics of the endangered Canarian endemic *Atractylis arbuscula* (Asteraceae): implications for taxonomy and conservation. *Plant Systematics and Evolution* **274**:99–109.

Caujapé-Castells J, Tye A, Crawford DJ, Santos-Guerra A, Sakai A, Beaver K, Lobin W, Vincent Florens FB, Moura M, Jardim R,

Gomes I, Kueffer C. 2010. Conservation of oceanic island floras: present and future global challenges. *Perspectives in Plant Ecology, Evolution and Systematics* **12**:107–129.

CBD. 2012. *Convention on biological diversity. Global strategy for plant conservation: 2011–2020.* Richmond, UK: Botanic Gardens Conservation International.

Costa H, Aranda SC, Lourenço P, Medeiros V, Azevedo EB, Silva L. 2012. Predicting successful replacement of forest invaders by native species using species distribution models: the case of *Pittosporum undulatum* and *Morella faya* in the Azores. *Forest Ecology and Management* **279**:90–96.

Costa H, Medeiros V, Azevedo EB, Silva L. 2013a. Evaluating the ecological-niche factor analysis as a modelling tool for environmental weed management in island systems. *Weed Research* **53**:221–230.

Costa H, Bettencourt MJ, Silva CMN, Teodósio J, Gil A, Silva L. 2013b. Invasive alien plants in the Azorean protected areas: invasion status and mitigation actions. In: Foxcroft LC, Pyšek P, Richardson DM, Genovesi P, eds. *Plant invasions in protected areas: patterns, problems and challenges.* Dordrecht: Springer, 375–394.

Crain BJ, Sánchez-Cuervo AM, White JW, Steinberg SJ. 2014. Conservation ecology of rare plants within complex local habitat networks. *Oryx*; doi:10.1017/S0030605313001245.

Cunha AG, Sobrinho LG. 1939. Estudos Botânicos no Arquipélago dos Açôres. *Revista da Faculdade de Ciências, Universidade de Lisboa* **1**:177–220.

Dias EF, Sardos J, Silva L, Maciel MGB, Moura M. 2014. Microsatellite markers unravel the population genetic structure of the Azorean *Leontodon*: implications in conservation. *Plant Systematics and Evolution* **300**:987–1001.

Don RH, Cox PT, Wainwright BJ, Baker K, Mattick JS. 1991. 'Touchdown' PCR to circumvent spurious priming during gene amplification. *Nucleic Acids Research* **19**:4008.

Donlan CJ, Croll DA, Tershy BR. 2003. Islands, exotic herbivores, and invasive plants: their roles in Coastal California restoration. *Restoration Ecology* **11**:524–530.

Doyle JJ, Dickson EE. 1987. Preservation of plant samples for DNA restriction endonuclease analysis. *Taxon* **36**:715–722.

Dubuis A, Giovanettina S, Pellissier L, Pottier J, Vittoz P, Guisan A. 2013. Improving the prediction of plant species distribution and community composition by adding edaphic to topo-climatic variables. *Journal of Vegetation Science* **24**:593–606.

Earl DA, von Holdt BM. 2012. STRUCTURE HARVESTER: a website and program for visualizing STRUCTURE output and implementing the Evanno method. *Conservation Genetics Resources* **4**:359–361.

Elith J, Leathwick JR. 2009. Species distribution models: ecological explanation and prediction across space and time. *Annual Review of Ecology, Evolution, and Systematics* **40**:677–697.

Engler R, Guisan A, Rechsteiner L. 2004. An improved approach for predicting the distribution of rare and endangered species from occurrence and pseudo-absence data. *Journal of Applied Ecology* **41**:263–274.

Evanno G, Regnaut S, Goudet J. 2005. Detecting the number of clusters of individuals using the software STRUCTURE: a simulation study. *Molecular Ecology* **14**:2611–2620.

Evans SM, Sinclair EA, Poore AGB, Steinberg PD, Kendrick GA, Vergés A. 2014. Genetic diversity in threatened *Posidonia australis* seagrass meadows. *Conservation Genetics* **15**:717–728.

Faircloth BC. 2008. MSATCOMMANDER: detection of microsatellite repeat arrays and automated, locus-specific primer design. *Molecular Ecology Resources* **8**:92–94.

Falush D, Stephens M, Pritchard JK. 2003. Inference of population structure using multilocus genotype data: linked loci and correlated allele frequencies. *Genetics* **164**:1567–1587.

Fordham DA, Resit Akçakaya H, Araújo MB, Elith J, Keith DA, Pearson R, Auld TD, Mellin C, Morgan JW, Regan TJ, Tozer M, Watts MJ, White M, Wintle BA, Yates C, Brook BW. 2012. Plant extinction risk under climate change: are forecast range shifts alone a good indicator of species vulnerability to global warming? *Global Change Biology* **18**:1357–1371.

Foxcroft LC, Pyšek P, Richardson DM, Genovesi P. 2013. *Plant invasions in protected areas: patterns, problems and challenges.* Dordrecht: Springer.

Garnock-Jones PI. 1976. Breeding systems and pollination in New Zealand *Parahebe* (Scrophulariaceae). *New Zealand Journal of Botany* **14**:291–298.

Garzón-Machado V, González-Mancebo JM, Palomares-Martínez A, Acevedo-Rodríguez A, Fernández-Palacios JM, Del-Arco-Aguilar M, Pérez-de-Paz PL. 2010. Strong negative effect of alien herbivores on endemic legumes of the Canary pine forest. *Biological Conservation* **143**:2685–2694.

Glenn TC, Schable NA. 2005. Isolating microsatellite DNA loci. *Methods in Enzymology* **395**:202–222.

Godefroid S, Piazza C, Rossi G, Buord S, Stevens A-D, Aguraiuja R, Cowell C, Weekley CW, Vogg G, Iriondo JM, Johnson I, Dixon B, Gordon D, Magnanon S, Valentin B, Bjureke K, Koopman R, Vicens M, Virevaire M, Vanderborght T. 2011. How successful are plant species reintroductions? *Biological Conservation* **144**:672–682.

Guisan A, Zimmermann NE. 2000. Predictive habitat distribution models in ecology. *Ecological Modelling* **135**:147–186.

Halbur MM, Sloop CM, Zanis MJ, Emery NC. 2014. The population biology of mitigation: impacts of habitat creation on an endangered plant species. *Conservation Genetics* **15**:679–695.

Hancock N, Hughes L. 2014. Turning up the heat on the provenance debate: testing the 'local is best' paradigm under heatwave conditions. *Austral Ecology* **39**:600–611.

Hirzel AH, Le Lay G. 2008. Habitat suitability modelling and niche theory. *Journal of Applied Ecology* **45**:1372–1381.

Hirzel AH, Hausser J, Chessel D, Perrin N. 2002. Ecological-niche factor analysis: how to compute habitat-suitability maps without absence data? *Ecology* **83**:2027–2036.

Hirzel AH, Le Lay G, Helfer V, Randin C, Guisan A. 2006. Evaluating the ability of habitat suitability models to predict species presences. *Ecological Modelling* **199**:142–152.

Hirzel AH, Hausser J, Perrin N. 2007. *Biomapper 4.0.* Lausanne, Switzerland: Laboratory for Conservation Biology, University of Lausanne.

Houston DB, Schreiner EG, Moorhead BB. 1994. *Mountain goats in Olympic National Park: biology and management of an introduced species.* Scientific Monograph NPS/NROLYM/NRSM-94/25. Port Angeles, WA: United States Department of the Interior.

Huang X, Madan A. 1999. CAP3: a DNA sequence assembly program. *Genome Research* **9**:868–877.

Jakobsson M, Rosenberg NA. 2007. CLUMPP: a cluster matching and permutation program for dealing with label switching and multimodality in analysis of population structure. *Bioinformatics* **23**:1801–1806.

King TJ. 1975. Inhibition of seed germination under leaf canopies in *Arenaria serpyllifolia*, *Veronica arvensis* and *Cerastum holosteoides*. *New Phytologist* **75**:87–90.

Kumar S, Skjæveland A, Orr RJS, Enger P, Ruden T, Mevik B, Burki F, Botnen A, Shalchian-Tabrizi K. 2009. AIR: a batch-oriented web program package for construction of supermatrices ready for phylogenomic analyses. *BMC Bioinformatics* **10**:357.

Lance SL, Light JE, Jones KL, Hagen C, Hafner JC. 2010. Isolation and characterization of 17 polymorphic microsatellite loci in the kangaroo mouse, genus *Microdipodops* (Rodentia: Heteromyidae). *Conservation Genetic Resources* **2**:139–141.

Latera P, Bazzalo ME. 1999. Seed-to-seed allelopathic effects between two invaders of burned pampa grasslands. *Weed Research* **39**:297–308.

Li Y-Y, Guan S-M, Yang S-Z, Luo Y, Chen X-Y. 2012. Genetic decline and inbreeding depression in an extremely rare tree. *Conservation Genetics* **13**:343–347.

Lomba A, Pellissier L, Randin C, Vicente J, Moreira F, Honrado J, Guisan A. 2010. Overcoming the rare species modelling paradox: a novel hierarchical framework applied to an Iberian endemic plant. *Biological Conservation* **143**:2647–2657.

Marcer A, Sáez L, Molowny-Horas R, Pons X, Pino J. 2013. Using species distribution modelling to disentangle realised versus potential distributions for rare species conservation. *Biological Conservation* **166**:221–230.

Martinez I, Carreño F, Escudero A, Rubio A. 2006. Are threatened lichen species well-protected in Spain? Effectiveness of a protected areas network. *Biological Conservation* **133**:500–511.

Martins JM, Moreira OCB, Rainha NFP, Baptista JAB, Silva L, Moura MMT. 2012. Morphophysiological dormancy and germination in seeds of the Azorean tree *Picconia azorica*. *Seed Science and Technology* **40**:163–176.

Martins JM, Moreira OCB, Sardos J, Maciel MGB, Silva L, Moura MMT. 2013. Population genetics and conservation of the Azorean tree *Picconia azorica*. *Biochemical Systematics and Ecology* **49**:135–143.

Martins J, Costa H, Moreira O, Azevedo EB, Moura M, Silva L. 2015. Distribution and conservation status of the endangered Azorean tree *Picconia azorica*. *International Journal of Biological Sciences and Applications* **2**:1–9.

Menges ES. 1990. Population viability analysis for an endangered plant. *Conservation Biology* **4**:52–62.

Milchunas DG, Noy-Meir I. 2002. Grazing refuges, external avoidance of herbivory and plant diversity. *Oikos* **99**:113–130.

Mir BA, Mir SA, Koul S. 2014. In vitro propagation and withaferin A production in *Withania ashwagandha*, a rare medicinal plant of India. *Physiology and Molecular Biology of Plants* **20**:357–364.

Moreira O, Martins J, Silva L, Moura M. 2012. Seed germination and seedling growth of the endangered Azorean cherry *Prunus azorica*. *HortScience* **47**:1222–1227.

Moreira O, Costa H, Martins J, Azevedo EB, Moura M, Silva L. 2014. Present and potential distribution of the endangered tree *Prunus lusitanica* subsp. *azorica*: implications in conservation. *International Journal of Biological Sciences and Applications* **1**:190–200.

Moreira OCB, Martins JM, Sardos J, Maciel MGB, Silva L, Moura MMT. 2013. Population genetic structure and conservation of the Azorean tree *Prunus azorica* (Rosaceae). *Plant Systematics and Evolution* **299**:1737–1748.

Moura M, Silva L. 2010. Seed germination of *Viburnum treleasei* Gand., an Azorean endemic with high ornamental potential. *Propagation of Ornamental Plants* **10**:129–135.

Moura M, Silva L, Caujapé-Castells J. 2013. Population genetics in the conservation of the Azorean shrub *Viburnum treleasei* Gand. *Plant Systematics and Evolution* **299**:1809–1817.

Moura M, Silva L, Dias EF, Schaefer H, Carine M. 2015. A revision of the genus *Leontodon* (Asteraceae) in the Azores based on morphological and molecular evidence. *Phytotaxa* **210**:24–46.

Ollerton J, Winfree R, Tarrant S. 2011. How many flowering plants are pollinated by animals? *Oikos* **120**:321–326.

Peakall R, Smouse PE. 2012. GenAlEx 6.5: genetic analysis in Excel. Population genetic software for teaching and research—an update. *Bioinformatics* **28**:2537–2539.

Pence VC. 2013. In vitro methods and the challenge of exceptional species for Target 8 of the Global Strategy for Plant Conservation. *Annals of the Missouri Botanical Garden* **99**:214–220.

Pereira JC, Schaefer H, Paiva J. 2002. New records of *Veronica dabneyi* Hochst. (Scrophulariaceae), an Azorean endemic plant not collected since 1938. *Botanical Journal of the Linnean Society* **139**:311–315.

Phillips SJ, Dudík M. 2008. Modeling of species distributions with Maxent: new extensions and a comprehensive evaluation. *Ecography* **31**:161–175.

Phillips SJ, Dudík M, Schapire RE. 2004. A maximum entropy approach to species distribution modeling. In: *Proceedings of the 21st International Conference on Machine Learning*, Banff, Canada.

Phillips SJ, Anderson RP, Schapire RE. 2006. Maximum entropy modeling of species geographic distributions. *Ecological Modelling* **190**:231–259.

Pritchard JK, Stephens M, Donnelly P. 2000. Inference of population structure using multilocus genotype data. *Genetics* **155**:945–959.

Rodríguez B, Siverio F, Siverio M, Barone R, Rodríguez A. 2015. Nectar and pollen of the invasive century plant *Agave americana* as a food resource for endemic birds. *Bird Study* **62**:232–242.

Rosenberg NA. 2004. Distruct: a program for the graphical display of population structure. *Molecular Ecology Notes* **4**:137–138.

Rozen S, Skaletsky H. 2000. Primer3 on the WWW for general users and for biologist programmers. *Methods in Molecular Biology* **132**:365–386.

Rychlik W. 1995. Selection of primers for polymerase chain reaction. *Molecular Biotechnology* **3**:129–134.

Schaefer H. 2003. Chorology and diversity of the Azorean flora. *Dissertationes Botanicae* **374**:1–130.

Schaefer H, Carine MA, Rumsey FJ. 2011. From European priority species to invasive weed: *Marsilea azorica* (Marsileaceae) is a misidentified alien. *Systematic Botany* **36**:845–853.

Schuelke M. 2000. An economic method for the fluorescent labeling of PCR fragments: a poor man's approach to genotyping for research and high-throughput diagnostics. *Nature Biotechnology* **18**:233–234.

Seubert M. 1844. *Flora Azorica quam ex collectionibus schedisque Hochstetteri patris et filii elaboravit et tabulis XV propria manu aeri incisis illustravit Mauritius Seubert*. Bonnae: Adolphum Marcum.

Silva L, Ojeda-Land E, Rodriguez-Luengo J-L. 2008. *Invasive terrestrial flora and fauna of Macaronesia. Top 100 in Azores, Madeira and Canaries*. Ponta Delgada: ARENA.

Silva L, Martins M, Maciel G, Moura M. 2009. *Azorean vascular flora. Priorities in conservation*. Ponta Delgada: Amigos dos Açores & CCPA.

Silva L, Elias RB, Moura M, Meimberg H, Dias E. 2011. Genetic variability and differentiation among populations of the Azorean endemic gymnosperm *Juniperus brevifolia*: baseline information for a conservation and restoration perspective. *Biochemical Genetics* **49**:715–734.

Sjögren J. 1973. Recent changes in the vascular flora and vegetation of the Azores Islands. *Memórias da Sociedade Broteriana* **22**:5–453.

Svenning J-C. 2001. On the role of microenvironmental heterogeneity in the ecology and diversification of neotropical rain-forest palms (Arecaceae). *The Botanical Review* **67**:1–53.

Václavík T, Meentemeyer RK. 2012. Equilibrium or not? Modelling potential distribution of invasive species in different stages of invasion. *Diversity and Distributions* **18**:73–83.

Watson HC. 1870. Botany of the Azores. In: Godman FC, ed. *Natural history of the Azores or Western Islands*. London: Taylor & Francis, 113–288.

Wu H, Qiang S, Peng G. 2010. Genetic diversity in *Veronica hederifolia* (Plantaginaceae), an invasive weed in China, assessed using AFLP markers. *Annales Botanici Fennici* **47**:190–198.

Yu S, Sternberg M, Kutiel P, Chen H. 2007. Seed mass, shape, and persistence in the soil seed bank of Israeli coastal sand dune flora. *Evolutionary Ecology Research* **9**:325–340.

6

Space, time and aliens: charting the dynamic structure of Galápagos pollination networks

Anna Traveset[1]*, Susana Chamorro[1,4], Jens M. Olesen[2] and Ruben Heleno[3]

[1] Laboratorio Internacional de Cambio Global (LINC-Global), Institut Mediterrani d'Estudis Avançats (CSIC-UIB), C/Miquel Marqués 21, 07190-Esporles, Mallorca, Balearic Islands, Spain
[2] Department of Bioscience, Aarhus University, Ny Munkegade 114, DK-8000 Aarhus C, Denmark
[3] Centre for Functional Ecology, Department of Life Sciences, University of Coimbra, Calçada Martim de Freitas, 3000-456 Coimbra, Portugal
[4] Present address: Universidad Internacional SEK, Facultad de Ciencias Ambientales, Calle Alberto Einstein y 5ta transversal, Quito, Ecuador

Guest Editor: Donald Drake

Abstract. Oceanic archipelagos are threatened by the introduction of alien species which can severely disrupt the structure, function and stability of native communities. Here we investigated the pollination interactions in the two most disturbed Galápagos Islands, comparing the three main habitats and the two seasons, and assessing the impacts of alien plant invasions on network structure. We found that the pollination network structure was rather consistent between the two islands, but differed across habitats and seasons. Overall, the arid zone had the largest networks and highest species generalization levels whereas either the transition between habitats or the humid habitat showed lower values. Our data suggest that alien plants integrate easily into the communities, but with low impact on overall network structure, except for an increase in network selectiveness. The humid zone showed the highest nestedness and the lowest modularity, which might be explained by the low species diversity and the higher incidence of alien plants in this habitat. Both pollinators and plants were also more generalized in the hot season, when networks showed to be more nested. Alien species (both plants and pollinators) represented a high fraction ($\sim 56\%$) of the total number of interactions in the networks. It is thus likely that, in spite of the overall weak effect we found of alien plant invasion on pollination network structure, these introduced species influence the reproductive success of native ones, and by doing so, they affect the functioning of the community. This certainly deserves further investigation.

Keywords: Alien plants; alien pollinators; biological invasions; global change; mutualistic interactions; oceanic islands.

* Corresponding author's e-mail address: atraveset@imedea.csic-uib.es

Introduction

Sexual reproduction is an essential step for the life cycle of most plant species, and is chiefly limited by the quantity and quality of pollen grains arriving to their stigmas (Ashman *et al.* 2004). Pollination is thus a critical step in plant reproduction, and many animals, mostly insects, have a vital role in facilitating this step in ~90 % of the worlds' plant species (Ollerton *et al.* 2011).

Islands harbour a disproportionate part of the worlds' biological diversity and are particularly rich in endemic and threatened species (Sax and Gaines 2008). Oceanic islands, in particular, are generally characterized by low insect diversity (Gillespie and Roderick 2002) and simplified pollination networks when compared with mainland systems (Olesen and Jordano 2002; Traveset *et al.* 2015b). This low abundance and diversity of pollinators on islands is likely to translate into reduced pollinator redundancy, potentially leading to highly vulnerable communities when faced with disturbances, e.g. El Niño Southern Oscillations (Traveset and Richardson 2006). Ecological networks offer a most valuable solution to evaluate the overall changes in community structure and function as a response to disturbances affecting species composition (Bascompte 2010; Heleno *et al.* 2014). For example, in order to survive in such low diversity ecosystems, some animal species that successfully colonize isolated islands tend to broaden their trophic niches, thus interacting with more species (mutualistic partners or prey) than their continental counterparts (Carlquist 1974; Olesen *et al.* 2002). This expansion of the feeding niche can characterize entire island communities, a phenomena coined 'interaction release' and that tends to have a stabilizing effect on insular interaction networks (Traveset *et al.* 2015a). Apart from their low diversity and high generalization when compared with continental communities, oceanic island interaction networks tend to be characterized by an increased nestedness, i.e. an ordered interaction distribution pattern where specialist species interact with specific sub-sets of the partners of most generalist species (Olesen and Jordano 2002; Padrón *et al.* 2009; Kaiser-Bunbury *et al.* 2010; Traveset *et al.* 2013). While increased generalization and nestedness may increase network stability (Sebastián-González *et al.* 2015), overall low biodiversity and the existence of small endemic populations suggest high species vulnerability at least to some specific sources of disturbance, such as invasive species (Berglund *et al.* 2009; Traveset and Richardson 2014).

In this study we focus on the impacts of alien plants on pollination networks. Biological invasions are a growing threat to the worlds' biodiversity (Lambertini *et al.* 2011) and particularly worrying on oceanic islands, where the arrival of alien species frequently triggers serious disruptive

effects on the intricate network of interactions established between native species throughout their shared evolutionary history (Kaiser-Bunbury *et al.* 2011; Traveset *et al.* 2014). Specifically, applying a network approach to frame biological invasions at the community level is particularly suitable for clarifying how invasive species can integrate into the existing interaction networks, the likely consequences for community structure, and the consequences for the most vulnerable species (Memmott *et al.* 2007; Bascompte 2009; Traveset and Richardson 2014). In fact, if pollination networks vary naturally in space and time, it is likely that opportunities for alien pollinators and plants to 'infiltrate' those networks will also vary in space and time. Thus, aliens may find particularly favourable biotic and abiotic conditions under which their integration into the native communities (and potential invasion) is more likely. Recent studies have begun to evaluate the temporal and spatial variability of pollination network structure (e.g. Olesen *et al.* 2008; Petanidou *et al.* 2008; Dupont *et al.* 2009); however, we are only starting to understand such patterns, how they are related to each other and, particularly, how spatio-temporal dynamics might affect the capacity of alien species to infiltrate into and impact pollination networks. For example, an invasion might be more likely during a particular season, year (e.g. during particularly wet years), or in certain habitats.

As in most archipelagos throughout the World, the number of alien species in the Galapagos began to accumulate even before the first permanent human settlement of the islands, increased exponentially over the last 50 years in step with increasing human pressure (Tye 2006) and currently forms over 60 % of the vascular flora (Jaramillo *et al.* 2014). Indeed, alien species, both plants and animals, are generally considered the main threat to the conservation of the unique Galápagos biodiversity (Bensted-Smith 2002), and predicting the effects of alien plants and pollinators on the reproduction of native vegetation is a major conservation and scientific goal (Tapia *et al.* 2009; Traveset *et al.* 2013). A recent compilation of plant–animal pollination interactions retrieved data from 38 studies published in the last 100 years in highly scattered literature (Chamorro *et al.* 2012). This study concluded that most interactions were documented by observations highly limited in space and time, and thus identified strong biases in the sampling effort dedicated to different islands, times of day, focal plants and functional groups of visitors, reducing our ability to derive solid generalizations from these incomplete datasets (Chamorro *et al.* 2012).

While alien invasive plants may have a direct negative effect on native plants due to direct competition for space (Magee *et al.* 2001), repercussions may also cascade throughout the entire network of biological interactions of an island or archipelago without necessarily leading to local extinction of native species (Jäger et al. 2009). Such a disturbance scenario can be better understood with a network approach (Traveset et al. 2013), such as the one we apply here.

Human pressure is not evenly distributed across the islands but is heavily concentrated on the two large central islands: Santa Cruz, the most populous island, and San Cristóbal, which holds the administrative capital of the archipelago. Human developments are restricted to a small proportion of each island's area; however, an extensive use of the transition and highland zones for agriculture boosted the number of alien plant species. Thus, these two islands offer suitable models to improve our understanding of the disruptive effect of alien species on the native interaction networks of the Galápagos and to forecast short- and mid-term impacts on the islands with low human presence (Isabela and Floreana), and long-term impacts on the most pristine uninhabited islands. The main objective of this study was thus to assess the spatio-temporal variation of pollination interactions in the two most disturbed Galápagos islands and to determine whether and how alien plants may modify such interaction patterns. In addition, we investigated if alien species (both animals and plants) differ from endemic and non-endemic natives in their integration into the pollination networks.

Methods

Study sites

The Galápagos archipelago lies at the Equator in the Pacific Ocean, 960 km west of mainland Ecuador (Fig. 1). The archipelago is currently formed by 13 islands larger than 10 km², which were formed by volcanic activity between 0.035 and 4.0 My ago (Poulakakis et al. 2012), some of them having been merged in the past due to sea level fluctuations (Ali and Aitchison 2014).

The Galápagos vegetation is marked by strong zonation associated with altitude, with 60 % of the islands' surfaces being markedly dry (Trueman and d'Ozouville 2010). This dry zone occupies the lowlands of all islands and holds most plant diversity and endemic species (McMullen 1999; Guézou et al. 2010). The humid zone is restricted to the highlands of the six islands higher than 600 m and is dominated by large patches of the endemic tree Miconia robinsoniana, and by woodlands of 16 endemic species of arboreal Asteraceae (Scalesia spp.) (McMullen 1999). These two zones are separated by a transition zone characterized by closed mixed forest dominated by several native trees and shrubs including Zanthoxylum fagara and Tournefortia spp.

The Galápagos climate is characterized by two seasons. The hot/wet season, from January to May, is associated with frequent rain throughout the islands (Trueman and d'Ozouville 2010), and is the time in which most plants flower (Traveset et al. 2013) and fruit (Heleno et al. 2013a). In contrast, the cold/dry season, between June and December, is characterized by virtually no precipitation in the lowlands (Ziegler 1995; Trueman and d'Ozouville 2010). A permanent drizzle generates an evergreen humid habitat in the highest part of the tallest islands, including the two islands included in this study.

This study was conducted at 12 sites on the islands of Santa Cruz and San Cristóbal (Fig. 1). These two islands are highly comparable in terms of area (986 vs. 558 km²), elevation (864 m vs. 735 m above sea level), and latitude (0°29′–0.46′S vs. 0°40′–0°56′S, respectively). San Cristóbal is older (max. age = 4.0 My) than Santa Cruz (2.3 My) and is also more isolated from other islands (Fig. 1). Due to the difficulty in finding pristine sites that are accessible from the populated areas in Galápagos, a classical paired experimental design, comparing invaded and uninvaded sites was not possible. Instead, we evaluated the impact of invasion level by quantifying the proportions of alien plants among the 12 selected sites. The invasion level was estimated as the proportion of alien flowers at each site, based on counts of all alien and native flowers at each site, and it ranged from 0 to 73 %. We did not select sites where the native vegetation has been completely replaced by invasive plants such as Psidium guava, Rubus nivaeus, Syzygium jambos, but rather tried to select sites as diverse as possible.

Pollination observations

Data were collected from 12 sites using a hierarchical design (Fig. 1) including the two most human-populated islands (Santa Cruz and San Cristóbal), and the two most widespread habitats (dry lowland and humid highland) and their transition zone. Sites were sampled during two seasons (wet/hot and dry/cold).

At each site, regular focal flower censuses were performed to quantify the contact of flying animals, mostly insects, with the reproductive organs of open flowers. Each site was visited every other week, when climatic conditions allowed, from March to May (rainy season) and from July to September (dry season) in 2010. In 2011, the 12 sites were re-sampled but only during the flowering peak (hot season); this year, sampling took place between January and May. On each sampling day, flower diversity and abundance were measured along a 500 × 6 m transect (i.e. 3000 m²), as follows: (i) all species with open flowers were identified, (ii) the number of individuals of each species was counted, (iii) for each species, the number of flowers on two individuals having an

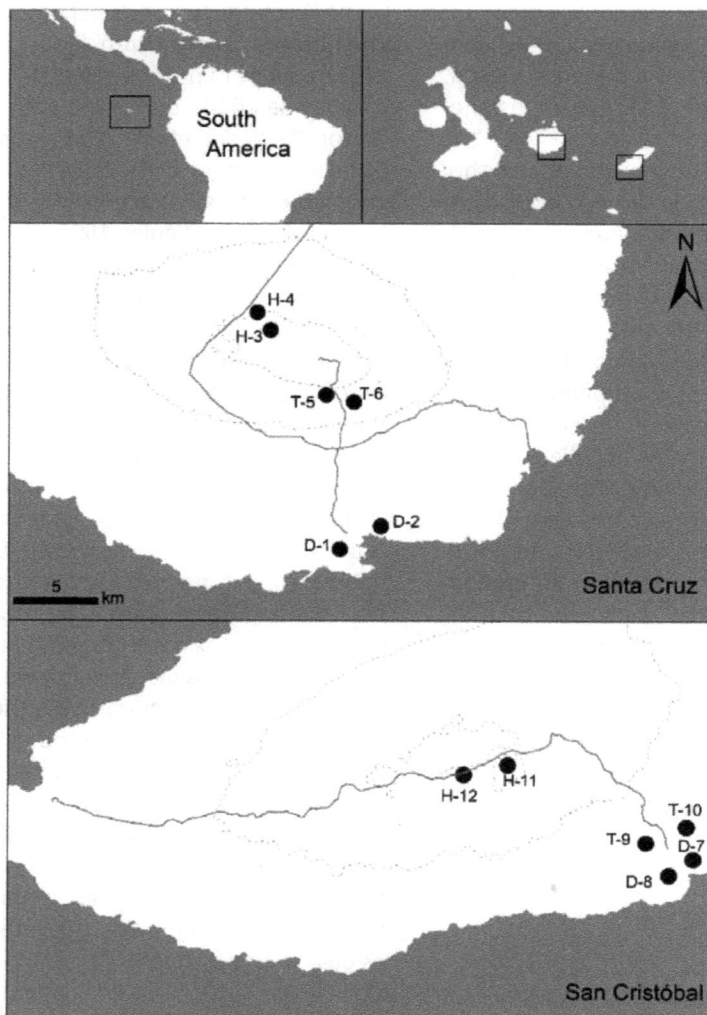

Island	Habitat	Map code	Name	UTM Coordinates
Santa Cruz	Dry zone	1	Puerto Ayora	90°19'26,566"W - 0°45'3,449"S
		2		90°18'6,347"W - 0°44'17,43"S
	Transition zone	3	Mina de Granillo Rojo	90°21'58,831"W - 0°37'6,254"S
		4		90°21'56,65"W - 0°37'20,302"S
	Humid zone	5	Media Luna	90°19'21,857"W 0°39'56,048"S
		6		90°19'30,907"W - 0°39'56,419"S
San Cristóbal	Dry zone	7	Puerto Chino	89°28'57,04"W - 0°53'38,342"S
		8		90°19'21,857"W - 0°39'56,048"S
	Transition zone	9	Galapaguera	90°21'58,831"W - 0°37'6,254"S
		10		89°26'8,394"W - 0°54'51,354"S
	Humid zone	11	El Junco	89°28'57,04"W - 0°53'38,342"S
		12		89°29'31,74"W - 0°53'48,545"S

Figure 1. The location of field sites on the islands of Santa Cruz and San Cristóbal in the Galápagos. Contour lines indicate the 300 m and 600 m isoclines. D, Dry lowland sites; T, transition habitat sites; and H, humid highland sites.

average flower display was counted and (iv) the number of open flowers for each species in the transect was extrapolated from the number of flowers per individual × number of individuals. For species with tightly clustered inflorescences (e.g. the capitula of Asteraceae) we scored each inflorescence as a flower, as this is the ecologically relevant unit visited by pollinators.

Flower-visitors were censused during periods of 10 min in front of target plants (~1 m away). On each census day, all species with open flowers (regardless of their anthesis stage and nectar production) were observed for at least two non-consecutive periods between 06:00 h (sunrise) and 22:00 h. Nocturnal censuses were made by means of red (low energy) l.e.d. headlights to avoid affecting insects. Species were arbitrarily selected for the different time periods to avoid censusing the same species always at similar times of the day. In each census period we recorded: (i) identity of the flowering plant species, (ii) number of open flowers observed on each individual plant (often only one branch was observed in the case of shrubs or trees), (iii) identity of each flower-visitor, (iv) number of individuals of each species visiting flowers and (v) number of flowers visited by each individual flower visitor. The sampling protocol resulted in 1145 h of flower visitation censuses (on a total of 283 287 flowers) of 119 plant species of which 36 are introduced in the Galápagos. Plant identifications followed McMullen (1999) and information available at the Charles Darwin Foundation Herbarium (Jaramillo et al. 2014). Insects that could not be identified in the field were collected for further identification by taxonomists at the Charles Darwin Entomological Collection (see Acknowledgements). Note that here we use the term 'pollinator' regardless of its effectiveness in the pollination process, as we do not know whether flower visitation results in pollination.

Network and statistical analyses

For each site (12 sites), season and year, we built a quantitative plant–pollinator interaction matrix. Thus, we ended up with a total of 36 matrices: 24 for 2010 and 12 (only hot season) for 2011. In each matrix, interactions were quantified by means of visitation frequency, expressed as the total number of visits to the flowers of each species per unit of time, standardized by the number of flowers observed in each census and by the overall flower abundance of each species (Castro-Urgal et al. 2012). From each matrix, we obtained 10 parameters commonly used to describe network structure. Seven are network-level parameters: species richness (S); connectance (C); interaction strength asymmetry (IAc), corrected for network asymmetry; interaction evenness (IE); complementary network specialization (H_2'); nestedness [weighted nestedness based on overlap and decreasing fill (WNODF)];

and modularity (M). The other three are species-level parameters: linkage level (L), species specialization (d') and species strength (st), for both pollinators and plants. Definitions of each of these parameters can be found in **Appendix S1 [see Supporting Information]**. All parameters were computed for the 36 networks using the package 'bipartite' v. 2.00 (Dormann et al. 2009) in R v. 3.1.0 (R Development Core Team 2014), the software NODF v. 2.0 (nestedness based on overlap and decreasing fill; Almeida-Neto and Ulrich 2011) (http://www.keib.umk.pl/nodf/) for the calculation of WNODF, and the software MODULAR (Marquitti et al. 2014) for the computation of modularity.

We used generalized linear models (GLMs) to test for a significant variation in the network level parameters between islands, habitats, seasons and in relation to invasion level. The fitted models, one for each parameter as a response variable, thus included all four predictors. Species richness (count data) followed a Poisson distribution, and was thus approached by a log link function (Zuur et al. 2009), whereas the rest of parameters (all continuous) were approached by the identity link function. Network size (N) was included as a covariate in all models, except for S (directly related to N) and H_2' (known to be independent of N; Blüthgen et al. 2007). For the species-level parameters, we fitted generalized linear mixed models (GLMMs), one model for each parameter as a response variable, and included site (i.e. network) as a random factor to prevent any possible effect of pseudo-replication (as species coexisting in the same site are not independent from each other). The linkage level was fitted to a Poisson distribution and approached by a log link function whereas d' and st followed normal distributions and were approached by the identity function. The dredge function in the MuMIn (Multi-model inference) package v. 1.10.5 in R v. 3.1.0 (R Development Core Team 2014) was used to select the best model, i.e. best random and fixed structure of the model for each metric, according to the corrected Akaike's Information Criterion (AIC) (Zuur et al. 2009). In order to determine the differences among species, either plants or animals, of different origin (aliens, endemic and non-endemic natives), separate tests were performed with a subset of the data excluding those species whose origin was unknown. For plants, only one unidentified species from the Fabaceae family was excluded from this analysis, whereas for animals (mostly insects), origin was unknown for 83 species (out of the 212 recorded on the flowers; **see Supporting Information—Tables S1 and S2**) and thus the dataset included the remaining 129 species. Origin of insect pollinators was obtained from the Charles Darwin Foundation database (http://www.darwinfoundation.org/datazone/checklists). All analyses were performed using functions lme and lmer implemented in package *lme4* in the R package v. 3.1.0 (R Development Core Team 2014).

Results

We recorded a total of 11 125 visits (one individual visitor visiting one flower) by 212 animal species (57 alien to the islands) to the flowers of 111 plant species (32 alien) **[see Supporting Information—Fig. S1 and Tables S1 and S2]**. Except for three species of birds (*Geospiza scandens, G. fuliginosa* and *Setophaga petechia*) and one species of lava lizard (*Microlophus bivittatus*), all flower visitors in our networks were insects. The insects belonged to the following taxonomic groups, in order of species richness: Diptera (63 spp.), Lepidoptera (52), Hymenoptera (41), Coleoptera (29), Hemiptera (16), Orthoptera (5), Odonata (1), Collembola (1) and Thysanoptera (1). Overall, 1214 unique, i.e. species-specific, interactions were recorded, of which the majority (43.8 %) corresponded to those between native plants and animals. One-third (33.7 %) of the interactions was found between native plants and alien insects, and alien plants were visited by native and alien insects on 13.5 and 9.0 % of the interactions recorded, respectively.

The proportion of alien plant species was greater on Santa Cruz (27 % overall average) than San Cristóbal (19 %). On Santa Cruz, the number of alien plant species was highest in the humid habitats (40 %), whereas in San Cristóbal most aliens were located in the transition zone (where they represented ~23 % of the plant species) and in the humid zone (19 %). The frequency of alien plant species was fairly consistent between the two seasons, on both islands and in all three habitats (Table 1). The

proportion of alien insect species recorded on the flowers was similar on the two islands, representing an average of 38 % of the total number of insect species. Unlike plants, the frequency of alien insects varied throughout the year and across habitats; on both islands, the highest fraction of alien insect species was found in the cold season in the arid and transition zones (Table 1).

Spatio-temporal patterns at the network level

Data from 2010 showed that species richness was somewhat higher in Santa Cruz than in San Cristóbal (Table 2), although differences were not significant ($\chi^2 = 18.26$, d.f. $= 3$, $P < 0.001$) (Fig. 2). There were significant differences among habitats in the number of species in the network ($\chi^2 = 27.17$, d.f. $= 4$, $P < 0.001$), arid zone showing higher number than either the transition or the humid zone, which did not differ from each other (Fig. 2). A significant interaction between island and habitat was found ($\chi^2 = 12.74$, d.f. $= 2$, $P < 0.01$), as differences among habitats were not consistent between Santa Cruz and San Cristóbal (Fig. 2). On both islands, networks were larger in the hot/rainy season, when most flowers are in bloom and more insects are flying, than in the cold season ($\chi^2 = 95.91$, d.f. $= 1$, $P < 0.001$). The level of plant invasion showed no effect on species richness ($\chi^2 = 0.52$, d.f. $= 1$, $P = 0.47$) and was not included in the best model.

The fraction of realized interactions out of all possible in the network (*connectance*) did not vary either between islands, habitats or seasons (Fig. 2), and it was not influenced by the level of invasion (all $P > 0.05$). The same result was found for *interaction asymmetry*, which indicates the difference in the dependence of animals on

Table 1. Frequency of alien plants and pollinators in the 12 study communities (networks).

Island	Habitat/zone	Season	Total plants	% alien plants	Total pollinators	% alien pollinators
Santa Cruz	Arid	Hot	29	3.45	50	42.00
Santa Cruz	Transition	Hot	26	23.08	57	42.11
Santa Cruz	Humid	Hot	26	38.46	46	41.30
Santa Cruz	Arid	Cold	9	11.11	17	58.82
Santa Cruz	Transition	Cold	11	18.18	10	40.00
Santa Cruz	Humid	Cold	15	46.67	10	30.00
San Cristóbal	Arid	Hot	18	5.56	50	44.00
San Cristóbal	Transition	Hot	15	20.00	24	41.67
San Cristóbal	Humid	Hot	14	21.43	29	34.48
San Cristóbal	Arid	Cold	8	0.00	18	55.56
San Cristóbal	Transition	Cold	11	27.27	15	60.00
San Cristóbal	Humid	Cold	6	16.67	10	20.00

Table 2. Network-level parameters of the 36 matrices corresponding to the first year of the study and the 12 matrices built for the second year, in which only the hot season was considered. None of the modularity values (M) showed to be significant (all P values >0.05). P, number of plants; A, number of animals (pollinators); S, total number of species in the network; C, connectance; IE, interaction evenness; H'_2, network specialization; IAc, corrected interaction asymmetry; WNODF, weighted nestedness (asterisks imply that it is significant); M, modularity; n_modules, number of modules in the network. **$P \leq 0.01$, *$P < 0.05$.

Year	Island	Season	Habitat	Invasion level	P	A	S	C	IE	H'_2	IAc	WNODF	M	n_modules
2010	San Cristóbal	Hot	Arid	7.59	28	29	57	0.12	0.28	0.82	0.01	23.66**	0.45	5
2010	San Cristóbal	Cold	Transition	72.75	11	15	26	0.17	0.37	0.89	0.12	11.25**	0.58	4
2010	San Cristóbal	Cold	Humid	0.65	6	8	14	0.35	0.48	0.18	0.07	44.88	0.37	4
2010	San Cristóbal	Cold	Humid	0.24	6	10	16	0.33	0.48	0.63	0.13	40.83**	0.38	3
2010	San Cristóbal	Hot	Arid	0.42	14	40	54	0.16	0.52	0.69	0.24	19.74**	0.45	5
2010	San Cristóbal	Hot	Transition	41.79	17	27	44	0.15	0.36	0.72	0.12	21.05**	0.41	7
2010	San Cristóbal	Hot	Transition	53.74	12	14	26	0.21	0.4	0.48	0.04	22.53**	0.41	6
2010	San Cristóbal	Hot	Humid	0.32	12	26	38	0.21	0.55	0.64	0.17	24.32**	0.37	5
2010	San Cristóbal	Hot	Humid	5.56	13	24	37	0.17	0.45	0.63	0.17	16.85**	0.49	5
2010	San Cristóbal	Cold	Arid	16.76	17	25	42	0.13	0.49	0.75	0.12	34.42*	0.53	6
2010	San Cristóbal	Cold	Arid	0	8	18	26	0.22	0.29	0.69	0.24	34.42*	0.43	5
2010	San Cristóbal	Cold	Transition	58.19	8	9	17	0.22	0.32	0.82	0.05	14.06**	0.53	5
2010	Santa Cruz	Hot	Arid	2.47	24	29	53	0.13	0.51	0.5	0.05	20.46**	0.42	6
2010	Santa Cruz	Cold	Transition	0.41	11	10	21	0.22	0.4	0.55	−0.03	14.27	0.51	6
2010	Santa Cruz	Cold	Humid	17.28	13	9	22	0.27	0.47	0.38	−0.09	44.15*	0.34	5
2010	Santa Cruz	Cold	Humid	8.16	15	10	25	0.21	0.49	0.38	−0.13	43.42	0.44	4
2010	Santa Cruz	Hot	Arid	0.69	24	31	55	0.12	0.43	0.71	0.07	17.06**	0.45	7
2010	Santa Cruz	Hot	Transition	61.91	20	37	57	0.15	0.36	0.67	0.12	17.2**	0.4	7
2010	Santa Cruz	Hot	Transition	18.08	22	39	61	0.16	0.46	0.68	0.1	22.86**	0.37	6
2010	Santa Cruz	Hot	Humid	16.81	14	26	40	0.18	0.46	0.66	0.14	21.44**	0.41	6
2010	Santa Cruz	Hot	Humid	27.01	21	31	52	0.14	0.4	0.71	0.09	21.72**	0.43	6
2010	Santa Cruz	Cold	Arid	0.01	8	14	22	0.23	0.2	0.72	0.15	18.15**	0.45	5
2010	Santa Cruz	Cold	Arid	0.01	9	17	26	0.2	0.27	0.2	0.19	25.15**	0.48	5
2010	Santa Cruz	Cold	Transition	11.68	9	23	32	0.22	0.37	0.56	0.24	15.16**	0.49	5
2011	San Cristóbal	Hot	Arid	8.25	15	19	34	0.13	0.48	0.69	0.09	9.55**	0.61	9
2011	San Cristóbal	Hot	Arid	2.29	10	22	32	0.17	0.37	0.31	0.27	14.01**	0.52	7

Continued

Table 2. Continued

Year	Island	Season	Habitat	Invasion level	P	A	S	C	IE	H'_2	IAc	WNODF	M	n_modules
2011	San Cristóbal	Hot	Transition	16.99	13	18	31	0.13	0.4	0.74	0.14	14.79**	0.63	7
2011	San Cristóbal	Hot	Transition	61.91	11	13	24	0.17	0.31	0.87	0.07	14.29**	0.61	5
2011	San Cristóbal	Hot	Humid	4.56	5	4	9	0.4	0.34	0.42	-0.08	56.25*	0.39	3
2011	San Cristóbal	Hot	Humid	1.03	10	14	24	0.19	0.52	0.58	0.12	19.49	0.5	6
2011	Santa Cruz	Hot	Arid	2.35	25	67	92	0.09	0.47	0.52	0.24	16.3**	0.49	6
2011	Santa Cruz	Hot	Arid	1.73	24	43	67	0.13	0.45	0.58	0.12	18.98**	0.39	6
2011	Santa Cruz	Hot	Transition	17.48	13	41	54	0.13	0.41	0.56	0.33	15.19**	0.53	7
2011	Santa Cruz	Hot	Transition	52.41	19	47	66	0.14	0.45	0.6	0.18	25.06**	0.38	5
2011	Santa Cruz	Hot	Humid	43.28	11	25	36	0.19	0.43	0.74	0.22	18.06**	0.46	5
2011	Santa Cruz	Hot	Humid	71.93	22	35	57	0.14	0.44	0.62	0.1	17.09**	0.41	5

plants and *vice versa*. Interaction evenness, which measures the uniformity in the distribution of inter-action frequencies differed only across habitats, i.e. habitat was the only factor included in the best model ($\chi^2 = 0.05$, d.f. $= 2$, $P = 0.02$). The humid zone showed a more even frequency of interactions than the arid and the transition zones (Fig. 2). Interaction evenness was also independent of invasion level ($\chi^2 = 1.90$, d.f. $= 1$, $P = 0.17$). In contrast, the best model for *complementary specialization* (H'_2) included only invasion level ($\chi^2 = 0.11$, d.f. $= 1$, $P = 0.057$); a high fraction of alien flowers in the community was positively associated with higher H'_2 (Fig. 3), i.e. with higher levels of selective-ness or niche differentiation, implying that species tended to visit (pollinators) or be visited (plants) by part-ners more frequently than expected from the relative abundances of the latter.

Habitat and season were the only variables included in the best model predicting *quantitative nestedness* (WNODF). Networks from the humid zone showed signi-ficantly higher levels of nestedness than those from the transition or the arid zone and ($\chi^2 = 890.5$, d.f. $= 2$, $P < 0.001$) (Table 2 and Fig. 2). Nestedness was higher in the cold than in the hot season ($\chi^2 = 347.1$, d.f. $= 1$, $P = 0.01$) (Fig. 2). The level of invasion did not affect the nested pattern of the networks, which was significant in 30 out of the 36 networks (Table 2). Finally, despite none of the networks was significantly modular (i.e. when com-pared to a null model), the degree of modularity (M) was slightly lower in the humid zone than in either the arid or the transition zone ($\chi^2 = 0.02$, d.f. $= 2$, $P = 0.04$) and was marginally higher in the cold than in the hot *season* ($\chi^2 = 0.01$, d.f. $= 1$, $P = 0.07$) (Fig. 2).

In 2011, the number of species in the networks was almost twice as high in Santa Cruz as in San Cristóbal ($\chi^2 = 93.13$, d.f. $= 1$, $P < 0.001$). Again, the humid zone showed the lowest species richness ($\chi^2 = 28.31$, d.f. $= 2$, $P < 0.001$). This year we found no significant differences in interaction evenness, nestedness or modularity across habitats [see Supporting Information—Fig. S2].

Spatio-temporal patterns at the species level

In 2010, pollinators in Santa Cruz had a higher *linkage level* (L_a) than in San Cristóbal ($\chi^2 = 6.13$, d.f. $= 1$, $P = 0.01$) and also tended to visit more plant species in the humid than in the arid zone ($\chi^2 = 9.11$, d.f. $= 4$, $P = 0.05$), and more in the hot than in the cold season ($\chi^2 = 20.98$, d.f. $= 3$, $P < 0.001$). There was a significant interaction between habitat and season ($\chi^2 = 9.11$, d.f. $= 2$, $P = 0.01$), as differences among habitats were not consistent between the two seasons [see Supporting Information—Fig. S3]. Results were consistent in 2011, except that this year L_a was positively influenced by invasion level

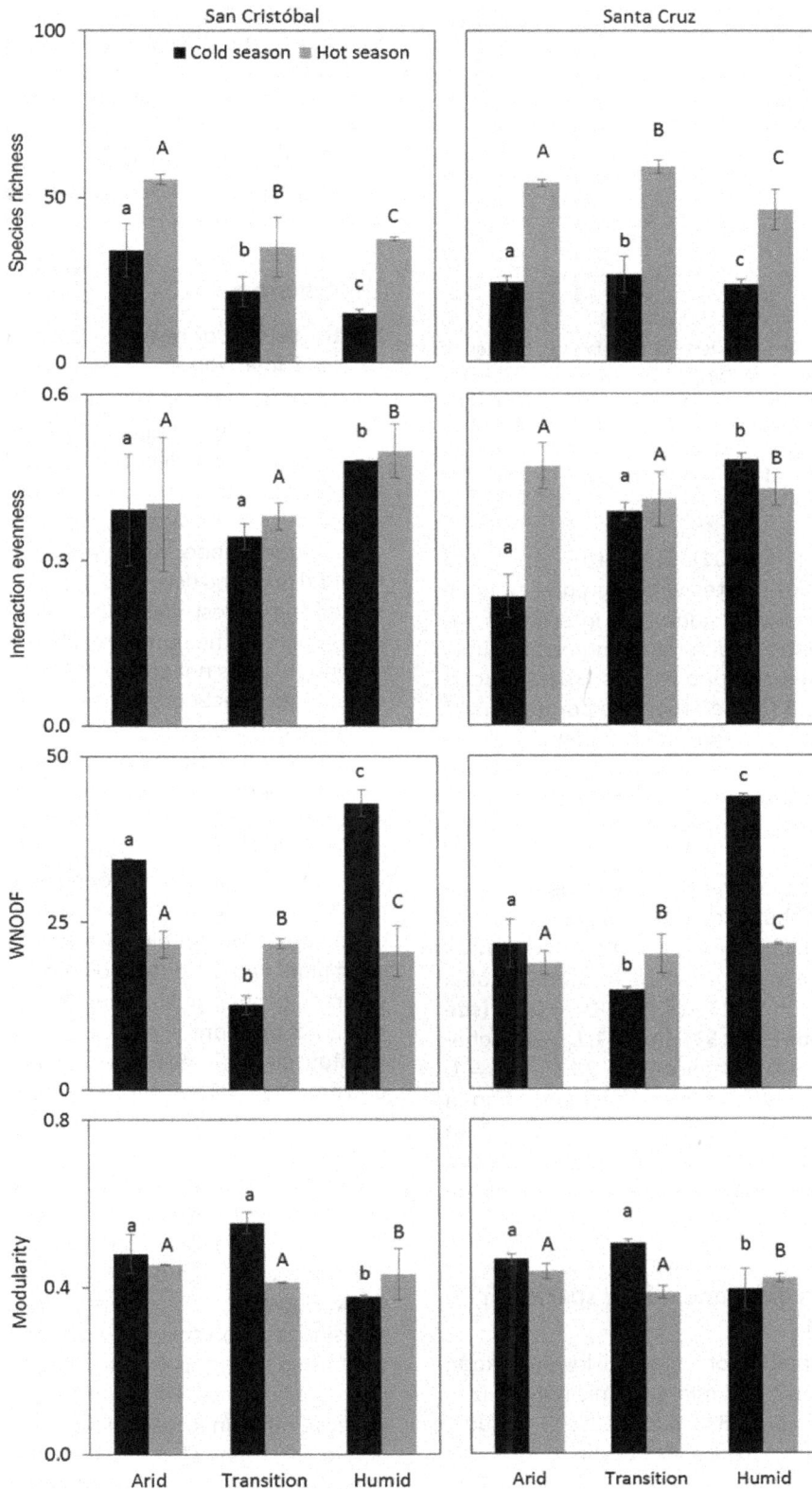

Figure 2. The mean ($\pm$ 1 SE) of the network parameters for each island, habitat and season. Data are from 2010. Only parameters that showed significant differences are shown. For each island and season, bars with the same letter indicate no differences across habitats ($P > 0.05$).

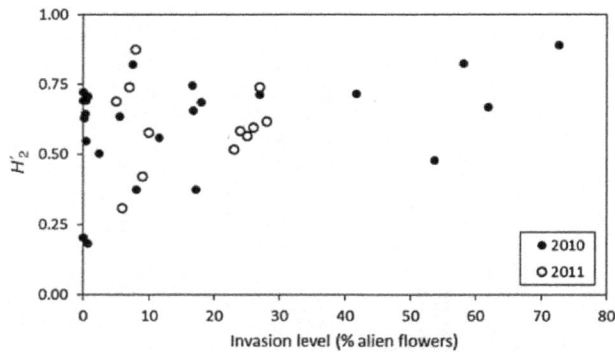

Figure 3. Relationship between the level of invasion (i.e. fraction of alien flowers out of all flowers in the site) and the level of network specialization H'_2 found during the 2 years of the study. Data from the two islands and the three habitats are pooled. The association is marginally significant in the two cases ($t = 1.9$, $P = 0.07$ and $t = 2.14$, $P = 0.06$, in 2010 and 2011, respectively).

($\chi^2 = 10.89$, d.f. $= 1$, $P < 0.001$); pollinators interacted with more plant species in sites with a greater fraction of alien flowers. There was a significant interaction between island and habitat, as differences among habitats varied slightly between the two islands **[see Supporting Information—Fig. S3]**. The other two parameters at the pollinator species level, *specialization level* (d') and *strength* (st), could not be predicted by any of the variables included in the models, i.e. they did not differ between islands, habitats, or seasons and were not influenced by invasion levels.

Regarding *plant linkage level* (L_p), the transition zone showed higher values than either the arid or the humid zone in 2010 ($\chi^2 = 16.39$, d.f. $= 6$, $P = 0.01$); differences among habitats were more marked in San Cristóbal than in Santa Cruz ($\chi^2 = 13.24$, d.f. $= 2$, $P = 0.001$) **[see Supporting Information—Fig. S3]**. In 2011, L_p was higher in Santa Cruz than in San Cristóbal ($\chi^2 = 29.35$, d.f. $= 1$, $P < 0.001$), and it was lower in the humid zone than in the other two habitats ($\chi^2 = 6.04$, d.f. $= 2$, $P < 0.05$). In contrast, neither d' nor st were significantly influenced by any of the predictor variables included in any of the models.

Differences between alien and native species in interaction patterns

In both years, alien pollinators showed lower linkage levels than both endemic and non-endemic natives ($z = 6.45$, d.f. $= 415$ and $z = 3.83$, d.f. $= 265$, $P < 0.001$, respectively), although in 2011 aliens and non-endemic natives did not differ significantly ($z = 0.40$, d.f. $= 265$, $P = 0.69$) (Fig. 4). Alien pollinators had lower d' and st values than endemic ones in 2010 ($t = 2.05$, d.f. $= 415$, $P = 0.04$ and $t = 4.74$, d.f. $= 415$, $P < 0.001$, respectively;

Fig. 4). In 2011, d' did not differ between the two groups but st was again significantly lower for alien than for endemic pollinators ($t = 2.56$, d.f. $= 265$, $P = 0.01$; Fig. 4).

On the other hand, in both years, alien plants showed lower L_p and st than endemic plants, whereas they did not differ significantly from non-endemic native species. No differences were found in d' depending upon plant species' origin (Fig. 4).

Discussion

Spatio-temporal network patterns and influence of plant invasion

Despite network size being larger in Santa Cruz than in San Cristóbal, especially in 2011 when it was twice as large, the overall pollination network structure was similar between the two islands. Strong spatial variation in network structure was detected, however, across habitats. The arid zone, which includes the vast majority of the land area and bears the highest species richness, supported the largest pollination networks. In contrast, the transition and the humid zone were more similar in size, though this was not consistent between islands or years. Flower and insect abundance are known to be influenced by abiotic conditions such as temperature or rainfall which can vary much spatially and temporally (Ziegler 1995; Trueman and d'Ozouville 2010). Alien plant species represented up to 40 % of the plants in some networks, particularly in the humid zone of Santa Cruz and in the transition zone in San Cristóbal. However, overall network size was not affected by the level of invasion—measured as the proportion of alien flowers—suggesting both that alien plants do not differ from natives with respect to the diversity of their pollinators and that aliens do not displace native plant species in the pollination networks. Habitats also differed in interaction evenness, nestedness and modularity. The uniformity in the distribution of interaction frequencies was higher in the humid habitat than in the two other habitats. In a previous study in the Galápagos (Traveset *et al.* 2013), a decrease in interaction evenness was observed along a gradient of invasion intensity at the island scale, being attributed to shifts in the proportion of strong and weak interactions in the network. However, the present work showed no effect of invasion level on this network parameter and, actually, the humid habitat is that bearing the highest fraction of alien species. Interaction evenness has been reported to increase after an invasion in one study on seed dispersal networks (Heleno *et al.* 2013b) but not in another (Heleno *et al.* 2013a). Hence, further data are needed to generalize about how this network parameter, known to be inversely related to network stability (Rooney and McCann 2012), is influenced by alien invasions. The humid habitat

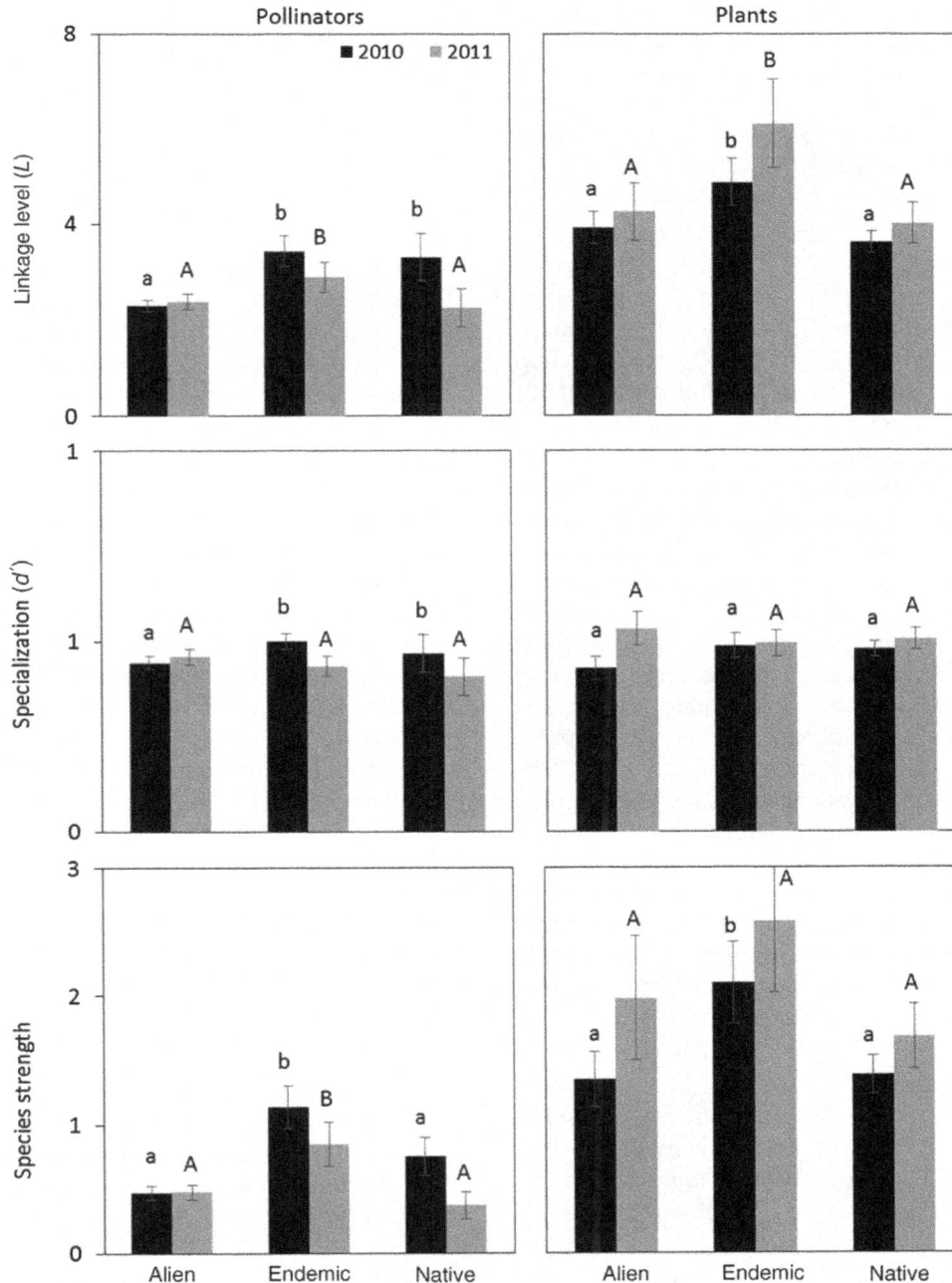

Figure 4. Mean (± 1 SE) of the species-level parameters analysed in this study, for both pollinators and plants, showing differences among species of different origin for the 2 years of the study. Data from the two islands, three habitats and two seasons were pooled here for simplification. For each year, bars with the same letter indicate no differences across habitats ($P > 0.05$).

showed the strongest nested pattern (in which specialist species link to a subset of species with which generalists also interact), which could also be attributed, at least partly, to its high level of invasion. The degree of nestedness has been found to increase with the integration of alien species (Padrón et al. 2009; Santos et al. 2012); this is because aliens tend to be generalist species and/or are linked to generalist species (Aizen et al. 2008; Traveset

et al. 2013). Thus, although the level of invasion overall was a poor predictor of nestedness, we cannot discard the possibility that a higher incidence of alien flowers enhances a nested pattern in a habitat. Modularity—another common parameter that informs us on how cohesive the network is and how vulnerable it can be to different types of disturbances (Olesen et al. 2007)—was lower in the humid zone, i.e. this zone had a weaker segregation of species into cores of strong interactions, than the arid and transition zones. The lower modularity in the humid zone might be associated with its lower plant and animal diversity compared with the transition and arid zones, and also with its relatively higher linkage levels (see below). The level of invasion has been documented to decrease modularity, and thus to enhance network cohesiveness in some studies (Santos *et al.* 2012; Albrecht *et al.* 2014). It is thus possible that the lower modularity in the humid habitat is partly due to its higher incidence of aliens. A low modularity has potential effects on network functioning, reciprocal selection regimes and the cascade of perturbations throughout the network (Albrecht *et al.* 2014).

Other network descriptors, such as connectance, interaction strength asymmetry and network complementary specialization (H'_2), did not vary much either in space or time. The level of network connectance, which is inversely related to network size, was both spatially and temporally consistent, despite species richness in each network varying across islands, habitats and seasons. This parameter is a measure, albeit crude, of network generalization level and, as expected from other island studies (Olesen and Jordano 2002; Traveset *et al.* 2013), we found relatively high values (~18 % on average, ranging from 12 to 40 %, across the 36 matrices analysed). No effect of invasion level on connectance was observed, which is consistent with previous studies (Forup and Memmott 2005; Heleno *et al.* 2012), although network rewiring can actually occur and, as a result, the number of interactions between native species can decrease (Aizen *et al.* 2008; Padrón *et al.* 2009; Kaiser-Bunbury *et al.* 2011). Besides being consistent in space and time, most values of interaction strength asymmetry were positive which indicates that animals are more dependent upon plants than *vice versa* (Blüthgen *et al.* 2007), a result commonly found in other oceanic archipelagos (Kaiser-Bunbury *et al.* 2010; Traveset *et al.* 2015*b*), and a pattern not found to be influenced by invasion level in this study. Finally, an interesting finding from our study was that H'_2 increased with the level of plant invasion, implying that species become more selective in their choice of mutualists by being compelled to interact with less abundant partners as invasion progresses. This finding contrasts with results from other studies which have reported a decrease in H'_2 after an invasion (Heleno *et al.* 2013*b*).

Regarding the species-level parameters, pollinators tended to visit more plant species in Santa Cruz than in San Cristóbal, what can be attributed to the higher plant species richness in the former. Pollinators were also more generalist in the humid zone even though here the number of plants is lower than in the other two zones. It is possible, thus, that the lower amount of

floral resources in the humid zone promotes insects visiting more plant species, as has been found in a number of island studies (Olesen *et al.* 2002; Kaiser-Bunbury *et al.* 2009; Padrón *et al.* 2009; Traveset *et al.* 2013). Interestingly, one of the years (2011), pollinators visited more plant species at sites with a greater fraction of alien flowers, suggesting that pollinators might be attracted to the new species which in turn would enhance their visitation to the other native plants in the community. Such 'facilitative' effects of alien plant species on native ones have been often reported in different systems (e.g. Moeller 2005; Jakobsson *et al.* 2009). Plant species, on the other hand, showed higher generalization levels in their pollination interactions in Santa Cruz than in San Cristóbal, at least in 2011 when more insect species were found on the former island. Plants were visited by less pollinator species in the humid zone, as the total number of pollinators is also lower in this zone compared with the other two zones. The other two parameters, species specialization d' and strength st, were highly consistent in space. Both pollinators and plants had a similar level of selectiveness in their flower or pollinator use, respectively, and were also equally important to the plant or pollinator communities, respectively, in the two islands and in the three habitats. A fairly constant value of d' for both plants and pollinators has been previously reported across five of the Galápagos Islands (Traveset *et al.* 2013). Moreover, those two parameters were not influenced by plant invasion level. In contrast, at least one study (of seed-dispersal networks) has reported the level of invasion to decrease species specialization d' of native species (Heleno *et al.* 2013*a*).

Except for a few differences between pollination networks of different habitats, our results were highly consistent between the 2 years of the study, which were both considered 'normal' years in terms of precipitation and sea surface patterns (FCD Weather report, data not shown), despite the usual fluctuations in flower production and flower-visitors' presence/abundance. Thus, we focus on the temporal differences observed between seasons. All pollination networks were larger during the hot rainy season, when more plant species are blooming and more insects are flying, than in the cold dry season. Both pollinators and plants actually showed higher linkage levels in the hot than in the cold season, given the greater availability of partner species in the former. Moreover, networks were more nested in the cold than the hot season after controlling for network size, which influences this parameter. Such temporal difference in the degree of nestedness suggests that the interactions in the hot season tend to be more specific, with specialist species interacting more than expected with each other and less so with generalists.

Integration of alien species on pollination networks

Alien pollinators were consistently found to visit fewer plant species than endemic pollinators and, at least one year, also visited fewer plant species than native pollinators, which suggests that these newly-arrived species are focusing flower visitation on species with particular traits. However, the fact that alien pollinators also showed lower levels of selectiveness than endemic pollinators implies that they tend to visit more abundant flower resources compared with endemic pollinators, which visit even rare flowers. Likewise, species strength was consistently lower for alien than endemic pollinators, indicating that the former are less important to plants. In a previous study focusing on the pollination networks of the arid zone in five Galápagos islands, we found that alien insects had more links than either endemics or non-endemic natives (Traveset et al. 2013), which suggests that the inclusion of the two other habitats, transition and humid, in the present study masks that pattern and/or that Santa Cruz and San Cristóbal are somewhat outliers in archipelago wide patterns, possibly due to the high level of disturbance.

Alien plants were also consistently more specialized than endemic plants, although they were similar to native species. In contrast to pollinators, plants showed similar selectiveness regardless of their origin, but again, endemic plant species were more important to the pollinator community than alien plants. These findings were consistent with our previous study (Traveset et al. 2013). It might be possible that aliens do not rely as much on pollinators as native species do. However, no data are currently available on the breeding system for the large majority of plants and, thus, future studies are needed to test this hypothesis.

Conclusions

The structure of pollination networks is highly consistent on the two most disturbed islands of the Galápagos archipelago. Differences in network structure exist across the main habitats. The most widespread arid habitat consistently bears the largest pollination networks and differs strongly from the humid habitat in descriptors such as interaction evenness, nestedness and modularity. The transition habitat between the arid and the humid zone shows pollination networks more similar in structure to those in the arid than in the humid areas. The humid habitat is also the most invaded by alien species and this could partly explain some of the differences in its network structure, such as its more nested pattern and its lower modularity level compared with the arid and the transition zones. Pollinators appear to interact with more plants in the humid habitat than in the arid one. The incidence of alien flowers might actually increase the level of pollinator generalization, although results are

inconclusive as this was observed in only one of the two study years. Overall, the level of invasion has a weak influence on pollination network structure and seems to be associated with only one metric, H'_2 which measures the level of selectiveness; thus, as invasion progresses, species in the network appear to become more selective in their choice of partners, interacting with less abundant species more than would be expected by chance.

Pollination networks are larger during the hot/rainy season, when most flowers are in bloom and more insects are flying, than in the cold/dry season. They are also more nested in the hot season, and thus probably more robust to disturbances. Pollinators visit more plant species, and plants are visited by more pollinator species, during the hot season. In the cold season, the number of insects is especially low in the humid zones and thus the number of pollinators visiting plants is also lower in that season and habitat. In contrast, both pollinator and plant selectiveness (d') and strength (st, importance to the plant and pollinator community, respectively) were spatially and temporally consistent and not influenced by alien plants.

Alien pollinators interacted with fewer plants, were less selective in their choice (i.e. tended to visit the most abundant species) and were less important to the plant community (i.e. showed lower species strength) than endemic and native pollinators. They, however, infiltrated the native communities of all habitats and in both seasons and currently represent over 40 % of all recorded pollination interactions. Alien plants, on the other hand, were visited by approximately the same number of pollinators as natives—but less than endemic plants—implying that they are also well integrated into the native communities. In this study, we found a rather feeble effect of alien plants on the structure of pollination networks. As previously mentioned in the methods, our study intentionally considered sites that are not completely disturbed by highly invasive species (e.g. Psidium guava, Rubus nivaeus, Syzygium jambos) which have displaced many native species in the invaded areas, mainly in the humid zones (Guézou et al. 2010). Hence, the overall weak effect we found does not imply a weak influence of plant invasions on the reproductive success of native species. The fact that alien plant species are present in all habitats and in both seasons and that they are involved in ~25 % of all pollination interactions, actually leads us to think that their effect on the functioning of native communities is far from negligible.

Sources of Funding

This research is part of a larger project funded by BBVA Foundation and is also framed within project CGL-2013-44386-P financed by the Spanish Government.

R.H. was supported by the Portuguese grant FCT-IF/00441/2013, and by the Marie Curie action FP7-2012-CIG-321794.

Contributions by the Authors

A.T., J.M.O. and R.H. designed the study, S.C. collected the data and did some preliminary analyses, and A.T. and R.H. performed the final analyses. A.T. led the writing and all authors contributed to the text. All authors read and approved the final manuscript.

Acknowledgements

We thank the staff at the Galápagos National Park and the Charles Darwin Foundation for allowing us to use their laboratories, the herbarium and the insect collections. We are especially grateful to the taxonomists who identified the insects, especially Ana Maria Ortega, Alejandro Mieles and Henri Herrera, and to different people that helped in the field, especially Rocío Castro-Urgal and Ana Carrión. Donald Drake and two anonymous reviewers made valuable suggestions to a previous version of the manuscript. Finally, we thank our project team in Galápagos for stimulating discussions, specifically Manuel Nogales, Pablo Vargas and Conley McMullen.

Supporting Information

The following additional information is available in the online version of this article –

Appendix S1. Definitions of the metrics used in this study to describe network structure.

Table S1. Complete list of flowering species observed in this study, including origin, overall number of observed visits per plant species and estimated flower abundance (calculated by multiplying the mean number of flowers on two individuals per transect by the total number of individuals counted along the transects). The eight plant species that have not received any recorded visit during the censuses are marked with an asterisk.

Table S2. Complete list of flower-visitors found in this study, including origin and the overall number of observed visits per animal species.

Figure S1. Illustration of the entire pollination network, comprising data from the two islands (Santa Cruz and San Cristóbal), the three habitats (arid, transition and humid zones) and the two seasons (hot and cold). Data from 2010 to 2011 are also pooled. Plant species are depicted at the bottom of the network, whereas pollinators are at the top. Alien (A) plants and their links are represented in red to illustrate the magnitude of the interactions in which they are involved, whereas endemic (Nze) and non-endemic natives (Nt) are represented in black and grey, respectively. Questionably native plants (Nq) are also shown in grey.

Figure S2. Mean (± 1 SE) of network metrics showing differences between the two seasons across habitats for each study island in the hot season of 2011. Data on IE are shown for comparison with data from Fig. 2 (2010 data), although differences across habitats were not significant this year. For each island, bars with the same letters indicate no differences across habitats ($P > 0.05$).

Figure S3. Mean (± 1 SD) of the species-level parameters analysed in this study, for both pollinators and plants, showing differences among species from different habitats for the two islands, and the two seasons of 2010. Bars with the same letters on each graph indicate no differences across habitats ($P > 0.05$).

Literature Cited

Aizen MA, Morales CL, Morales JM. 2008. Invasive mutualists erode native pollination webs. *PLoS Biology* **6**:e31. doi:10.1371/journal.pbio.0060031

Albrecht M, Padrón B, Bartomeus I, Traveset A. 2014. Consequences of plant invasions on compartmentalization and species' roles in plant-pollinator networks. *Proceedings of the Royal Society B: Biological Sciences* **281**:20140773.

Ali JR, Aitchison JC. 2014. Exploring the combined role of eustasy and oceanic island thermal subsidence in shaping biodiversity on the Galápagos. *Journal of Biogeography* **41**:1227–1241.

Almeida-Neto M, Ulrich W. 2011. A straightforward computational approach for measuring nestedness using quantitative matrices. *Environmental Modelling & Software* **26**:173–178.

Ashman TL, Knight TM, Steets JA, Amarasekare P, Burd M, Campbell DR, Dudash MR, Johnston MO, Mazer SJ, Mitchell RJ, Morgan MT, Wilson WG. 2004. Pollen limitation of plant reproduction: ecological and evolutionary causes and consequences. *Ecology* **85**: 2408–2421.

Bascompte J. 2009. Disentangling the Web of Life. *Science* **325**: 416–419.

Bascompte J. 2010. Structure and dynamics of ecological networks. *Science* **329**:765–766.

Bensted-Smith R. 2002. *A biodiversity vision for the Galapagos Islands.* Puerto Ayora, Galápagos: Fundación Charles Darwin para las islas Galápagos y Fondo Mundial para la Naturaleza.

Berglund H, Järemo J, Bengtsson G. 2009. Endemism predicts intrinsic vulnerability to nonindigenous species on islands. *The American Naturalist* **174**:94–101.

Blüthgen N, Menzel F, Hovestadt T, Fiala B, Blüthgen N. 2007. Specialization, constraints, and conflicting interests in mutualistic networks. *Current Biology* **17**:341–346.

Carlquist S. 1974. *Island biology.* Columbia: Columbia University Press.

Castro-Urgal R, Tur C, Albrecht M, Traveset A. 2012. How different link weights affect the structure of quantitative flower–visitation networks. *Basic and Applied Ecology* **13**:500–508.

Chamorro S, Heleno R, Olesen JM, McMullen CK, Traveset A. 2012. Pollination patterns and plant breeding systems in the Galápagos: a review. *Annals of Botany* **110**:1489–1501.

Dormann CF, Fründ J, Blüthgen N, Gruber B. 2009. Indices, graphs and null models: analyzing bipartite ecological networks. *The Open Ecology Journal* **2**:7–24.

Dupont YL, Padrón B, Olesen JM, Petanidou T. 2009. Spatio-temporal variation in the structure of pollination networks. *Oikos* **118**: 1261–1269.

Forup ML, Memmott J. 2005. The restoration of plant-pollinator interactions in hay meadows. *Restoration Ecology* **13**:265–274.

Gillespie RG, Roderick GK. 2002. Arthropods on islands: colonization, speciation, and conservation. *Annual Review of Entomology* **47**: 595–632.

Guézou A, Trueman M, Buddenhagen CE, Chamorro S, Guerrero AM, Pozo P, Atkinson R. 2010. An extensive alien plant inventory from the inhabited areas of Galapagos. *PLoS ONE* **5**:e10276. doi: 10.1371/journal.pone.0010276

Heleno R, Devoto M, Pocock M. 2012. Connectance of species interaction networks and conservation value: is it any good to be well connected? *Ecological Indicators* **14**:7–10.

Heleno R, Garcia C, Jordano P, Traveset A, Gómez JM, Blüthgen N, Memmott J, Moora M, Cerdeira J, Rodríguez-Echeverría S, Freitas H, Olesen JM. 2014. Ecological networks: delving into the architecture of biodiversity. *Biology Letters* **10**:20131000.

Heleno RH, Olesen JM, Nogales M, Vargas P, Traveset A. 2013a. Seed dispersal networks in the Galápagos and the consequences of alien plant invasions. *Proceedings of the Royal Society B: Biological Sciences* **280**:20122112.

Heleno RH, Ramos JA, Memmott J. 2013b. Integration of exotic seeds into an Azorean seed dispersal network. *Biological Invasions* **15**: 1143–1154.

Jäger H, Kowarik I, Tye A. 2009. Destruction without extinction: long-term impacts of an invasive tree species on Galapagos highland vegetation. *Journal of Ecology* **97**:1252–1263.

Jakobsson A, Padrón B, Traveset A. 2009. Competition for pollinators between invasive and native plants: effects of spatial scale of investigation. *Ecoscience* **16**:138–141.

Jaramillo P, Guézou A, Mauchamp A, Tye A. 2014. CDF checklist of Galapagos flowering plants. In: Bungartz F, Herrera H, Jaramillo P, Tirado N, Jímenez-Uzcategui G, Ruiz D, Guézou A, Ziemmeck F, eds. *Charles Darwin Foundation Galapagos Species Checklist.* Puerto Ayora, Galapagos: Charles Darwin Foundation. http://checklists.datazone.darwinfoundation.org/.

Kaiser-Bunbury CN, Memmott J, Müller CB. 2009. Community structure of pollination webs of Mauritian heathland habitats. *Perspectives in Plant Ecology, Evolution and Systematics* **11**:241–254.

Kaiser-Bunbury CN, Traveset A, Hansen DM. 2010. Conservation and restoration of plant-animal mutualisms on oceanic islands. *Perspectives in Plant Ecology, Evolution and Systematics* **12**:131–143.

Kaiser-Bunbury CN, Valentin T, Mougal J, Matatiken D, Ghazoul J. 2011. The tolerance of island plant-pollinator networks to alien plants. *Journal of Ecology* **99**:202–213.

Lambertini M, Leape J, Marton-Lefèvre J, Mittermeier RA, Rose M, Robinson JG, Stuart SN, Waldman B, Genovesi P. 2011. Invasives: a major conservation threat. *Science* **333**:404–405.

Magee J, McMullen CK, Reaser JK, Spitzer E, Struve S, Tufts C, Tye A, Woodruff G. 2001. Green invaders of the Galapagos islands. *Science* **294**:1279–1280.

Marquitti FMD, Guimarães PR, Pires MM, Bittencourt LF. 2014. MODU-LAR: software for the autonomous computation of modularity in large network sets. *Ecography* **37**:221–224.

McMullen CK. 1999. *Flowering plants of the Galapagos.* Cornell: Comstock Publishing Associates.

Memmott J, Gibson R, Carvalheiro L, Henson K, Heleno R, Lopezaraiza M, Pearce S. 2007. The conservation of ecological interactions. In: Stewart AA, New TR, Lewis OT, eds. *Insect conservation biology.* London: The Royal Entomological Society.

Moeller DA. 2005. Pollinator community structure and sources of spatial variation in plant–pollinator interactions in *Clarkia xantiana* ssp. *xantiana. Oecologia* **142**:28–37.

Olesen JM, Jordano P. 2002. Geographic patterns in plant-pollinator mutualistic networks. *Ecology* **83**:2416–2424.

Olesen JM, Eskildsen LI, Venkatasamy S. 2002. Invasion of pollination networks on oceanic islands: importance of invader complexes and endemic super generalists. *Diversity and Distributions* **8**: 181–192.

Olesen JM, Bascompte J, Dupont YL, Jordano P. 2007. The modularity of pollination networks. *Proceedings of the National Academy of Sciences of the USA* **104**:19891–19896.

Olesen JM, Bascompte J, Elberling H, Jordano P. 2008. Temporal dynamics in a pollination network. *Ecology* **89**:1573–1582.

Ollerton J, Winfree R, Tarrant S. 2011. How many flowering plants are pollinated by animals? *Oikos* **120**:321–326.

Padrón B, Traveset A, Biedenweg T, Díaz D, Nogales M, Olesen JM. 2009. Impact of alien plant invaders on pollination networks in two archipelagos. *PLoS ONE* **4**:e6275. doi:10.1371/journal.pone.0006275

Petanidou T, Kallimanis AS, Tzanopoulos J, Sgardelis SP, Pantis JD. 2008. Long-term observation of a pollination network: fluctuation in species and interactions, relative invariance of network structure and implications for estimates of specialization. *Ecology Letters* **11**:564–575.

Poulakakis N, Russello M, Geist D, Caccone A. 2012. Unravelling the peculiarities of island life: vicariance, dispersal and the diversification of the extinct and extant giant Galápagos tortoises. *Molecular Ecology* **21**:160–173.

R Development Core Team. 2014. *R: a language and environment for statistical computing.* Vienna, Austria: R Foundation for Statistical Computing.

Rooney N, McCann KS. 2012. Integrating food web diversity, structure and stability. *Trends in Ecology and Evolution* **27**:40–46.

Santos GMD, Aguiar CML, Genini J, Martins CF, Zanella FCV, Mello MAR. 2012. Invasive Africanized honeybees change the structure of native pollination networks in Brazil. *Biological Invasions* **14**: 2369–2378.

Sax DF, Gaines SD. 2008. Species invasions and extinction: the future of native biodiversity on islands. *Proceedings of the National Academy of Sciences of the USA* **105**:11490–11497.

Sebastián-González E, Dalsgaard B, Sandel B, Guimarães PR. 2015. Macroecological trends in nestedness and modularity of seed-dispersal networks: human impact matters. *Global Ecology and Biogeography* **24**:293–303.

Tapia W, Ospina P, Quiroga D, González JA, Montes C. 2009. *Ciencia para la sostenibilidad en Galápagos: el papel de la investigación científica y tecnológica en el pasado, presente y futuro del archipiélago.* Quito, Parque Nacional Galápagos: Universidad Andina Simón Bolívar, Universidad Autónoma de Madrid y Universidad San Francisco de Quito.

Traveset A, Richardson DM. 2006. Biological invasions as disruptors of plant reproductive mutualisms. *Trends in Ecology and Evolution* **21**:208–216.

Traveset A, Richardson DM. 2014. Mutualistic interactions and biological invasions. *Annual Review of Ecology, Evolution, and Systematics* **45**:89–113.

Traveset A, Heleno R, Chamorro S, Vargas P, McMullen CK, Castro-Urgal R, Nogales M, Herrera HW, Olesen JM. 2013. Invaders of pollination networks in the Galápagos Islands: emergence of novel communities. *Proceedings of the Royal Society B: Biological Sciences* **280**:20123040.

Traveset A, Heleno RH, Nogales M. 2014. The ecology of seed dispersal. In: Gallagher RS, ed. *Seeds: the ecology of regeneration in plant communities*. Oxfordshire, UK: CABI.

Traveset A, Olesen JM, Nogales M, Vargas P, Jaramillo P, Antolín E, Trigo MM, Heleno R. 2015a. Bird-flower visitation networks in the Galápagos unveil a widespread interaction release. *Nature Communications* **6**:6376.

Traveset A, Tur C, Trøjelsgaard K, Heleno R, Castro-Urgal R, Olesen JM. 2015b. Global patterns of mainland and insular pollination networks. *Global Ecology and Biogeography*. In press.

Trueman M, d'Ozouville N. 2010. Characterizing the Galapagos terrestrial climate in the face of global climate change. *Galapagos Research* **67**:26–37.

Tye A. 2006. Can we infer island introduction and naturalization rates from inventory data? Evidence from introduced plants in Galápagos. *Biological Invasions* **8**:201–215.

Ziegler W. 1995. El Archipiélago de las Galápagos. Ubicación clima, condiciones atmosféricas y origen geológico. In: Zizka G, Klemmer K, eds. *Flora y Fauna de las Islas Galápagos: Origen, Investigación, Amenazas y Protección*. Frankfurt, Germany: Palmengarten der Stadt.

Zuur A, Ieno EN, Walker N, Saveliev AA, Smith GM. 2009. *Mixed effects models and extensions in ecology with R*. New York, USA: Springer.

7

Introduced birds incompletely replace seed dispersal by a native frugivore

Liba Pejchar*

Department of Fish, Wildlife and Conservation Biology, Colorado State University, Fort Collins, CO 80523, USA

Guest Editor: Donald Drake

Abstract. The widespread loss of native species and the introduction of non-native species has important consequences for island ecosystems. Non-native species may or may not functionally replace the role of native species in ecological processes such as seed dispersal. Although the majority of Hawaii's native plants require bird-mediated seed dispersal, only one native frugivore, Omao (*Myadestes obscurus*), persists in sufficient numbers to fill this functional role. Omao are restricted to less than half their original range, but two introduced frugivores are abundant throughout Hawaii. Given large-scale extinctions on islands, it is important to understand whether introduced birds serve as functional replacements or whether the absence of native frugivores alters plant communities. To assess seed dispersal by native and introduced birds, seed rain, vegetation characteristics, bird diet, density and habitat use were measured at three sites with Omao and three sites without Omao on Hawaii Island. The diet of native and introduced birds overlapped substantially, but Omao dispersed a variety of native species ($n = 6$) relatively evenly. In contrast, introduced birds dispersed an invasive species and fewer native species ($n = 4$), and $>90\%$ of seeds dispersed by introduced birds were from two ubiquitous small-seeded species. Seed rain was significantly greater and more species rich at sites with Omao. These findings suggest that patterns of seed dispersal are altered following the local extinction of a native island frugivore. To more directly evaluate the relative roles of native and introduced frugivores in ecological processes, future studies could include reintroducing Omao to a suitable habitat within its historic range, or novel introductions to nearby islands where closely related species are now extinct. In an era of widespread extinction and invasion of island ecosystems, understanding the consequences of novel animal assemblages for processes like seed dispersal will be critical for maintaining diverse and self-regenerating plant communities.

Keywords: Ecological processes; forest regeneration; frugivory; Hawaiian islands; invasive species; mutualism; plant–animal interactions; seed rain.

* Corresponding author's e-mail address: liba.pejchar@colostate.edu

Introduction

Ecosystem processes are increasingly disrupted by the loss and functional extinction of ecologically important species (Anderson *et al.* 2011; Dirzo *et al.* 2014). For example, the decline of fruit-eating birds is altering seed dispersal dynamics (McConkey *et al.* 2012; Kurten 2013), with diverse consequences for plant communities. The absence of vertebrate dispersers can lead to dispersal failure (Chimera and Drake 2010), decrease plant density and diversity (Traveset and Riera 2005; Sharam *et al.* 2009; Traveset *et al.* 2012), shift community composition towards small-seeded generalists and abiotically dispersed plants (Terborgh *et al.* 2008) or a few dominant species (McConkey *et al.* 2012; Kurten 2013) and limit the resilience of plant and animal communities to global change (Elmqvist *et al.* 2003).

In some cases, introduced birds, which are increasingly well established in ecosystems worldwide (Blackburn *et al.* 2009), are the sole dispersers of native plants and may fill the functional role of lost native species (Foster and Robinson 2007), but more often they appear to provide limited benefits or generate negative effects by preferentially dispersing seeds of invasive plants (Kelly *et al.* 2006; Williams 2006; Chimera and Drake 2010; Aslan *et al.* 2012). Understanding the extent to which introduced birds ecologically replace native species is important because seed dispersal is believed to play a crucial role in structuring and maintaining plant communities (Howe and Smallwood 1982; Terborgh *et al.* 2008) through several well-established mechanisms. For example, bird-mediated dispersal enhances recruitment by moving seeds away from parent plants to suitable microsites, allowing seedlings to escape competition and density-dependent mortality (Janzen 1970; Connell 1971; Wotton and Kelly 2011). Thus, if introduced species are inadequate substitutes for native frugivores, the rapid decline and extinction of native frugivores is even more likely to dramatically alter plant assemblages (Bond 1994; Sekercioglu *et al.* 2004; Traveset and Richardson 2006) and reduce overall native plant diversity (Webb and Peart 2001).

The ecological role filled by native frugivores may be disproportionately important on islands such as Hawaii, where most woody plants have historically been dispersed by birds (Carlquist 1974). This relationship between native birds and plants is threatened by the loss or functional extinction of much of Hawaii's avifauna from habitat loss, disease and predation by introduced mammals (Pratt *et al.* 2009). In particular, all crows (*Corvus* spp.) and thrushes (*Myadestes* spp.) are extinct in the wild with the exception of a critically endangered thrush (Puaiohi; *Myadestes palmeri*) on Kauai Island and Omao (*Myadestes obscurus*) on the Island of Hawaii. The Omao is locally common and believed to be stable, but globally listed as vulnerable to extinction (Birdlife International 2012) because it is restricted to a fraction of its historic range (25–30 %) on a single Island (Scott *et al.* 1986; Wakelee and Fancy 1999). Reasons for this range contraction are not well understood (Ralph and Fancy 1994), but could include historic habitat degradation (Scott *et al.* 1986) and fragmentation (Wu *et al.* 2014), predation by introduced mammals (Kilpatrick 2006) or exposure to diseases such as avian pox or malaria (Atkinson *et al.* 2001).

Concurrent with the extinction and decline of Hawaii's native birds, the archipelago has received more introduced birds than anywhere else in the world (>58 species established; Moulton and Pimm 1983; Pyle 2002). Two of these birds, Japanese White-eye (*Zosterops japonicus*) and Red-billed Leiothrix (*Leiothrix lutea*), introduced from east Asia and the Indian subcontinent, respectively, have been highly successful in colonizing native forest communities. Both of these species are small bodied and small gaped (Japanese White-eye: weight 10–12 g, culmen 10–11 mm; Red-billed Leiothrix: 20–23 g, 10–12 mm). Japanese White-eye are generalists that consume insects, fruit and nectar; Red-billed Leiothrix consume fruit as well as insects (Mountainspring and Scott 1985; Male *et al.* 1998). In contrast, Omao are larger-bodied (weight 46–49 g, culmen 12–16 mm), but like Red-billed Leiothrix, they are primarily frugivorous and also consume insects (Ralph and Fancy 1994; Wakelee and Fancy 1999). Each of these species belong to a different family within the order Passeriformes.

Because introduced frugivorous species are abundant throughout Hawaii's forests, in the absence of native frugivores, they have the potential to play a positive role in maintaining native plant communities (Foster and Robinson 2007). Alternatively, Japanese White-eye and Red-billed Leiothrix may only disperse a subset of native seeds dispersed by Omao, or they may facilitate the spread of undesirable invasive plants (Traveset and Richardson 2006).

To address the role of native and introduced birds in seed dispersal and the regeneration of Hawaii's plant communities, this study poses the following specific questions. (i) Do introduced birds disperse the same seeds as the native Omao? (ii) Do landscapes with Omao receive a higher abundance and diversity of seed rain from native plants relative to landscapes without Omao? (iii) Does habitat use by Omao differ from introduced frugivorous birds?

These questions are highly relevant to several emerging topics in applied ecology, including whether or not to embrace novel ecosystems (areas where non-native species are established and inter-mingle with native species), and under what circumstances to pursue refaunation as a conservation strategy. Refaunation is defined as the restoration of ecologically important species either within or beyond their historic range. Making informed decisions about refaunation requires understanding the relationship between native and introduced animal communities and ecosystem processes (Armstrong and Seddon 2008), particularly in the increasing fraction of the world that is dominated by novel ecosystems (Hobbs *et al.* 2009).

Methods

Study sites

This study took place from 1 June 2006 to 20 August 2006 and was located on the Island of Hawaii at three sites where Omao are present and three sites where Omao occurred historically but are now locally extinct. The three sites with Omao are the Pua Akala tract of Hakalau Forest National Wildlife Refuge, Keauhou Ranch, which is owned and managed by Kamehameha Schools, and The Nature Conservancy of Hawaii's Kaiholena Preserve. The three sites without Omao are Puu Waawaa Bird Sanctuary, managed by the Hawaii Department of Natural Resources, Honaunau Forest Reserve, owned and managed by Kamehameha Schools, and The Nature Conservancy of Hawaii's Kona Hema Preserve.

The sites with and without Omao are spatially segregated due to the contraction of the former range of this species. All sites, however, can be characterized as mesic rainforest with precipitation ranging from ~900–2000 mm year^{-1} (Juvik and Juvik 1998). These study sites were located at elevations between 975 and 1900 m **[see Supporting Information]**. The vegetation communities are dominated by ohia (*Metrosideros polymorpha*) and koa (*Acacia koa*) in the canopy and diverse sub-canopy and understory plants, many of which are woody fleshy-fruited species.

Each of these sites is recovering from a varied history of grazing by cattle and other introduced mammals, but management for these species did not vary systematically between sites with and without Omao at the time of the study. The sites with Omao have all experienced a long history of herbivory by cattle and feral pigs (*Sus scrofa*) either within or immediately adjacent to the forest. Feral pigs were present at all three of these sites during the time of this study, and management activities to remove or reduce feral pig densities was occurring at two of the three sites. Similarly, the sites without Omao also have a long history of disturbance by the same non-native mammals, and with the exception of the Kona Hema preserve, all contained feral pigs at the time of this study. Active management to remove feral pigs was occurring at the Puu Waawaa bird sanctuary.

A square grid was established at each site that consisted of 16 points, 200 m apart. All seed rain traps,

vegetation surveys and point counts were conducted at these points and all collection of faecal samples and bird habitat use observations occurred within these grids.

Bird diet

In order to determine whether Omao and introduced frugivorous birds disperse seeds from different sets of plants, faecal samples were collected and analyzed from each target bird species. Faecal samples were obtained by using mistnets to capture birds. Nets ($n = 6–12$) were in use for 3–4 days/month at each site throughout the duration of the field season for a total of 3160 net hours. The nets were opened at approximately 0630 (1 h after sunrise) and closed at 12 pm, or earlier in the case of inclement weather. All potential frugivores [Omao, Red-billed Leiothrix, Japanese White-eye, Hawaii Amakihi (*Hemignathus virens*) and Northern Cardinal (*Cardinalis cardinalis*)] were extracted from the nets and placed in clean cotton bags, at which point the bird would usually defecate. The faecal sample was then removed from the bag, placed in a plastic vial with 70 % ethanol and stored in a cool location. Metal and colour bands were placed on each bird, and weight, wing length, tail length, tarsus length, bill length and sex and age (when possible), were recorded. Other research teams working at the same sites during the same time period, occasionally collected faecal samples from these study species and donated them to this study. All of the faecal samples were later sorted under a light microscope. Seeds were counted and identified to species using a reference collection of seeds collected from the same sites.

Vegetation

To test for potential differences in plant species diversity and density among sites with and without Omao, per cent cover of all plant species in 10 m radius plots was recorded. Using ocular methods, per cent cover was estimated at three canopy levels at all sites: overstory (>10 m), understory (1–10 m) and ground cover (<1 m). The vegetation plots were located at all odd-numbered grid points at each site, resulting in 8 plots/site or a total surface area of ~2500 m^2/site. These data were used to calculate per cent cover (m^2/ha) of ground cover types, fleshy-fruited plant species and canopy species at each site. Each month of the study ($n = 3$) fruiting phenology was documented by recording the extent to which each individual fleshy-fruited plant species within 10 m of the plot centre was fruiting. The per cent of all surface area of a particular species that was in fruit was estimated using the following categories: 0, 1–33, 34–66 and 67–100 %. Only ripe fruits were included in this estimate.

Seed rain

To test for differences in seed dispersal by birds at sites with and without Omao, 40 seed traps were established at each site (total traps = 240). Groups of 10 traps were clustered at four randomly chosen grid points. Half of the traps at each point were placed under randomly chosen wind-dispersed canopy trees (*M. polymorpha*) and the remaining half were placed under randomly chosen fleshy-fruited understory plants. Each trap was composed of a 33-inch tomato cage sunk into the ground with a seven gallon plastic plant pot sitting firmly on top of the cage. The funnel-shaped cage was designed to make the traps largely inaccessible to seed predation by rats and mice. Rat faeces were observed in the traps only two times and these faeces may have fallen from the tree canopy. The plant pots had holes on the bottom for drainage and were covered in fine mosquito mesh fitted so that the mesh formed a concave bowl within the pot. A rock was placed on top of the mesh to prevent wind from disturbing the concave/funnel shape of the mesh. All bird-dispersed seeds and other organic material were thus caught within this concave mesh funnel.

Once each month all of the traps were checked, large pieces of litter (e.g. branches) were brushed off above the trap and the remaining material was transferred to a quart size zip-lock bag. The content of these bags were then examined under a light microscope and all seeds were counted and identified to species. Only seeds without a fruit coating were counted. In addition, because decomposition could result in fruit removal from seeds, seeds of the same species as the tree and shrub species above the trap were excluded from the analysis.

Frugivorous bird density

Point counts were conducted at all six sites to determine the density of each frugivorous bird species. Following standard procedures for 8-min variable circular point counts (Reynolds *et al.* 1980), the following information was recorded: distance to each bird, the method of detection (visual or aural) and several weather parameters (cloud cover, rain, wind). All point counts took place from 0630 to 1100 during June 2006, which is within the breeding season for all of the focal bird species.

Habitat use

Habitat use by frugivorous birds could influence patterns of seed dispersal. To explore this hypothesis, at least 20 individuals of each focal species were observed at all study sites. Birds were located by systematically traversing each study site and searching for individuals of each focal species (Vanderwerf 1994; Pejchar *et al.* 2005). Observation times varied from 1 to 8 min, and terminated when visual contact with the bird was lost. The plant species in which the bird was foraging, perching or vocalizing was recorded, as well as the total time spent in each canopy or understory species.

Data analysis

Potential differences in the diet of Omao and introduced birds were assessed by calculating the relative proportion of seed species dispersed across all faecal samples for each bird species. Site-level per cent cover (m^2/ha) of canopy, understory plants and ground cover and the extent of fruiting (proportion of plant surface area) were calculated by averaging values across the eight vegetation plots at each site. To assess potential differences in plant community composition among sites, Student's *t*-tests (two-tailed, unequal variance) were used to compare per cent cover of each plant species, per cent cover of all fleshy-fruited plants, extent of fruiting and species richness of fleshy-fruited plant species at sites with and without Omao.

Prior to comparing seed rain at sites with and without Omao, seed rain was adjusted to account for differences in the number of seed trap days and the per cent cover and extent of fruiting of bird-dispersed plants at each site. First, seed rain was summed across all traps at each site and divided by 50 to determine seed rain/50 days per site. Second, using the vegetation data collected for each site (described above), seed rain was re-calculated as a function of site-specific per cent cover for species that appeared in seed rain (e.g. the sum of the number of seeds per 50 trap days/% cover of each plant species), and the extent of fruiting (sum of the number of seeds per 50 trap days/proportion of canopy in fruit). The difference in seed rain from all plant species among sites with and without Omao was assessed using a full factorial two-way ANOVA with site type (with or without Omao) and trap location (under fleshy-fruited or wind-dispersed plants) as the main effects and site type × trap location as an interaction effect.

To achieve sufficient replication for statistical analysis, only seed species that occurred in seed rain at two or more sites with and without Omao met the criteria for species-level comparisons. To compare seed rain of these species among sites, Student's *t*-tests were used to compare the mean ± SD seed rain adjusted for vegetation cover at sites with and without Omao. The same test was used to compare species richness of seed rain between sites with and without Omao. This approach, incorporating sources of variation (number of trap days and vegetation cover) into the response variable (seed rain), was adopted because the limited number of data points in this study ($n = 6$) did not provide sufficient degrees of freedom for a statistical model that included habitat and/or site characteristics as covariates.

To compare bird density (birds/ha) at each site, detections were truncated at a radius of 50 m at each point prior to analysis. Student's *t*-tests were used to compare mean (± SD) bird density of all frugivorous birds, introduced frugivorous birds and each frugivorous bird species at sites with and without Omao. The relative proportion of each species in the frugivorous bird community at sites with and without Omao was also calculated (e.g. density of species X/overall bird density). To explore differences in habitat use by Omao and introduced birds, all observations for each bird species were summed across sites. These data were used to calculate the relative time spent by Omao, Japanese White-eye and Red-billed Leiothrix each plant species (e.g. total time observed in plant species X/total observation time).

Results

Bird diet

Faecal samples ($n = 93$) were collected from five bird species known to consume fruit (Omao, Japanese White-eye, Red-billed Leiothrix, Hawaii Amakihi and Northern Cardinal. These samples contained a total of 714 seeds from eight plant species. Two of the bird species had few (Hawaii Amakihi = 1) or no (Northern Cardinal = 0) seeds in their faecal samples, suggesting that they are unlikely to be important seed dispersers in this system. Thus, the following analyses focus on the three bird species which do disperse large numbers of seeds: Omao, Japanese White-eye and Red-billed Leiothrix (Table 1).

The three seed species that occurred most commonly in all faecal samples were *Vaccinium calycinum* (43.8 %), *Rubus hawaiensis* (36.7 %) and *Cheirodendron trigynum* (6.6 %; Table 2). These species were dispersed by all three bird species. The remaining seed species were less common in the faecal samples and were dispersed only by the introduced bird species or only by Omao. *Perrottetia sandwicensis* (0.3 %) and *Rubus rosifolius* (3.9 %), an introduced plant, were dispersed only by the introduced birds. *Ilex anomala* (6.9 %), *Leptecophylla tameiameiae* (0.4 %) and *Psychotria* spp. (1.4 %) were dispersed only by Omao. The large majority of the seeds dispersed by introduced birds were either *R. hawaiensis* or *V. calycinum*; Japanese White-eye = 90.4 %, Red-billed Leiothrix = 92.1 %). In contrast Omao dispersed seeds more evenly (*I. anomala* = 34.0 %, *V. calycinum* = 22.2 %, *C. trigynum* = 20.1 %, *R. hawaiensis* = 14.6 %; Table 2).

Vegetation

There was no difference in the per cent cover of the primary canopy tree (*M. polymorpha*) or any ground cover type (fern, grass, rock, wood, bare ground, introduced herb, moss) between sites with and without Omao, but sites with Omao did have higher canopy cover of *A. koa* **[see Supporting Information]**. There was also no difference in the species richness of fleshy-fruited plant species

Table 1. The proportion of faecal samples from native and introduced birds that contained seeds, and the richness of native and introduced seeds dispersed by each species.

Bird species	Per cent of samples with seeds	Native seed richness	Introduced seed richness
Omao *Myadestes obscurus* (n = 19)	84.2	6	0
Japanese White-eye *Zosterops japonicus* (n = 33)	69.7	4	1
Red-billed Leiothrix *Leiothrix lutea* (n = 19)	84.2	4	1

Table 2. The per cent of seeds from each plant species found in the diet samples of the native bird (Omao), the two introduced birds (Japanese White-eye and Red-billed Leiothrix) and all bird species combined. Seed sizes range from 0.5 to 4 mm in length. * indicates that the bird or plant species is introduced.

Plant species	Seed size—length (mm)	Per cent of diet			
		Omao	Japanese White-eye*	Red-billed Leiothrix*	All bird species
R. hawaiensis	3	14.6	85.9	22.7	36.7
Vaccinium calycinum	0.5	22.2	4.5	69.4	43.8
C. trigynum	4	20.1	6.2	1.8	6.6
I. anomala	2	34.0	0	0	6.9
Styphelia tameiameiae	3.5	2.1	0	0	0.4
Psychotria spp.	3	6.9	0	0	1.4
P. sandwicensis	1.2	0	0.6	0.3	0.3
*R. rosifolius**	1	0	2.8	5.7	3.9

(sites with Omao = 11.7 ± 0.4; sites without Omao = 12 ± 0.4), the richness of plant species in fruit during the study season (sites with Omao = 7.7 ± 1.5; sites without Omao = 7.7 ± 1.1) or the per cent cover or extent of fruiting in the most common understory plants between sites with and without Omao **[see Supporting Information]**. Three exotic fleshy-fruited plant species (*Passiflora mollisima*, *Rubus argutus* and *R. rosifolius*) were observed at one or more sites both with and without Omao, and two exotic fruiting species (*Passiflora edulis* and *Psidium cattleianum*) were observed only at one or more sites without Omao.

Seed rain

A total of 1020 seeds from 11 plant species were collected in 240 traps over the collection period ($n = 12 150$ trap days). Just over half of the traps (52.1 %) contained seeds at some point during the study. Four out of the 11 seed species collected in seed traps (*R. hawaiensis*, *V. calycinum*, *C. trigynum* and *I. anomala*) were dispersed in numbers great enough to allow comparisons between sites with and without Omao. The remaining seven species (*Psychotria* spp., *P. sandwicensis*, *Myoporum sandwicense*, *P. mollisima*, *Clermontia* spp., *L. tameiameiae* and *Myrsine* spp.) were collected and/or were present at only one or two sites with or without Omao, which provided insufficient data for statistical analysis.

Overall, sites with Omao had significantly higher species richness of seed dispersed into seed traps (sites with Omao: 5.6 ± 0.5; sites without Omao: 4.3 ± 0.5; $t = -2.8$; df = 4; $P = 0.04$). These sites also had significantly higher numbers of seeds dispersed even after correcting for the per cent cover of understory plants (sites with Omao = 59.2 ± 9.7; sites without Omao = 12.2 ± 11.5; $t = 5.4$; df = 4; $P = 0.006$) and the per cent cover of fleshy-fruited plants at each site (sites with Omao = 232.7 ± 74.3; sites without Omao = 64.3 ± 73.0; $t = 2.8$; df = 4; $P = 0.05$). Native seed species richness was 5.5 ± 0.9 at sites with Omao and 3.3 ± 1.2 at sites without Omao. No exotic species appeared in the seed rain at sites with Omao and only a single exotic species (*P. mollisima*) was observed in the seed rain at sites without Omao.

There was significantly more *R. hawaiensis* and *C. trigynum* seeds dispersed (relative to per cent cover of each plant species) at sites with Omao (*R. hawaiensis*: $t = 7.7$; df = 3; $P = 0.004$; *C. trigynum*: $t = 3.5$; df = 3; $P = 0.04$; Fig. 1). *Vaccinium calycinum* seeds were only dispersed at sites with Omao; this species was not observed in fruit at non-Omao sites during the study season **[see Supporting Information]**. *Ilex anomala* seeds were only dispersed at the sites with Omao (Fig. 1), despite no difference in *I. anomala* per cent cover or

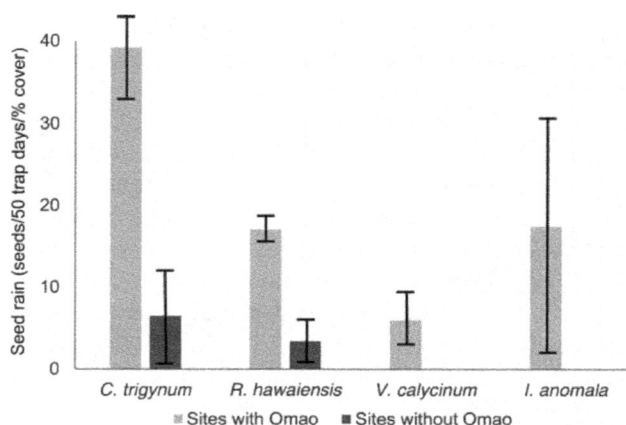

Figure 1. Seed rain from four fleshy-fruited species (number of seeds/50 trap day adjusted for per cent cover of plant species; mean ± SD) at sites with and without Omao.

fruiting among the sites with or without Omao **[see Supporting Information]**.

The two-way analysis of variance yielded differences in seed rain among sites, trap locations and the interaction between those two main effects ($F(3,8) = 22.6$; $P < 0.0003$). Seed rain differed by site type (sites with and without Omao; $P < 0.0003$), and trap location (under bird-dispersed fleshy-fruited plants or wind-dispersed canopy trees—*M. polymorpha*; $P < 0.0128$). The interaction effect (site type × trap location) was also significant ($P < 0.0017$); more seeds were collected under *M. polymorpha* at sites with Omao (Fig. 2).

Frugivorous bird density

The density (birds/ha) of introduced frugivores was significantly higher at sites without Omao (sites with Omao = 3.1 ± 0.1; sites without Omao = 4.4 ± 0.8; $t = -0.3$; df = 2; $P = 0.05$), but overall density of frugivores, and density of individual frugivores did not differ between sites with and without Omao **[see Supporting Information]**. On average, Omao consisted of 46 % of the frugivorous bird population at sites with Omao. At these sites, Japanese white-eyes were 35 %, and Red-billed Leiothrix were 19 % of all frugivorous birds. At sites without Omao, Japanese White-eye were 56 % and Red-billed Leiothrix were 44 % of the frugivorous bird population. The absence of Omao from a site was thus associated with a 1.6× increase in Japanese White-eye and a 2.3× increase in Red-billed Leiothrix (Fig. 3).

Habitat use

The three focal bird species were observed for a total of 535 min [Japanese White-eye 105 min ($n = 36$ individuals); Omao 352 min ($n = 55$ individuals); Red-billed Leiothrix 66 min ($n = 20$ individuals)] throughout the

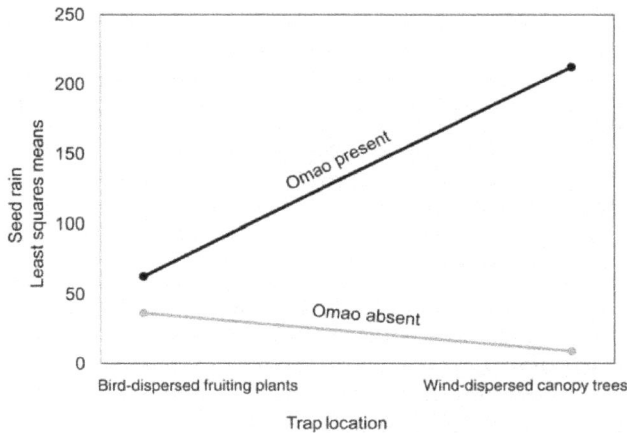

Figure 2. Seed rain (least squares means) under a wind-dispersed canopy tree (*M. polymorpha*) and under fleshy-fruited plants at sites with and without Omao.

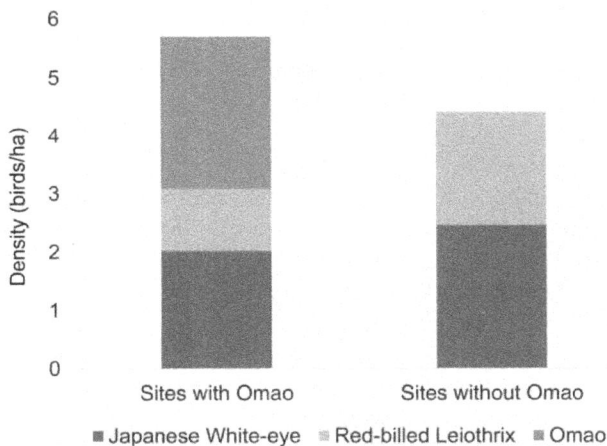

Figure 3. Mean density (birds/ha) of frugivorous bird species at sites with and without Omao.

study period. Omao spent most of their time perching and calling from *M. polymorpha* (71 % of total observation time), secondarily on *C. trigynum* (14 %) and sparingly in five other species **[see Supporting Information]**. Japanese White-eye also spent most of their time in *M. polymorpha* (64 % of total observation time), and were observed in 12 other species relatively evenly. In contrast, Red-billed Leiothrix were observed mostly in understory plants. This species spent most time in either *R. hawaiensis* (55 % of total observation time) or *V. calycinum* (29 %), venturing into *M. polymorpha* <5 % of the total observation time **[see Supporting Information]**.

Discussion

The decline and extinction of native frugivorous birds could have important consequences for plant communities

by altering patterns of seed dispersal (Loiselle and Blake 2002; Traveset and Richardson 2006). Alternatively, introduced birds may functionally replace ecological services provided by native species (Foster and Robinson 2007; Aslan *et al.* 2014). Understanding which of these is true is critical to knowing whether to prioritize conservation and refaunation of native dispersers within their historic range, to consider introducing ecologically important species outside their native ranges as proxies for extinct species, or to embrace some introduced species as fulfilling valuable ecological functions. These options are not mutually exclusive and each may be more or less desirable depending on context-specific conservation objectives.

In this exploratory study, Omao and introduced frugivorous birds dispersed similar seed species on Hawaii Island, but in different proportions. Faecal samples of introduced birds were strongly (>90 %) dominated by two common native species, while Omao dispersed a greater variety of seeds in more equal proportions. Only the introduced birds dispersed seeds from an introduced plant. Seed rain was greater and more diverse at sites with Omao, even after accounting for small differences in per cent cover of fruiting plants. Finally, more seeds were dispersed under wind-dispersed canopy trees than understory fleshy-fruited plants at sites with Omao. This finding is consistent with observed patterns of habitat use. Omao spent most of their time in the canopy; in their absence, a greater proportion of the frugivore community was active in the understory. Although this study measured only bird-mediated seed dispersal, this result suggests that not only is seed rain less frequent, less species rich and biased towards smaller seeded plants, but without native frugivores, seeds may also be dispersed into different microclimate conditions because of differences in habitat use among native and introduced frugivores.

In comparison to Omao, both Japanese White-eye and Red-billed Leiothrix are small-bodied generalists. Japanese White-eye in particular have a narrow, piercing bill and may remove flesh from fruit rather than consuming seeds. The smaller gape and more omnivorous diet of the introduced birds together with their diet, as observed in this study and others (Wu *et al.* 2014), suggest that these species are an incomplete ecological substitute for Omao. In contrast to this study, which found substantial dietary overlap between Omao and introduced birds, Wu *et al.* (2014), also working on Hawaii Island but in a single location, found that Japanese White-eye only disperse two seed species. These species were *R. hawaiensis* and *V. calycinum*, which also dominated the diet of introduced birds in this study.

Previous research has suggested introduced birds play an important ecological role in Hawaii, particularly in the absence of native frugivores. Foster and Robinson (2007) studied seed dispersal on the Island of Maui, where all native frugivores are presumed extinct. They reported a very different composition of seeds in faecal samples of introduced birds, despite working in a plant community dominated by similar species as this study. These differences may be partially explained by the length of their field season; they collected faecal samples over a full year, which may have included fruit that was not available or less abundant during the shorter timeframe of this study.

Common to these studies and others (Chimera and Drake 2010) is the concern that introduced birds could exacerbate the spread of invasive species. Foster and Robinson (2007) found that introduced birds disperse *Hedychium gardnerianum* and *R. argutus* and this study found evidence of regular dispersal of *R. rosifolius*—these species are all either considered noxious weeds in Hawaii (USDA 2015) and/or are listed in the IUCN's global invasive species database (IUCN 2015). Wu *et al.* (2014) demonstrated that Japanese White-eye can disperse seeds further than Omao. Collectively, these findings suggest introduced birds may contribute to primary succession, as well as spread small-seeded invasive species, particularly in fragmented landscapes. Additionally, larger seeded plants may be more susceptible to dispersal failure in the absence of the Omao and Hawaii's largest extant frugivore, the Hawaiian Crow (*Corvus hawaiiensis*), which persists only in captivity (Culliney *et al.* 2012).

The decline in the frequency and diversity of seed rain at sites without Omao supports the prediction that the loss of native frugivores alters patterns of seed dispersal. Bird-mediated dispersal, however, is just one component of the maintenance and regeneration of plant communities. Although measuring seedling recruitment and seedling survival was beyond the scope of this study, understanding whether plants are dispersal-limited or constrained by other factors (e.g. competition, seed predation or seedling herbivory by other introduced species) is critical (Denslow *et al.* 2006; Pender *et al.* 2013). Introduced rats, for example, have been shown to interact with native and invasive plants through both seed predation and dispersal (Shiels 2011; Shiels and Drake 2011).

Although the phenology of fruiting did not differ substantially among sites for most species [see Supporting Information], this may have contributed to the seed rain patterns observed, particularly since the sites with and without Omao were spatially disparate. For example, *V. calycinum* was only dispersed at sites with Omao. Although this species was present at all sites, it was not observed fruiting at sites without Omao during the study season. In contrast, *I. anomala* was also only dispersed at sites with Omao, but there was no difference in the

per cent cover or extent of fruiting in this plant among site with and without Omao (Fig. 1). Further, because many bird-dispersed species in Hawaii have long fruiting seasons, and not all of these overlapped with the period of this exploratory study, year-round and multi-year studies of the temporal and spatial patterns of bird diet and seed rain are warranted.

In the absence of Omao, the frugivorous bird community appears to shift towards greater dominance of Red-billed Leiothrix. It is unclear whether this change is due to relaxed competition with Omao or due to fundamental differences in site and habitat characteristics that were not measured in this study. Understanding the degree to which introduced species compete with native species is a ripe area for further inquiry in Hawaii and other island ecosystems (Moulton and Pimm 1983). Preliminary observations of habitat use in this study suggest that Red-billed Leiothrix forage largely in the understory and rarely venture into the canopy. This pattern is potentially consistent with the higher seed rain documented under *M. polymorpha* at sites with Omao compared with sites without Omao (Fig. 2). If the gap left empty by Omao is filled disproportionally by a bird that spends time in the understory rather than the canopy, seeds are likely to be dispersed into a different microclimate, which could have important implications for germination success (Janzen 1983). Bird-dispersal that originates in *M. polymorpha* may result in higher germination rates if the seeds are deposited within the canopy (plants that grow epiphytically are protected from introduced mammalian herbivores; Drake and Pratt 2001), or under the canopy, which could shade out introduced grasses and forbs, thus reducing competition for resources (Cordell *et al.* 2009). This prediction, that the absence of Omao leads not only to less seed dispersal overall but also to less seed rain into suitable microhabitat, warrants future research.

In addition to the short time frame and limited observations of habitat use by frugivorous birds, this study has several other important limitations. The sites with and without Omao are necessarily spatially segregated due to the current distribution of this species. Thus, it is possible that factors other than the presence or absence of Omao (e.g. land use history, density of seed predators, patterns of precipitation and primary productivity) could contribute to differences in the rate and magnitude of seed rain and density of introduced birds among the two groups of sites. For example, sites with Omao tend to receive more precipitation than sites where Omao are now absent [see Supporting Information]. Despite these limitations, this study's findings are consistent with previous information on the degree of diet specialization of the focal species (Wu *et al.* 2014). Given the restricted range of the Omao and the arguable functional extinction of all other native Hawaiian frugivores (Wakelee and Fancy 1999; Culliney *et al.* 2012), measuring seed rain before and after an experimental reintroduction may be the most effective means of assessing the link between native frugivorous birds, seed rain and plant communities in island systems transformed by anthropogenic activities.

Conclusions

Introduced birds are imperfect substitutes for native species on Hawaii Island. Although introduced birds provide some seed dispersal services, the relative abundance and proportion of seeds they disperse differs from native species, as does their habitat use, which could have implications for seed germination and seedling survival. Research priorities include understanding which plant species are truly dispersal limited in the absence of native frugivores (Denslow *et al.* 2006; Inman-Narahari *et al.* 2013) and whether these species have shared characteristics on islands with introduced avifauna (e.g. large-seeded plant species; Meehan *et al.* 2002). This requires assessing the importance of reduced or altered seed dispersal services relative to other factors (e.g. seed predation and herbivory) that effect recruitment.

Omao offer a unique opportunity to measure changes in seed rain over time following reintroduction or recolonization. This bird remains common within its current range, and this range may be expanding naturally (Judge *et al.* 2012). If past and ongoing threats are identified and alleviated, active reintroduction of Omao to ensure the persistence of this species and to restore ecological processes may become a priority (Atkinson *et al.* 2001; Fancy *et al.* 2001).

Maintaining the link between plants and seed dispersers is critical on islands such as Hawaii where frugivory is an important ecological process for the majority of woody plants. Given large scale habitat fragmentation and other extinction risks to native frugivores, conserving existing populations should be a priority. Increasingly, island biologists are also considering reintroducing ecologically important species to suitable habitat within their historic range and to neighbouring islands with depauperate communities of vertebrate seed dispersers. Although novel introductions should always be approached with caution, such actions have precedent (Seddon *et al.* 2014) and could offer an exciting opportunity to experimentally evaluate the ecological role of native frugivores. In lieu of handing over the fate of Hawaii's plants to introduced birds, conservation strategies should focus on protecting and recovering native bird species to ensure the maintenance and regeneration of diverse island plant communities.

Sources of Funding

Funding was provided by Stanford University's Center for Conservation Biology and the Stanford Woods Institute for the Environment.

Contributions by the Authors

L.P. designed the study, collected the data, conducted the analysis and wrote the manuscript.

Acknowledgements

I thank the following colleagues for their help in conceiving the study: G. Daily, P. Ehrlich, J. Goldstein, B. Brosi, R. Pringle and J. Ranganathan, and I am grateful to P. Taylor for excellent assistance in the field. I am indebted to the staff of the US Fish and Wildlife Service at Hakalau Forest NWR, The Land Assets Division of Kamehameha Schools, the Nature Conservancy of Hawaii and the State of Hawaii Division of Forestry and Wildlife for supporting this research on their lands.

Supporting Information

The following additional information is available in the online version of this article –

Table S1. Provides the presence/absence and/or mean ($\pm$ SD) per cent cover of canopy and understory plant species and ground cover types at sites with and without Omao.

Table S2. Provides the presence/absence and/or mean ($\pm$ SD) extent of fruiting on fleshy-fruited understory plant species at sites with and without Omao.

Table S3. Lists the mean ($\pm$ SD) density of frugivorous birds at sites with and without Omao.

Table S4. Reports relative habitat use of canopy and understory plant species by the three frugivorous bird species.

Table S5. Reports the elevation and average annual precipitation for each study site.

Literature Cited

Anderson SH, Kelly D, Ladley JJ, Molloy S, Terry J. 2011. Cascading effects of bird functional extinction reduce pollination and plant density. *Science* **331**:1068–1071.

Armstrong DP, Seddon PJ. 2008. Directions in reintroduction biology. *Trends in Ecology and Evolution* **23**:20–25.

Aslan CE, Zavaleta ES, Croll D, Tershy B. 2012. Effects of native and non-native vertebrate mutualists on plants. *Conservation Biology* **26**:778–789.

Aslan CE, Zavaleta ES, Tershy B, Croll D, Robichaux RH. 2014. Imperfect replacement of native species by non-native species as pollinators of endemic Hawaiian plants. *Conservation Biology* **28**:478–488.

Atkinson CT, Lease JK, Drake BM, Shema NP. 2001. Pathogenicity, serological responses, and diagnosis of experimental and natural malarial infections in native Hawaiian thrushes. *The Condor* **103**:209–218.

Birdlife International. 2012. *Myadestes obscurus*. In: *IUCN 2012. IUCN red list of threatened species*. Version 2012.2.

Blackburn TM, Lockwood JL, Cassey PB. 2009. *Avian invasions: the ecology and evolution of exotic birds*. New York: Oxford University Press.

Bond WJ. 1994. Do mutualisms matter? Assessing the impact of pollinator and disperser disruption on plant extinction. *Philosophical Transactions of the Royal Society B: Biological Sciences* **344**:83–90.

Carlquist S. 1974. *Island biology*. New York: Columbia University Press.

Chimera CG, Drake DR. 2010. Patterns of seed dispersal and dispersal failure in a Hawaiian dry forest having only introduced birds. *Biotropica* **42**:493–502.

Connell JH. 1971. On the role of natural enemies in preventing competitive exclusion in some marine animals and in rain forest trees. In: den Boer PJ, Gradwell GR, eds. *Dynamics of populations*. Wageningen, The Netherlands: Centre for Agricultural Publishing and Documentation, 298–312.

Cordell S, Ostertag R, Rowe B, Sweinhart L, Vasquez-Radonic L, Michaud J, Cole TC, Schulten JR. 2009. Evaluating barriers to native seedling establishment in an invaded Hawaiian lowland wet forest. *Biological Conservation* **142**:2997–3004.

Culliney S, Pejchar L, Switzer R, Ruiz-Gutierrez V. 2012. Seed dispersal by a captive corvid: the role of the 'Alalā (*Corvus hawaiiensis*) in shaping Hawaii's plant communities. *Ecological Applications* **22**:1718–1732.

Denslow JS, Uowolo AL, Hughes RF. 2006. Limitations to seedling establishment in a mesic Hawaiian forest. *Oecologia* **148**:118–128.

Dirzo R, Young HS, Galetti M, Ceballos G, Isaac NJB, Collen B. 2014. Defaunation in the anthropocene. *Science* **345**:401–406.

Drake DR, Pratt LW. 2001. Seedling mortality in Hawaiian rain forest: the role of small-scale physical disturbance. *Biotropica* **33**:319–323.

Elmqvist T, Folke C, Nyström M, Peterson G, Bengtsson J, Walker B, Norberg J. 2003. Response diversity, ecosystem change, and resilience. *Frontiers in Ecology and the Environment* **1**:488–494.

Fancy SG, Nelson JT, Harrity P, Kuhn J, Kuhn M, Kuehler M, Giffin JG. 2001. Reintroduction and translocation of Omao: a comparison of methods. *Studies in Avian Biology* **22**:347–353.

Foster JT, Robinson SK. 2007. Introduced birds and the fate of Hawaiian rainforests. *Conservation Biology* **21**:1248–1257.

Hobbs RJ, Higgs E, Harris JA. 2009. Novel ecosystems: implications for conservation and restoration. *Trends in Ecology and Evolution* **24**:599–605.

Howe HF, Smallwood J. 1982. Ecology of seed dispersal. *Annual Review of Ecology and Systematics* **13**:201–228.

Inman-Narahari F, Ostertag R, Cordell S, Giardina CP, Nelson-Kaula K, Sack L. 2013. Seedling recruitment factors in low-diversity Hawaiian wet forest: towards global comparisons among tropical forests. *Ecosphere* **4**:art24. http://dx.doi.org/10.1890/ES12-00164.1.

IUCN. 2015. Invasive species specialist group: global invasive species database. http://www.issg.org/ (1 March 2015).

Janzen DH. 1970. Herbivores and the number of tree species in tropical forests. *The American Naturalist* **104**:501.

Janzen DH. 1983. Dispersal of seeds by vertebrate guts. In: Futuyma DJ, Slatkin M, eds. *Coevolution*. Sunderland: Sinauer Associates, 232–262.

Judge SW, Gaudioso JM, Gorresen PM, Camp RJ. 2012. Reoccurrence of 'Ōma'o in leeward woodland habitat and their distribution in alpine habitat on Hawai'i Island. *Wilson Journal of Ornithology* **124**:675–681.

Juvik SP, Juvik JO. 1998. *Atlas of Hawaii*, 3rd edn. Honolulu: University of Hawaii Press.

Kelly D, Robertson AW, Ladley JJ, Anderson SH. 2006. Relative (Un)importance of introduced animals as pollinators and dispersers of native plants. In: Allen RB, Lee WG, eds. *Biological invasions in New Zealand*. Berlin: Springer.

Kilpatrick AM. 2006. Facilitating the evolution of resistance to avian malaria in Hawaiian birds. *Biological Conservation* **128**:475–485.

Kurten EL. 2013. Cascading effects of contemporaneous defaunation on tropical forest communities. *Biological Conservation* **163**:22–32.

Loiselle BA, Blake JG. 2002. Potential consequences of extinction of frugivorous birds for shrubs of a tropical wet forest. In: Levey DJ, Silva WR, Galetti M, eds. *Seed dispersal and frugivory: ecology, evolution and conservation*. New York: CABI, 397–405.

Male TD, Fancy SG, Ralph CJ. 1998. Red-billed Leiothrix (*Leiothrix lutea*). In: Poole A, Gill F, eds. *The birds of North America, No. 359*. Philadelphia: The Birds of North America, Inc.

McConkey KR, Prasad S, Corlett RT, Campos-Arceiz A, Brodie JF, Rogers H, Santamaria L. 2012. Seed dispersal in changing landscapes. *Biological Conservation* **146**:1–13.

Meehan HJ, McConkey KR, Drake DR. 2002. Potential disruptions to seed dispersal mutualisms in Tonga, Western Polynesia. *Journal of Biogeography* **29**:695–712.

Moulton MP, Pimm SL. 1983. The introduced Hawaiian avifauna: biogeographic evidence for competition. *The American Naturalist* **121**:669–690.

Mountainspring S, Scott JM. 1985. Interspecific competition among Hawaiian forest birds. *Ecological Monographs* **55**:219–239.

Pejchar L, Holl KD, Lockwood JL. 2005. Hawaiian honeycreeper home range size varies with habitat: implications for native *Acacia koa* forestry. *Ecological Applications* **15**:1053–1061.

Pender RJ, Shiels AB, Bialic-Murphy L, Mosher SM. 2013. Large-scale rodent control reduces pre- and post-dispersal seed predation of the endangered Hawaiian lobeliad, Cyanea superba subsp. superba (Campanulaceae). *Biological Invasions* **15**:213–223.

Pratt TK, Atkinson CT, Banko PC, Jacobi JD, Woodworth BL (eds). 2009. *Conservation biology of Hawaiian forest birds*. New Haven: Yale University Press.

Pyle RL. 2002. Checklist of the birds of Hawaii. *Elepaio* **62**:137–148.

Ralph CJ, Fancy SG. 1994. Demography and movements of the Omao (*Myadestes obscurus*). *The Condor* **96**:503–511.

Reynolds RT, Scott JM, Nussbaum RA. 1980. A variable circular-plot method for estimating bird numbers. *The Condor* **82**:309–313.

Scott JM, Mountainspring S, Ramsey FL, Kepler CP. 1986. Forest bird communities of the Hawaiian Islands. Studies in Avian Biology, no. 9. Lawrence, KS: Allen Press.

Seddon PJ, Griffiths CJ, Soorae PS, Armstrong DP. 2014. Reversing defaunation: restoring species in a changing world. *Science* **345**:406–412.

Sekercioglu CH, Daily GC, Ehrlich PR. 2004. Ecosystem consequences of bird declines. *Proceedings of the National Academy of Sciences of the USA* **101**:18042–18047.

Sharam GJ, Sinclair ARE, Turkington R. 2009. Serengeti birds maintain forests by inhibiting seed predators. *Science* **325**:51.

Shiels AB. 2011. Frugivory by introduced black rats (*Rattus rattus*) promotes dispersal of invasive plant seeds. *Biological Invasions* **13**:781–792.

Shiels AB, Drake DR. 2011. Are introduced rats (*Rattus rattus*) both seed predators and dispersers in Hawaii? *Biological Invasions* **13**:883–894.

Terborgh J, Nuñez-Iturri G, Pitman NCA, Valverde FHC, Alvarez P, Swamy V, Pringle EG, Paine CET. 2008. Tree recruitment in an empty forest. *Ecology* **89**:1757–1768.

Traveset A, Richardson DM. 2006. Biological invasions as disruptors of plant reproductive mutualisms. *Trends in Ecology and Evolution* **21**:208–216.

Traveset A, Riera N. 2005. Disruption of a plant-lizard seed dispersal system and its ecological effects on a threatened endemic plant in the Balearic Islands. *Conservation Biology* **19**:421–431.

Traveset A, Gonzalez-Varo JP, Valido A. 2012. Long-term demographic consequences of a seed dispersal disruption. *Proceedings of the Royal Society B: Biological Sciences* **279**:3298–3303.

USDA. 2015. Hawaii State-listed Noxious Weeds. http://plants.usda.gov/java/noxious?rptType=State&statefips=15 (29 April 2015).

VanderWerf EA. 1994. Intraspecific variation in Elepaio foraging behavior in Hawaiian forests of different structure. *The Auk* **111**:917–932.

Wakelee KM, Fancy SG. 1999. Omao (*Myadestes obscurus*), Olomao (*Myadestes lanaiensis*), Kamao (*Myadestes myadestinus*), Amaui (*Myadestes woahensis*). In: Poole A, ed. *The birds of North America*. Ithaca: Cornell of Ornithology.

Webb CO, Peart DR. 2001. High seed dispersal rates in faunally intact tropical rain forest: theoretical and conservation implications. *Ecology Letters* **4**:491–499.

Williams PA. 2006. The role of blackbirds (*Turdus merula*) in weed invasion in New Zealand. *New Zealand Journal of Ecology* **30**:285–291.

Wotton DM, Kelly D. 2011. Frugivore loss limits recruitment of large-seeded trees. *Proceedings of the Royal Society B: Biological Sciences* **278**:3345–3354.

Wu JX, Delparte DM, Hart PJ. 2014. Movement patterns of a native and non-native frugivore in Hawaii and implications for seed dispersal. *Biotropica* **46**:175–182.

8

Long-distance dispersal to oceanic islands: success of plants with multiple diaspore specializations

Pablo Vargas[1]*, Yurena Arjona[1,2], Manuel Nogales[2] and Ruben H. Heleno[3]

[1] Real Jardín Botánico de Madrid (RJB-CSIC), 28014 Madrid, Spain
[2] Island Ecology and Evolution Research Group (IPNA-CSIC), 38206 La Laguna, Tenerife, Canary Islands, Spain
[3] Centre for Functional Ecology, Department of Life Sciences, University of Coimbra, 3000-456 Coimbra, Portugal

Guest Editor: Donald Drake

Abstract. A great number of scientific papers claim that angiosperm diversification is manifested by an ample differentiation of diaspore traits favouring long-distance seed dispersal. Oceanic islands offer an ideal framework to test whether the acquisition of multiple sets of diaspore traits (syndromes) by a single species results in a wider geographic distribution. To this end, we performed floristic and syndrome analyses and found that diplochorous species (two syndromes) are overrepresented in the recipient flora of the Azores in contrast to that of mainland Europe, but not to mainland Portugal. An additional analysis of inter-island colonization showed a general trend of a higher number of islands colonized by species with a single syndrome (monochorous) and two syndromes than species with no syndrome (unspecialized). Nevertheless, statistical significance for differences in colonization is meagre in some cases, partially due to the low proportion of diplochorous species in Europe (244 of ~10 000 species), mainland Portugal (89 of 2294 species), and the Azores (9 of 148 species), Canaries (17 of 387 lowland species) and Galápagos (18 of 313 lowland species). Contrary to expectations, this first study shows only a very marginal advantage for long-distance dispersal of species bearing multiple syndromes.

Keywords: Anemochorous; diplochorous traits; endozoochorous; epizoochorous; insular colonization; thalassochorous.

Introduction

The evolutionary acquisition of fruits by angiosperms offered the opportunity to have one more dispersal-related structure (the fruit) subject to modification and natural selection (Vargas 2014). In particular, infructescences, fruits and seeds (reproductive diaspores) have been evolving specific traits involved in specializations that favour long-distance dispersal (hereafter LDD) and thus higher colonization success (Valcárcel and Vargas 2014). After more than 180 million years of evolution, angiosperms have spawned multiple evolutionary avenues, including a great diversity of fruit and seed types.

* Corresponding author's e-mail address: vargas@rjb.csic.es

This is interpreted as a result of selection on diaspore traits assisting the colonization of new territories and environments (Valcárcel and Vargas 2014). But, how effective have LDD syndromes actually been? The comparison of the floras of Europe and the Azores shows that some sets of diaspore traits (syndromes), chiefly specializations for floatation and survival in seawater (thalassochorous traits), appear to have been favoured in the colonization of the archipelago (Heleno and Vargas 2015). However, Europe (continental source) shows a majority of species with unspecialized diaspores ($\sim$63 %) that parallels the poor specialization ($\sim$63 %) of the flora of the Azores islands (recipient flora) (Heleno and Vargas 2015). Acquisition of diaspore and floral syndromes has been very dynamic in the course of evolution, and thus any flora constitutes a complex assemblage of specialized and generalized species (Whitney 2009). Most floras have some species adapted towards accumulating diaspore traits that can facilitate dispersal by more than one mechanism; for example, palatable fleshy fruits that can float and survive after immersion in saltwater can be dispersed internally by animals and/or by oceanic currents (Ridley 1930). This type of species is named diplochorous (see Vander Wall and Longland 2004). The question remains as to whether diaspores with multiple specializations (more than one set of diaspore traits) have been particularly favoured for LDD. Indeed, the theory of island biogeography predicts that oceanic archipelagos are disharmonic inasmuch as they represent a sample of the mainland diaspore pool, i.e. dispersal (diaspore specializations) and further establishment (habitat availability) act as filters to colonization (Whittaker and Fernández-Palacios 2007). Therefore, our working hypothesis is that plants with multiple diaspore specializations have been particularly favoured in the process of island colonization.

Diplochory has been defined as seed dispersal by a sequence of two or more steps or phases, each involving a different dispersal agent (Vander Wall and Longland 2004). Nevertheless, it is extremely difficult to document multiple dispersal events of new colonizer species on islands, and virtually impossible to reconstruct the colonization processes over long periods of time (millions of years). An alternative approach is to evaluate whether species having multiple syndromes (i.e. more than one set of traits related to wind, sea currents or diaspore dispersal by animals—both internally and externally) have been particularly favoured for LDD over species with a single or no LDD syndrome. Although distribution frequencies of single LDD syndromes between floras of distant territories, such as Tasmania–New Zealand (Jordan 2001) and Europe–Azores (Heleno and Vargas 2015), have already been analysed, the additional

value of bearing multiple syndromes on a single species remains unexplored. Species spectra of insular floras offer different datasets to contrast the colonization success of species bearing no, one, two or more diaspore syndromes. Comparative analyses of LDD syndrome distributions provide a general framework to test this and other explicit evolutionary hypotheses about dispersal traits potentially favoured for island colonization within the theory of island biogeography (Heleno and Vargas 2015). For instance, a long-standing hypothesis posited for Hawaiian plants proposed loss of dispersability during the process of evolution on islands (Carlquist 1966). The most suitable approach to test dispersability changes is the phylogenetic method because it evaluates shifts of diaspore syndromes related to LDD (reconstruction of ancestral characters) (Vargas 2007). Alternatively, species spectra that contrast traits of endemic vs. indigenous (Vargas *et al.* 2014) and congeneric species pairs (Vazačová and Münzbergová 2014) have already been used to test loss of dispersability.

Oceanic islands emerge lifeless from the sea floor and receive all their species by LDD. That is why oceanic archipelagos provide an ideal spatio-temporal system in which to analyse plant dispersal traits related to current distributions of species across islands. The present study evaluates the success of species with multiple dispersal syndromes in island colonization, by analysing the floras of the European continent (including mainland Portugal) and the Azores. In addition, the success of multiple dispersal syndromes in inter-island colonization is analysed for the floras of the Azores, Canaries and Galápagos. Specific questions are as follows. (i) How are multiple sets of LDD traits distributed within each species and within each flora? (ii) Are multiple dispersal syndromes overrepresented in the recipient flora of the Azores, relative to the source floras of Europe and mainland Portugal? (iii) How well distributed are species with no, one and multiple dispersal syndromes within each archipelago? (iv) Is there evidence of loss or acquisition of multiple dispersal syndromes during speciation on islands?

Methods

Flora of Europe and Azores

Contrast analyses were performed on full spectra of LDD syndromes from the native Azorean flora (recipient) and those of the source floras of Europe and mainland Portugal (see Heleno and Vargas 2015). These analyses are largely complementary because the European flora includes the closest related continental flora available for colonization of the Azores, while the subset formed by the mainland Portuguese flora represents a more comparable territory in terms of habitat similarity, notably maximum elevation, latitude range, coastal length and historical climate (see Patiño et al. 2015). The lists of plant species native to Europe (10 792 species) and mainland Portugal (~2294 species) were retrieved from Flora Europaea (Tutin et al. 1980) (see Heleno and Vargas 2015); the Azorean flora, comprising 148 native species, was retrieved from Schaefer et al. (2011). To contrast the proportions of species syndromes between the Azores and the two mainland floras, we excluded endemic and introduced species from the analysis, using only shared native species.

Archipelago floras

The relative importance of LDD syndromes within the floras of the Galápagos (Pacific Ocean) and the Azores (North Atlantic Ocean) has been recently evaluated (Heleno and Vargas 2015; Vargas et al. 2014). One more North Atlantic archipelago (the Canary Islands) has partially been analysed for syndrome categorization (Arjona et al. 2015) and is herein used to evaluate the effect of multiple dispersal syndromes on island LDD and colonization. The species used in this study share similar habitats (typically lowland habitats) that occur on all the islands of each archipelago (Aranda et al. 2014; Vargas et al. 2014). In other words, we excluded medium- and high-altitude species from the analysis of the Galápagos and the Canaries, as their habitats are not present on all islands; thus inter-island connectivity could be tested because all the islands share similar lowland conditions for establishment. All the Azores species have been included because of their low number; altitudinal zonation is not critical there and habitat limitation is an unlikely barrier to colonization. As a result, the three archipelagos display different numbers of species: the Azores (148 native species, Schaefer et al. 2011), Canary Islands (387 lowland of 703 native species, Acebes Girovés et al. 2010) and Galápagos (313 lowland of 403 native species, Vargas et al. 2014). The distribution of species across islands of each archipelago (i.e. number of islands where present) is used as a proxy for diaspore LDD capacity. Therefore, the study contrasts the distribution of species across islands under similar ecological conditions and considering two or more dispersal syndromes (vs. the distribution of species with a single syndrome or none). Each archipelago was initially analysed independently, after which general patterns were then compared among islands.

Number of islands

The three archipelagos included in this study have different numbers of large islands (>10 km^2): the Azores (9), Canaries (9) and Galápagos (12). However, fluctuating sea level (eustasy) and volcanic activity imply that some present islands might have been connected by land bridges in the past (Ali and Aitchison 2014). Thus, colonization across long-standing islands is better studied based on the number of islands that have been isolated since emergence (palaeo-islands). Accordingly, we used the number of palaeo-islands pre-dating the last glaciations for each archipelago, namely seven in the Galápagos (Ali and Aitchison 2014), eight in the Azores (Rijsdijk et al. 2014) and six in the Canaries (Fernández-Palacios et al. 2011).

Diaspore traits and syndrome assignment

Long-distance dispersal is herein understood in a biogeographical sense, i.e. plant connections between the mainland and the Azores and among islands within the same archipelago (Azores, Canaries, Galápagos). We classified diaspores into five classes according to the presence/absence of specialized morphological traits favourable to particular LDD vectors: anemochory (wings, plumes or hairs promoting dispersal by wind), thalassochory (floatation and survival favouring dispersal by oceanic currents), endozoochory (nutritive tissues promoting internal dispersal by animals), epizoochory (hooks, hairs or adhesive substances aiding external dispersal by animals) and unspecialized (no LDD traits). For a detailed guide to syndrome categorization see Appendices in Heleno and Vargas (2015). As we intentionally do not consider information regarding 'actual dispersal' for syndrome assignment, the categorization includes four sets of diaspore traits that potentially provide an evolutionary advantage for LDD, namely anemochorous, thalassochorous, endozoochorous and epizoochorous traits (see Van der Pijl 1982). Therefore, as an important difference regarding other studies, we did not analyse actual dispersal for inter-island colonization (i.e. actual vectors) or categorization into specific and often fuzzy dispersal mechanisms (e.g. mud dispersal; see Nogales et al. 2012). Alternatively, our approach tests the likely success of particular morphologies (diaspore specializations) acquired in the evolutionary history of angiosperms. Previous studies have already implemented this syndrome approach for plants from the Galápagos (Vargas et al. 2014) and Azores (Heleno and Vargas 2015). The same analysis has been performed for part of the flora of the Canaries (Arjona et al. 2015).

Multiple syndromes in single species

Combining two means of diaspore dispersal can increase the probability of seed dispersal reaching suitable habitats and reducing seed mortality (Vander Wall and Longland 2004; Valcárcel and Vargas 2014). This typically includes multiple adaptations of different plant dispersal units (seeds, fruits or infructescences) related to the four LDD syndromes; for instance *Astydamia latifolia* of the Canary Islands has a winged fruit (anemochorous) with seeds that survive a long time in sea water (thalassochorous) (Y. Arjona *et al.* unpubl. data). The two syndromes can be displayed on different diaspore parts, as in the case of *A. latifolia* (above), or on the same diaspore part, as in *Corema alba* from the Azores that displays fleshy fruits promoting both endozoochory and thalassochory (C. F. Esteves *et al.* unpubl. data). In some other cases the same species can have intra-individual variation in the way that the same plant part has two different diaspore types; for instance, Asteraceae and some other families have inflorescence heterocarpy—that is two types of fruits such as adhesive achenes on the capitulum periphery and plumed achenes in the capitulum centre (Sorensen 1986).

Irrespective of the origin of multiple diaspore adaptations within a single species, three groups of species were considered: with unspecialized diaspores, with one syndrome (monochorous) and with two or more syndromes (multichorous). Datasets for the five floras (Europe, Portugal, Azores, Canaries and Galápagos) were taken from our previous studies and reanalysed after careful examination of one or more syndromes on single species. We also studied multichorous species to find out if they could be grouped into diplochorous (two-syndrome) and triplochorous (three-syndrome) groups. Accordingly, this classification also suitably incorporates the evolutionary advantage of seed specializations that potentially favour travelling by multiple, consecutive dispersal vectors.

Statistical analyses

To evaluate if species with more than one LDD syndrome were particularly favoured in the colonization of the Azores from mainland diaspores, we used a contingency analysis. A likelihood ratio test (*G* test) was used to compare the proportion of insular species with multichorous diaspores vs. those in continental Europe and continental Portugal. To find any signal of loss of dispersability, proportions of diaspore specializations from non-endemic natives (expressing speciation within the continent) and archipelago endemics (speciation on islands) were also analysed using a likelihood ratio test.

The effect of having no, one or multiple LDD syndromes (categorical predictor) on the distribution of plant species within each archipelago was evaluated with generalized linear models, considering the number of palaeo-islands (Poisson distributed error) as the dependent variable. When an effect was detected, multiple comparisons between each flora type (diaspores with unspecialized, monochorous and multichorous traits) were performed by a Tukey post hoc test. In order to look for an overall effect of unspecialized, monochorous and diplochorous diaspores on plant distribution across all datasets, we used a generalized linear mixed-effects model, including 'archipelago' as a random variable and assuming a Poisson distribution for the number of palaeo-islands. This test was followed by a Tukey multiple comparisons test. All analyses were performed using the packages Base, lme4, multcomp and Deducer in R v. 3.1.0 (R Development Core Team 2012).

Results

The results shown here are based on the codification of diaspore traits into LDD syndromes in ~11 000 angiosperm species from the floras of Galápagos, Azores, Canaries and continental Europe, including mainland Portugal.

Distribution of multichorous traits between floras

None of the species analysed had more than three syndromes and even three syndromes were identified in only five species (<0.01 % of the species screened). These five species shared traits related to anemochory, epizoochory and thalassochory, and were only present in mainland Europe: two *Limonium* (*L. lobatum* and *L. sinuatum*), two *Vulpia* (*V. alopecuros* and *V. fasciculata*) and one *Armeria* (*A. maritima*). We did not find any insular species with three or more sets of LDD syndromes. Species with two syndromes are much more frequent than those with three syndromes and represent a low proportion of the floras of Europe (244 of ~10 000 species; 2.4 %) and mainland Portugal (89 of 2294 species; 3.9 %) (Fig. 1A). Insular floras also displayed a low representation of diplochorous traits (considering species with potential habitat on all islands—see Methods): Azores (9 of 148 species; 6.1 %), Canaries (17 of 387 lowland species; 4.4 %) and Galápagos (18 of 313 lowland species; 5.8 %).

As a whole, the three archipelagos displayed 44 species with two syndromes, with approximately half of them (21 species; 48 %) having both anemochorous and thalassochorous traits (Table 1). In the flora of Europe, the highest proportions are found when combining anemochorous and epizoochorous (68.5 %), followed by anemochorous and thalassochorous (23.0 %) traits. Only in Europe we found three species (*Adoxa moschatellina*, *Viscum album* and *Viscum cruciatum*) with both epizoochorous and endozoochorous traits, and one species (*Corema alba*) with both thalassochorous and endozoochorous traits. The only syndrome pair not found in any of the studied floras is that including both anemochorous and endozoochorous traits (Table 1C).

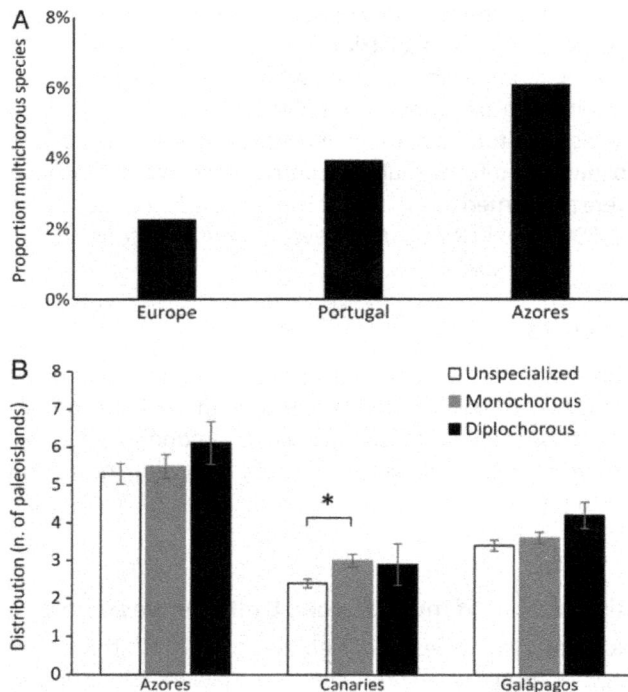

Figure 1. (A) Proportion of multichorous (diplochorous) species in mainland (Europe, Portugal) and the Azores. (B) Mean distribution of plant species with one, two and no LDD syndromes (monochorous, diplochorous and unspecialized, respectively) within the palaeo-islands of the Azores, Canaries and Galápagos. Error bars indicate the standard error of the mean. Significant differences (to $\alpha = 0.05$) are marked with an asterisk.

Diplochorous traits within and among islands

The proportion of species with more than one LDD syndrome is significantly higher in the Azores than in the whole flora of Europe ($G = 6.19$, df $= 1$, $P = 0.013$) and also higher, but not significantly so, than that of Portugal ($G = 1.37$, df $= 1$, $P = 0.248$), which is a more comparable territory.

Overall, there is a weak tendency for species bearing more LDD syndromes to occur on more islands (Fig. 1B, Table 2). Nevertheless, this tendency is non-significant (Azores $\chi^2 = 1.09$, df $= 2$, $P = 0.59$; Galápagos $\chi^2 = 2.85$, df $= 2$, $P = 0.25$; Table 3) with the exception of a broader distribution for species with one syndrome than species with no syndromes in the Canary Islands (Canaries $\chi^2 = 12.35$, df $= 2$, $P = 0.0022$; Tukey's test $z = -3.45$, $P = 0.0016$) (Tables 3 and 4). Figure 1 also shows that the Azores archipelago clearly harbours a wider distribution of species, irrespective of the presence or absence of syndromes, followed by Galápagos and then the Canaries. When considering the three archipelagos together, plant distribution is significantly affected by the presence and number of LDD syndromes ($\chi^2 = 10.82$,

df $= 4$, $P = 0.0045$). Post hoc comparisons revealed that the distribution of plants with one and two syndromes was higher than that of plants with unspecialized diaspores ($\chi^2 = -2.76$, $P = 0.014$, and $z = -2.40$, $P = 0.014$, respectively); however, there was no difference between the distribution of species with one and multiple syndromes ($z = 1.04$, $P = 0.539$; Table 5).

Diplochorous traits in endemic vs. indigenous species

We failed to find significant differences between the proportion of plants with diplochorous traits between the endemic and non-endemic native flora of the Azores ($G = 0.29$, df $= 1$, $P = 0.59$), Canaries ($G = 0.88$, df $= 1$, $P = 0.35$) and Galápagos ($G = 0.54$, df $= 1$, $P = 0.46$). Accordingly, island species displayed neither gain nor loss of diaspore traits related to LDD.

Discussion

Diaspore specialization of European angiosperms for LDD appears to be poor, inasmuch as the majority of the species from the two insular and two mainland floras show a high level of unspecialized diaspores (54–67 %) (see Heleno and Vargas 2015; Arjona et al. 2015). A similar figure was found for the flora of Galápagos (55.6 %; Vargas et al. 2014). These unspecialized plants were successful long-distance dispersers despite the apparent disadvantage of lacking specialized adaptations for LDD. Indeed, we know from landscape studies that particular sets of diaspore traits can increase both dispersal success and seedling establishment (Schupp et al. 2010). To the best of our knowledge, the hypothesis that plants displaying multiple sets of diaspore traits adapted for abiotic (sea, wind) and biotic (animal) LDD dispersal result in greater success in colonizing islands has, however, not been tested yet.

Multiple LDD syndromes in continental and insular floras

In a recent study, Heleno and Vargas (2015) found that only plants displaying diaspores adapted for sea dispersal (thalassochory) showed evidence of overrepresentation in the flora of Azores with respect to the mainland European and Portuguese floras. Here we show that diplochorous traits are also overrepresented in the Azores with respect to the flora of mainland Europe (but not to that of Portugal). Given that bioclimatic conditions (habitat similarity, maximum elevation, latitude range, coastal length, climate) of the flora of the Azores rather parallel conditions of mainland Portugal (see Heleno and Vargas 2015; Patiño et al. 2015), the contrast between Europe and Azores should be considered with caution.

Table 1. Frequency of two shared LDD syndromes on the same species of the following floras: (A) the Azores, mainland Portugal and Europe; (B) the Azores, Canaries and Galápagos and (C) the three archipelagos (altogether). The three archipelagos only have species with diplochorous traits, while Europe and Portugal also include a few cases of triplochorous species.

(A)

	Endozoochorous			Epizoochorous			Thalassochorous		
	Azores	Portugal	Europe	Azores	Portugal	Europe	Azores	Portugal	Europe
Epizoochorous	0	2	3	–	–	–	–	–	–
Thalassochorous	1	1	1	2	7	18	–	–	–
Anemochorous	0	0	0	4	57	161	2	16	54

(B)

	Endozoochorous			Epizoochorous			Thalassochorous		
	Azores	Canaries	Galápagos	Azores	Canaries	Galápagos	Azores	Canaries	Galápagos
Epizoochorous	0	0	3	–	–	–	–	–	–
Thalassochorous	1	0	7	2	1	1	–	–	–
Anemochorous	0	0	0	4	3	1	2	13	6

(C)

	Endozoochorous	Epizoochorous	Thalassochorous
Epizoochorous	3	–	–
Thalassochorous	8	4	–
Anemochorous	0	8	21

Certain combinations of two syndromes are predominant in temperate floras. Indeed, the majority of the diplochorous species in the floras of the Azores, Canaries and mainland Europe have anemochorous coupled with epizoochorous or thalassochorous traits. The predominance of these two pairs of syndromes in the three temperate floras leads us to hypothesize a primary role of either evolutionary constraints of angiosperms (as a whole) in the process of acquisition of co-occurring morphologies or the existence of ecological conditions favouring certain syndrome pairs depending on biogeographic areas. It is a fact that diplochorous traits in the tropical flora of Galápagos result from endozoochorous coupled with epizoochorous or thalassochorous traits, which are very rare in temperate floras (only four cases in Europe). This favours the hypothesis of differential adaptation of fleshy fruits in tropical areas (Ridley 1930; Moles *et al.* 2007). Indeed, in woody species, endozoochorous seeds are more frequent in neo- and palaeotropical (>70 %) than in temperate forests (<44 %) (Jordano 2014). The association between syndrome combination and latitude needs to be further explored given the scarce knowledge of dispersal syndromes in some biogeographic areas (see Fig. 3 in Moles *et al.* 2007).

Inter-island colonization within the Azores, Canaries and Galápagos

There is a low number of species displaying multiple syndromes in the floras of Azores (9 species), Canaries (17 species) and Galápagos (18 species). Nevertheless, an analysis of all diplochorous species from the three archipelagos (44 species) shows statistical significance for a general pattern of a higher number of islands colonized by species with two syndromes (diplochorous) than species with no syndromes (unspecialized). The same is true for a single syndrome (monochorous) vs. unspecialized

Table 4. Summary information of the multiple comparisons performed with the Tukey post hoc test, exploring differences on plant distribution across the Canaries palaeo-islands, according to the number of dispersal syndromes present on their diaspores (categorical variable: 0, unspecialized; 1, one dispersal syndrome; ≥ 2, two or more dispersal syndromes).

	Estimate	SE	Z value	P value
≥ 2 vs. 1 syndrome	−0.031	0.149	−0.207	0.97546
0 vs. 1 syndrome	−0.220	0.064	−3.455	0.00163
0 vs. ≥ 2 syndromes	−0.148	0.150	−1.280	0.38932

Table 5. Summary information of the multiple comparisons performed with the Tukey post hoc test, exploring differences on plant distribution across the palaeo-islands of three archipelagos (Azores, Canaries and Galapagos, as random factor) taken together, according to the number of dispersal syndromes present on plant diaspores (categorical variable: 0, unspecialized; 1, one dispersal syndrome; ≥ 2, multiple dispersal syndromes). *Significant differences at $\alpha = 0.05$.

	Estimate	SE	Z value	Adj. Pr(>|z|)
≥ 2 vs. 1 syndrome	0.082	0.080	1.039	0.5386
0 vs. 1 syndrome	−0.106	0.038	−2.761	0.0144*
0 vs. ≥ 2 syndromes	−0.188	0.078	−2.396	0.0401*

Table 2. Distribution of species displaying multiple (multichorous), single (monochorous) and unspecialized traits related to LDD within the Azores, Canaries and the Galápagos, measured as the mean number of palaeo-islands.

Diaspore type	Azores	Canaries	Galapagos
Total number of palaeo-islands (potential distribution)	8	6	7
Unspecialized	5.3 ± 2.4	2.4 ± 1.7	3.4 ± 1.7
Monochorous	5.5 ± 2.4	3.0 ± 2.1	3.6 ± 1.8
Multichorous	6.1 ± 1.6	2.9 ± 2.2	4.2 ± 1.4

Table 3. ANOVA table for the three generalized linear models explaining the distribution of plant species within Azores, Canaries and Galápagos (number of palaeo-islands where each species occurs; Poisson distributed) by the number of dispersal syndromes present on their diaspores (categorical variable: unspecialized; one dispersal syndrome; two or more (multiple) dispersal syndromes). One model was constructed for each archipelago.

		df	Deviance	Residual df	Residual deviance	χ^2	P value
Azores	Null			147	180.04		
	Number of syndromes	2	1.062	145	178.98	1.092	0.5881
Canaries	Null			386	504.73		
	Number of syndromes	2	12.253	384	495.48	12.352	0.00218
Galápagos	Null			298	273.56		
	Number of syndromes	2	2.741	296	270.82	2.859	0.254

species from the three archipelagos. Overall, we confirmed the expected advantage for intra-island colonization of species with either one or two syndromes when compared with unspecialized species. However, we failed to detect any measurable advantage of species bearing any combination of two syndromes when compared with monochorous species.

Vander Wall and Longland (2005) proposed two models of evolutionary transition for species with multiple syndromes: co-occurrence of either two competing or two sequential modes of primary dispersal over evolutionary periods. To the best of our knowledge, there is no study that tests these evolutionary hypotheses of multiple LDD syndromes in a biogeographic sense. Nevertheless, our study sheds some light on the importance of evolutionary changes in diaspores over time (speciation). Two recent papers using endemic species did not find loss of dispersability in the Galápagos (Vargas et al. 2014) or the Canaries (Vazačová and Münzbergová 2014), as historically proposed for the flora of Hawai'i (Carlquist 1966). Our results revealed that the proportion of species bearing multiple dispersal syndromes is not lower among endemics (adapted to new island conditions) than non-endemic species (more related to mainland colonizers), as predicted by the historical hypothesis of loss of dispersability (see Vargas 2014). Instead, the proportion of multichorous species appears to have been maintained during the process of speciation in the endemic plants of the Azores (5 endemics out of 9 native species), Canaries (8 endemics among 17 native species) and Galápagos (6 endemics among 18 native species). The question remains as to whether the loss of dispersability hypothesis is only restricted to certain archipelagos such as Hawai'i and Samoa (Carlquist 1966). Phylogenetic studies based on sister species and estimates of divergence times are essential to test evolutionary transitions, including loss of dispersability during the speciation process and the two models of evolutionary transition over time.

Conclusions

The emergence of angiosperms more than 180 million years ago is characterized by the acquisition of three dispersal units (seed, fruit, infructescence) subject to evolutionary change, and thus to favour morphological differentiation into multiple LDD syndromes. Nevertheless, only a few species bear two or more syndromes (<5 % of any flora tested herein). In addition, not all syndrome combinations appear to have been similarly acquired during this long period of time. Some syndrome pairs (e.g. anemochorous/endozoochorous) are absent, whereas others (e.g. anemochorous/thalassochorous) are more common. The floras of Europe, Azores, Canaries

and Galápagos also aid us in interpreting that the combination of some diplochorous traits could be the result of both evolutionary and ecological constraints. For instance, endozoochorous/thalassochorous traits appear to be more common in tropical areas, which supports a pattern of latitudinal variation. In a nutshell, our results show that the presence of any dispersal syndrome confers a colonization advantage; however, species having more than one dispersal syndrome possess only a meagre, and not always statistically traceable, improvement in island colonization by plants.

Sources of Funding

This study is framed within a biogeographic project (CGL2012-C02-01) financed by the *Ministerio de Economía y Competitividad* (Spain). Y.A. was financed by a pre-PhD contract from the *Ministerio de Ciencia y Tecnología* (Spain). R.H.H. was funded by the FCT grant IF/00441/2013 (Portugal) and the Marie Curie Action CIG-321794 (European Union).

Contributions by the Authors

P.V. designed the study; R.H.H. and Y.A. analysed the data; R.H.H., Y.A., M.N. and P.V. collected the data; P.V. led the writing; all authors read the text and provided significant improvement.

Acknowledgements

The editor (Don Drake) and two anonymous referees provided useful comments and suggestions that helped to improve this manuscript. We thank Guido Jones for his help in editing the text.

Literature Cited

Acebes Ginovés JR, León Arencibia C, Rodríguez Navarro ML, del Arco Aguilar M, García Gallo A, Pérez de Paz PL, Rodríguez Delgado O, Martín Osorio VE, Wildpret de la Torre W. 2010. Pteridophyta, Spermatophyta. In: Arechavaleta M, Rodríguez S, Zurita N, García A, coords. *Lista de especies silvestres de Canarias. Hongos, plantas y animales terrestres*. Santa Cruz de Tenerife: Gobierno de Canarias Publisher.

Ali JR, Aitchison JC. 2014. Exploring the combined role of eustasy and oceanic island thermal subsidence in shaping biodiversity on the Galápagos. *Journal of Biogeography* 41:1227–1241.

Aranda SC, Gabriel R, Borges PAV, Santos AMC, de Azevedo EB, Patiño J, Hortal J, Lobo JM. 2014. Geographical, temporal and environmental determinants of bryophyte species richness in the Macaronesian Islands. *PLoS ONE* 9:e101786.

Arjona Y, Heleno R, Nogales M, Vargas P. 2015. Do dispersal syndromes matter? Inter-island plant dispersal across the Canary Islands. In: *FloraMac 2015*, March 24–27, Las Palmas de Gran Canaria, Spain.

Carlquist S. 1966. The biota of long-distance dispersal. III. Loss of dispersibility in the Hawaiian flora. *Brittonia* **18**:310–335.

Fernández-Palacios JM, De Nascimento L, Otto R, Delgado JD, Garcia-Del-Rey E, Arévalo JR, Whittaker RJ. 2011. A reconstruction of Palaeo-Macaronesia, with particular reference to the long-term biogeography of the Atlantic island laurel forests. *Journal of Biogeography* **38**:226–246.

Heleno R, Vargas P. 2015. How do islands become green? *Global Ecology and Biogeography* **24**:518–526.

Jordan GJ. 2001. An investigation of long-distance dispersal based on species native to both Tasmania and New Zealand. *Australian Journal of Botany* **49**:333–340.

Jordano P. 2014. Fruits and frugivory. In: Gallagher RS, ed. *Seeds: the ecology of regeneration of plant communities*. Wallingford: CABI Publishing, 18–61.

Moles AT, Ackerly DD, Tweddle JC, Dickie JB, Smith R, Leishman MR, Mayfield MM, Pitman A, Wood JT, Westoby M. 2007. Global patterns in seed size. *Global Ecology and Biogeography* **16**:109–116.

Nogales M, Heleno R, Traveset A, Vargas P. 2012. Evidence for overlooked mechanisms of long-distance seed dispersal to and between oceanic islands. *New Phytologist* **194**:313–317.

Patiño J, Carine M, Mardulyn P, Devos N, Mateo RG, González-Mancebo JM, Shaw AJ, Vanderpoorten A. 2015. Approximate Bayesian Computation reveals the crucial role of oceanic islands for the assembly of continental biodiversity. *Systematic Biology* **64**:579–589.

R Development Core Team. 2012. *R: a language and environment for statistical computing*. Vienna, Austria: R Foundation for Statistical Computing.

Ridley HN. 1930. *The dispersal of plants throughout the World*. Kent: L. Reeve & Co.

Rijsdijk KF, Hengl T, Norder SJ, Otto R, Emerson BC, Ávila SP, López H, van Loon EE, Tjørve E, Fernández-Palacios JM. 2014. Quantifying surface-area changes of volcanic islands driven by Pleistocene sea-level cycles: biogeographical implications for the Macaronesian archipelagos. *Journal of Biogeography* **41**:1242–1254.

Schaefer H, Hardy OJ, Silva L, Barraclough TG, Savolainen V. 2011. Testing Darwin's naturalization hypothesis in the Azores. *Ecology Letters* **14**:389–396.

Schupp EW, Jordano P, Gómez JM. 2010. Seed dispersal effectiveness revisited: a conceptual review. *New Phytologist* **188**:333–353.

Sorensen AE. 1986. Seed dispersal by adhesion. *Annual Review of Ecology and Systematics* **17**:443–463.

Tutin TG, Heywood VH, Burges NA, Moore DM. 1980. *Flora Europaea (vols I–V)*. Cambridge: Cambridge University Press.

Valcárcel V, Vargas P. 2014. Tracheophytes: land conquest by vascular plants. In: Vargas P, Zardoya R, eds. *The tree of life: evolution and classification of living organisms*. Sunderland, MA: Sinauer Associates, 97–108.

Van der Pijl L. 1982. *Principles of dispersal in higher plants*. Berlin: Springer.

Vander Wall SB, Longland WS. 2004. Diplochory: are two seed dispersers better than one? *Trends in Ecology and Evolution* **19**:155–161.

Vander Wall SB, Longland WS. 2005. Diplochory and the evolution of seed dispersal. In: Forget P-M, Lambert JE, Hulme PE, Vander Wall SB, eds. *Seed fate: predation, dispersal and seedling establishment*. Cambridge, MA: CABI Publishing, 297–314.

Vargas P. 2007. Are Macaronesian islands refugia of relict plant lineages?: a molecular survey. In: Weiss SJ, Ferrand N, eds. *Phylogeography in southern European refugia: evolutionary perspectives on the origins and conservation of European biodiversity*. Dordrecht: Springer, 297–314.

Vargas P. 2014. Evolution on islands. In: Vargas P, Zardoya R, eds. *The tree of life: evolution and classification of living organisms*. Sunderland, MA: Sinauer Associates, 577–594.

Vargas P, Nogales M, Jaramillo P, Olesen JM, Traveset A, Heleno R. 2014. Plant colonization across the Galápagos Islands: success of the sea dispersal syndrome. *Botanical Journal of the Linnean Society* **174**:349–358.

Vazačová K, Münzbergová Z. 2014. Dispersal ability of island endemic plants: What can we learn using multiple dispersal traits? *Flora* **209**:530–539.

Whitney KD. 2009. Comparative evolution of flower and fruit morphology. *Proceedings of the Royal Society of London B* **276**:2941–2947.

Whittaker RJ, Fernández-Palacios JM. 2007. *Island biogeography: ecology, evolution, and conservation*. Oxford: Oxford University Press.

9

Environmental correlates for tree occurrences, species distribution and richness on a high-elevation tropical island

Philippe Birnbaum[1,2]*, Thomas Ibanez[2], Robin Pouteau[2,3], Hervé Vandrot[2], Vanessa Hequet[3], Elodie Blanchard[2] and Tanguy Jaffré[3]

[1] CIRAD, UMR 51 AMAP, 34398 Montpellier, France
[2] Laboratory of Applied Botany and Plant Ecology, Institut Agronomique néo-Calédonien (IAC), Diversité biologique et fonctionnelle des écosystèmes terrestres, 98848 Noumea, New Caledonia
[3] Laboratory of Applied Botany and Plant Ecology, Institut de Recherche pour le Développement (IRD), UMR 123 AMAP, 98848 Noumea, New Caledonia

Guest Editor: Christoph Kueffer

Abstract. High-elevation tropical islands are ideally suited for examining the factors that determine species distribution, given the complex topographies and climatic gradients that create a wide variety of habitats within relatively small areas. New Caledonia, a megadiverse Pacific archipelago, has long focussed the attention of botanists working on the spatial and environmental ranges of specific groups, but few studies have embraced the entire tree flora of the archipelago. In this study we analyse the distribution of 702 native species of rainforest trees of New Caledonia, belonging to 195 genera and 80 families, along elevation and rainfall gradients on ultramafic (UM) and non-ultramafic (non-UM) substrates. We compiled four complementary data sources: (i) herbarium specimens, (ii) plots, (iii) photographs and (iv) observations, totalling 38 936 unique occurrences distributed across the main island. Compiled into a regular 1-min grid (1.852 × 1.852 km), this dataset covered ~22 % of the island. The studied rainforest species exhibited high environmental tolerance; 56 % of them were not affiliated to a substrate type and they exhibited wide elevation (average 891 ± 332 m) and rainfall (average 2.2 ± 0.8 m year^{-1}) ranges. Conversely their spatial distribution was highly aggregated, which suggests dispersal limitation. The observed species richness was driven mainly by the density of occurrences. However, at the highest elevations or rainfalls, and particularly on UM, the observed richness tends to be lower, independently of the sampling effort. The study highlights the imbalance of the dataset in favour of higher values of rainfall and of elevation. Projected onto a map, under-represented areas are a guide as to where future sampling efforts are most required to complete our understanding of rainforest tree species distribution.

Keywords: Area effect; biodiversity hotspot; α-diversity; island; species richness; tropical mountains; ultramafic substrate.

* Corresponding author's e-mail address: birnbaum@cirad.fr

Introduction

Despite an ever increasing amount of data, the geographical distribution of most plant species remains incomplete and biased, particularly for the most diverse taxonomic groups and regions (Whittaker *et al.* 2005). This so-called Wallacean shortfall frequently originates from the difficulty of evaluating the distribution of species diversity across large, heterogeneous areas where biodiversity is high but collecting efforts have been insufficient or inadequately planned (Schmidt-Lebuhn *et al.* 2012). However, when correctly assessed and accounted for, sampling biases do not entirely prevent identification of the mechanistic determinants of species distribution, which remains a central question in ecology and biogeography (Hortal *et al.* 2007).

Many hypotheses have been proposed to explain the geographic variability of species diversity. It is widely accepted that species richness decreases poleward (Hillebrand 2004). However, the relationship between species richness and elevation is more complex and highly dependent on the organism being considered (Lomolino 2001; Nogués-Bravo *et al.* 2008; McCain and Grytnes 2010; Kessler *et al.* 2011).

Environmental conditions (air temperature, precipitation and solar radiation) change with elevation and the available area tends to decrease as elevation increases, due to the conical shape of mountains, thus affecting the total number of species (McCain 2007; Barry 2008). After accounting for the decreasing availability of area with increasing elevation, many organisms show a hump-shaped pattern of richness with elevation. In a meta-analysis involving 204 elevational transects, Rahbek (2005) found ~80 % of hump-shaped richness patterns but only a small proportion of monotonic patterns. A number of explanations have been examined for these hump-shaped richness patterns (e.g. climate-derived productivity, source–sink dynamics, intermediate disturbance and mid-domain effect) but at this stage the dominant contributing factors remain unclear (McCain and Grytnes 2010).

Mountainous tropical islands are ideally suited for examining species distribution along complex environmental transects by virtue of their exaggerated climatic gradients and their complex topography. These combine to create a wide variety of habitats within a relatively small area. In addition to exhibiting steep environmental gradients, islands host biota with sharp variations in environmental tolerance and, paradoxically, harbour a greater proportion of narrow-range endemics when compared with continents (Carlquist 1974; Caujapé-Castells *et al.* 2010).

New Caledonia, an archipelago located in southwest Pacific Ocean, hosts a rich (more than 3200 species) and unique vascular flora (75 % of endemism) distributed within a remarkable mosaic of habitats (Morat *et al.* 2012). First considered as a vicariant Gondwanan refugium (Morat 1993), geologists have recently demonstrated the entire submersion by subduction before a final re-emersion of the main island 'Grande Terre' ~37 My ago coated with a fragment of oceanic crust at the origin of the ultramafic (UM) substrates (Cluzel and Chiron 1998; Neall and Trewick 2008). The entire submersion thus argues for a secondary colonization origin for the entire New Caledonian flora. Sherwin Carlquist was one of the first to support this view and stated in his book 'Island Biology' that there is 'no reason why New Caledonian flowering plants cannot be hypothesised to have arrived via long-distance dispersal' (Carlquist 1974). Several phylogenetic studies have provided evidence supporting the thesis that the flora of New Caledonia indeed originates from recent (<37 My) long-distance dispersal colonizations and diversifications (Murienne *et al.* 2005; Grandcolas *et al.* 2008; Murienne 2009; Espeland and Murienne 2011; Pillon 2012).

Such biogeographical mysteries have long focussed the attention of local and international botanists over more fundamental studies of rainforest structure, floristic composition and ecology. Most earlier studies examining plant distribution patterns in New Caledonia have focussed on particular groups: endangered species (Herbert 2006; Kumar and Stohlgren 2009), emblematic species (Pintaud *et al.* 1999; Jaffré *et al.* 2010; Poncet *et al.* 2013), narrow-range endemic species (Wulff *et al.* 2013) or montane species (Nasi *et al.* 2002). To our knowledge, few studies have embraced the entire tree flora of the archipelago.

The few plot-based studies, that have investigated an exhaustive number of tree species, point to a consensus that the species assemblage is of similar interest to the unusual biogeographical history of the archipelago (e.g. Jaffré and Veillon 1990; Read *et al.* 2000; Ibanez *et al.* 2014). In particular, the New Caledonian rainforests seem to be characterized by very high stem density, low α-diversity (low local richness) and high β-diversity (high between-plot dissimilarity) without significant correlations with either elevation or rainfall (Grytnes and Felde 2014). However, it is fair to question whether the apparent independence of diversity from the environment arises from an unusually high environmental tolerance of the New Caledonian tree species or from a failure of experimental design to capture the species α-diversity (Grytnes and Felde 2014).

In this study, we used a new-occurrences dataset to test whether the environment is the main force driving New Caledonian rainforest tree species diversity and individual species distribution. In other words, does the species assemblage in rainforests reflect a deterministic model (i.e. an environment or niche-based one) or a stochastic model (i.e. a null one)?

We draw up a starting-point statement of knowledge of rainforest tree species distribution in New Caledonia by compiling tree occurrences from four distinct sources: (i) herbarium specimens, (ii) plot inventories, (iii) photographs and (iv) other observations collected over several decades. First, we evaluated the geographic and environmental representativeness of this occurrence dataset. Second, we analysed how tree species richness found on the main island (i.e. the γ-diversity) is distributed along altitudinal and rainfall gradients on the two main substrate types. Last, we examined whether the spatial distribution of rainforest tree species is more driven by the environment than by dispersal ability. If species exhibit environmental specialisation, their distribution was hypothesized to be mainly driven by deterministic processes. Furthermore, if species exhibit aggregative distributional patterns, we then concluded that their distribution is mainly shaped by stochastic processes and controlled by dispersal ability.

Methods

Study site

New Caledonia is located slightly north of the Tropic of Capricorn (20–23°S, 164–167°E), ~1500 km east of Australia and 2000 km north of New Zealand. The main island 'Grande Terre', is long (400 km), narrow (40 km) and accounts for nearly 85 % of the archipelago area. It lies roughly SE : NW and is crossed by a central mountain chain, where the highest peaks reach 1628 m in the north (Mont Panié) and 1618 m in the south (Mont Humboldt).

Most of the UM substrates are located in the southeast of Grande Terre in a main massif called the Grand Massif du Sud while 12 other smaller UM massifs are scattered along the northwest coast (Fritsch 2012). The UM substrates provide a variety of soils with somewhat unusual characteristics—always a deficiency of phosphorus, potassium and calcium; frequently a high concentration of magnesium; often a low water retention and potentially phytotoxic levels of some metals including nickel, manganese, chromium and cobalt. Obviously, plants must be able to tolerate these soil conditions if they are to establish here (Jaffré 1980). Overall, New Caledonian trees can be classified into three balanced edaphic groupings: UM specialists, non-ultramafic (non-UM) specialists (on volcano-sedimentary or acidic substrates) and substrate-generalists (Ibanez et al. 2014).

Rainfalls range widely from 0.6 to 4.5 m year^{-1} and are generally lower on the leeward lowlands of the west coast and higher on the windward mountain slopes of the east-coast due to oro-topography and the eastern trade winds

(Météo-France 2007; Terry et al. 2008). A combination of rainfall and elevation is commonly used to classify the vegetation into the following: rainforest, dry sclerophyll forest, scrubland called 'maquis', savannah, secondary thickets and mangroves (Jaffré 1993; Jaffré et al. 2012). Rainforests are the richest vegetation type (more than 2000 native vascular species) and covers ~3800 km^2, with 1800 km^2 on non-UM, 1100 km^2 on UM and 900 km^2 on calcareous substrates, mainly located in the Loyalty Islands.

Selection of tree species

We focussed on the distribution of 702 woody tree species (i.e. excluding lianas, tree ferns and palms) reaching a diameter at breast height (DBH at 1.3 m) of 10 cm, at least once in our set of 37 597 trees in the New Caledonian Plant Inventory and Permanent Plot Network (NC-PIPPN). Intraspecific ranks (subspecies and varieties) were merged at the specific rank to ensure uniformity of identification across datasets. The nomenclature of tree species followed the New Caledonian taxonomic name reference Florical (Morat et al. 2012).

The NC-PIPPN inventory covers a surface area of ~15 ha and comprises: (i) 220 plots of 0.04 ha (20 × 20 m), with 30 221 inventoried trees (DBH >5 cm) located across rainforests of Grande Terre on both UM (111 plots) and non-UM substrates (89 plots) along a wide range of elevations (5–1292 m on UM and 105–1187 m on non-UM) and rainfalls (1.6–3.5 m year^{-1} on UM and 1.8–3.4 m year^{-1} on non-UM) (see Ibanez et al. 2014) and (ii) six recently established 1 ha plots (100 × 100 m), with 7376 inventoried trees (DBH >10 cm) located in rainforests of the Northern Province on non-UM substrates at mid-elevations (240–780 m) and mid-rainfalls (1.3–3.0 m year^{-1}).

Compilation of tree occurrences

Occurrences of the 702 tree species were compiled from four datasets: (i) the NC-PIPPN inventory (29 409 occurrences), (ii) herbarium specimens (22 715 specimens) compiled from the database of the herbarium of the IRD Centre of Noumea (NOU, http://herbier-noumea.plantnet-project.org), (iii) other observations (44 227 observations) from different unpublished inventories used for assessing the flora of areas under consideration for mining exploration or for conservation measures (**[see Supporting Information]**) and (iv) photographs acquired in the field (4326 photographs).

We then checked datasets and removed inaccurate geolocations. We only retained herbarium specimens that were: (i) georeferenced with a Global Positioning System (GPS), (ii) collected by Hugh S. MacKee (the principal contributor to the NOU Herbarium with ~45 000 specimens) and estimated to have a horizontal accuracy of

<500 m according to his online gazetteer (http://phanero.novcal.free.fr) or (iii) labelled with an elevation matching with a difference of ± 50 m the elevation extracted from a digital elevation model (DEM).

From the 100 677 initial occurrences, we compiled a dataset of 38 936 (~40 %) unique occurrences combining species with accurate geolocation: 11 845 from herbarium, 7420 from plots, 18 390 observations and 1281 photographs, used thereafter to describe the known distribution of tree species in New Caledonia. A total of 15 285 occurrences (i.e. <40 % of the whole dataset) were found in rainforests while the remainder occurred in other vegetation types.

Environmental features

Substrate types (i.e. UM and non-UM) were extracted from a UM substrate map downloaded from the Geographic Portal of New Caledonia (http://www.georep.nc/, DIMENC/SGNC-BRGM 2010). Elevation was extracted from a 50-m resolution digital elevation model (DEM-DTSI 2012) and rainfall from an interpolation model with a resolution of 1 km using mean annual rainfall compiled from 1990 to 2010 (AURELHY model, METEO-FRANCE). Finally, vegetation types were extracted from the vegetation map of the Atlas of New Caledonia (Bonvallot *et al.* 2012). This vegetation map is a broad-scale digitalization (scale of 1/1 600 000) based on aerial photographs in which only the largest rainforest units covering ~3276 km^2 were delineated (Jaffré *et al.* 2012).

Statistical analysis

Tree occurrences dataset. First, we subdivided the main island into 5546 cells of 1 min^2 resolution (1.852 × 1.852 km) and computed in each cell the number of occurrences (i.e. the occurrence density) and the number of species (i.e. the species richness or α-diversity). We then analysed the relative contributions of the four datasets and examined the occurrence and species patterns they produced.

Geographical and environmental distributions of *γ-diversity.* We then focussed on occurrences found exclusively in rainforests, and analysed their distribution as a function of environmental features including (i) the substrate type (UM or non-UM), (ii) elevation and (iii) mean annual rainfall. For each substrate, elevation and rainfall class (bands of 100 m and 0.25 m year^{-1}, respectively), we computed the number of occurrences found in rainforests, the observed γ-diversity and the theoretical γ-diversity (calculated by considering that a species occurs between the minimal and maximal class where it had been recorded). We then compared the class distribution of available rainforest area, occurrences,

observed and theoretical γ-diversity using linear log–log models of correlation. Finally, we computed species rarefaction curves and Fisher's α diversity index at low, mid and high classes of elevation (<400, 400–800 and >800 m) and rainfall (<2.5, 2.5–3.0 and >3.0 m year^{-1}) to avoid sampling bias (see McCain and Grytnes 2010). Occurrence-based rarefaction curves with 95 % confidence intervals were computed using 1000 random permutations (e.g. Gotelli and Colwell 2001).

Geographical and environmental distributions of species. Finally, we calculated a basic index of aggregation as the ratio between the number of occurrences and the number of grid-cells intercepted by each dataset (i.e. the mean density of occurrences per grid-cell). At the species level, we computed the Morisita index of aggregation (I_{mor}) in grid-cells as well as in the entire plots dataset (before removing species duplicates) and used the standardized I_{mst} which ranges from -1 to 1. Species with $I_{mst} \leq -0.5$ are equally distributed across grid-cells or plot inventories while species with $I_{mst} \geq 0.5$ are aggregated in some grid-cells or plots.

Geographic data processing was performed using Quantum GIS 2.6.0-Brighton (Quantum GIS Development Team, 2014) and statistical analyses using R 2.15.2 (R Foundation for Statistical Computing, Vienna, Austria), including the *vegan* package (Oksanen *et al.* 2013).

Results

Tree occurrences dataset

The 702 tree species selected belong to 195 accepted genera and 80 families. At the family level, Myrtaceae were the most common, followed in decreasing order by Cunoniaceae, Sapindaceae, Araliaceae, Sapotaceae, Clusiaceae, Rubiaceae, Lauraceae, Primulaceae, Rutaceae and Apocynaceae. Together these represented 50 % of the total number of occurrences **[see Supporting Information]**. At the species level, the 124 species most-represented (92–491 occurrences per species) contributed to 50 % of the total occurrences. In contrast, the 434 species least-represented (1–50 occurrences per species) contributed to <25 % of the total number of occurrences.

Tree occurrences were distributed into 1213 cells of a regular 1-min grid, which covers ~22 % of New Caledonia's main island (Fig. 1). Occurrence density ranged from 1 to 1405 per grid-cell (32 on average, ± 2 standard error). Occurrences from the plot inventory dataset were the most aggregated data, occurring in only 68 cells with an average of 109 occurrences ± 18 per grid-cell. The next most aggregated dataset was of observations (340 grid-cells and 54 occurrences ± 5 per grid-cell). This was

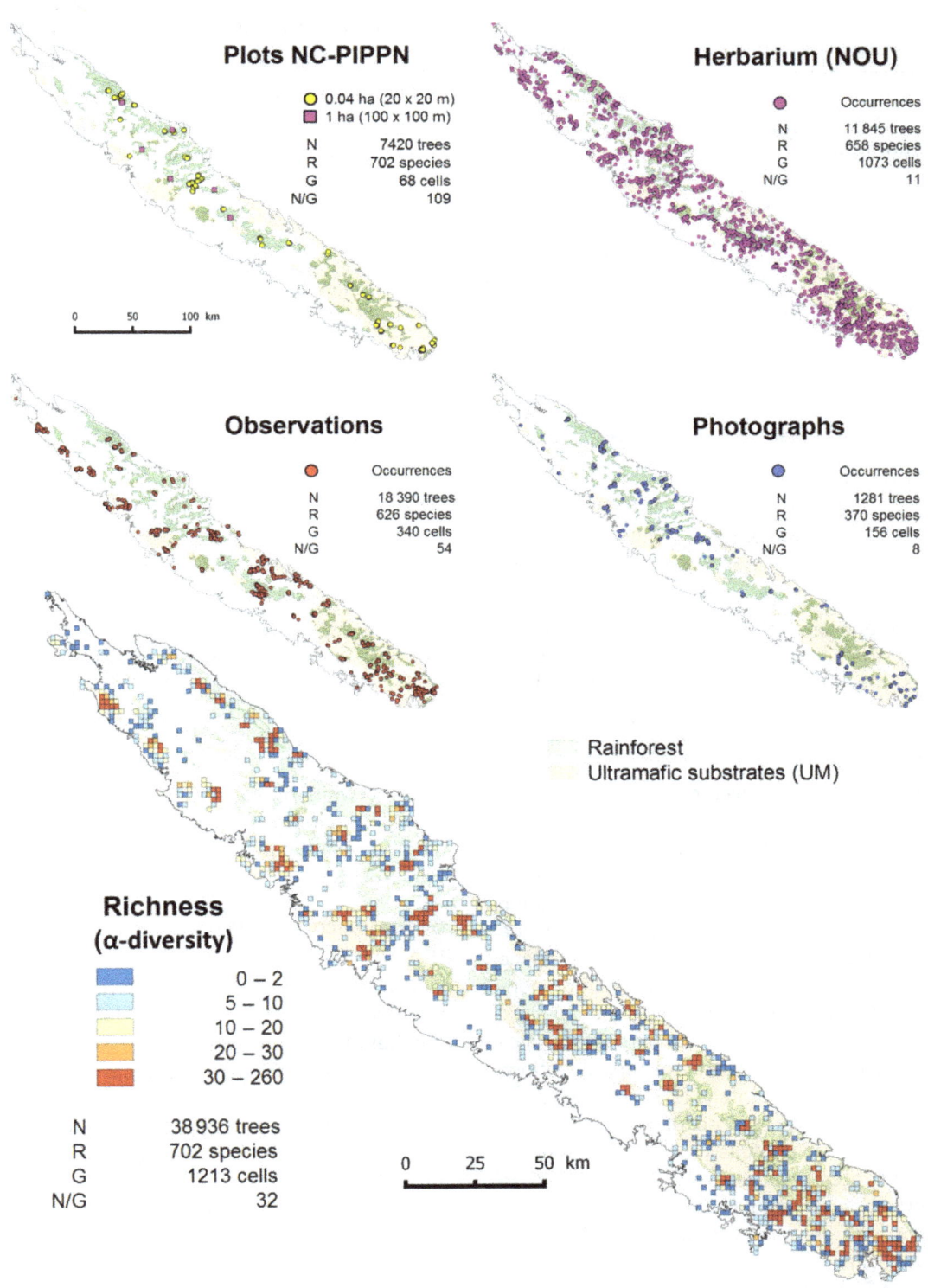

Figure 1. Distribution of the number of tree occurrences (*N*), the number of species or γ-diversity (*R*), number of intercepted cells on a 1 min-resolution grid (*G*) and the occurrences/cells ratio (*N/G*) within each dataset and the resulting α-diversity computed for overall occurrences by 1-min cell (1.852 × 1.852 km).

followed by herbarium specimens (1073 grid-cells and 11 occurrences ± 1 per grid-cell) and, last, photographs (156 grid-cells and 8 occurrences ± 1 per grid-cell).

Tree occurrences were substantially imbalanced among substrate types (Table 1). More than two-thirds of the occurrences were found on UM substrates while these

Table 1. Distribution of land area and tree occurrences in the whole *Grande Terre* (All) and in the outline of rainforests (Forest) for both UM and non-UM substrates. Italic values represent the relative contribution of each classes.

	All			Forest		
	UM	Non-UM	Total	UM	Non-UM	Total
Area (km²)	5805	11 469	17 274	1190	2086	3276
	33.6 %	*66.4 %*		*36.3 %*	*63.7 %*	
Occurrences (#)	26 340	12 596	38 936	7009	8276	15 285
	67.6 %	*32.4 %*		*45.9 %*	*54.1 %*	
Herbarium	6626	5219	11 845	2670	3204	5874
	55.9 %	*44.1 %*		*45.5 %*	*54.5 %*	
Plot	3831	3589	7420	780	3074	3854
	51.6 %	*48.4 %*		*20.2 %*	*79.8 %*	
Photograph	498	783	1281	192	599	791
	38.9 %	*61.1 %*		*24.3 %*	*75.7 %*	
Observation	15 385	3005	18 390	3367	1399	4766
	83.7 %	*16.3 %*		*70.6 %*	*29.4 %*	

substrates cover only one-third of the island (Fig. 1). This over-sampling was particularly high in the observation dataset. Thus, the number of occurrences for a given species was strongly correlated with its abundance on UM substrates (Pearson's correlation test $R^2 = 0.94$, P-value <0.001).

Overall, tree occurrences covered most of the elevation—rainfall combinations available on both UM and non-UM substrates (Fig. 2). As a result, we observed a significant correlation between the number of occurrences and the available rainforest area along the elevation gradient ($R^2 = 0.71$ on UM and 0.64 on non-UM, P-value <0.001 in both cases). This correlation was weaker along the rainfall gradient ($R^2 = 0.45$ on UM and 0.34 on non-UM, P-value <0.01 and 0.05, respectively, Table 2).

However, Fig. 2 reveals that occurrences on UM substrates were particularly under-represented at mid-elevations (500–800 m) and for mid-rainfalls (2.3–3.0 m year^{-1}), while they were over-represented at high elevations (above 1200 m) and for rainfalls (above 4.0 m year^{-1}). On non-UM substrates, occurrences were concentrated in a narrow elevation range (500–700 m) and also over-represented at high elevations (above 800 m) and for rainfalls (above 3.0 m year^{-1}).

Figure 3A and B shows under- and over-represented regions of the environmental space. On both substrates, below 500 m elevation and 2 m year^{-1} rainfall, occurrences were under-represented with regard to the relative rainforest surfaces. When projected on a map, under-represented areas covered 1873 km² of rainforests (i.e. 57 % of the total rainforest area) including 975 km²

(52 %) on UM substrates (Fig. 3C). The number of occurrences in under-represented areas was null in two-thirds of the grid-cells (no-data cells) and ranged from 1 to 1405 in the other third. Among major orographic massifs with no-data cells, those presenting an over-represented environment include 'Colnett', 'Me Maoya' or 'Saint Vincent' while those with an under-represented environment include 'Mandjelia/Balade', 'Tonine/Gaitada' 'Forêt plate', 'Source Neaoua', 'Karagreu/Boreare' and 'Kouakoué'.

Geographical and environmental distributions of γ-diversity

Overall, the α-diversity (i.e. the total number of species per 1-min cell) was strongly correlated with the total number of occurrences (Pearson's correlation test, $R^2 = 0.86$, P-value <0.001). The observed γ-diversity was strongly log-correlated with the number of occurrences, regardless of the substrate type or the environmental gradient ($R^2 > 0.90$ and P-value <0.001 in all cases, Table 2). Furthermore, the observed γ-diversity was also strongly log-correlated with the theoretical γ-diversity ($R^2 > 0.97$, P-value <0.001) except along the elevation gradient on non-UM substrates, where the correlation was weaker ($R^2 = 0.82$, P-value <0.001, Table 2). Nevertheless we observed important differences between observed and theoretical γ-diversity at mid-elevation and rainfall (Fig. 4). We note that along the elevation gradient, the highest γ-diversity on UM substrates was recorded at low elevation despite a low density of occurrences. Conversely, at high elevations and high rainfalls,

Area

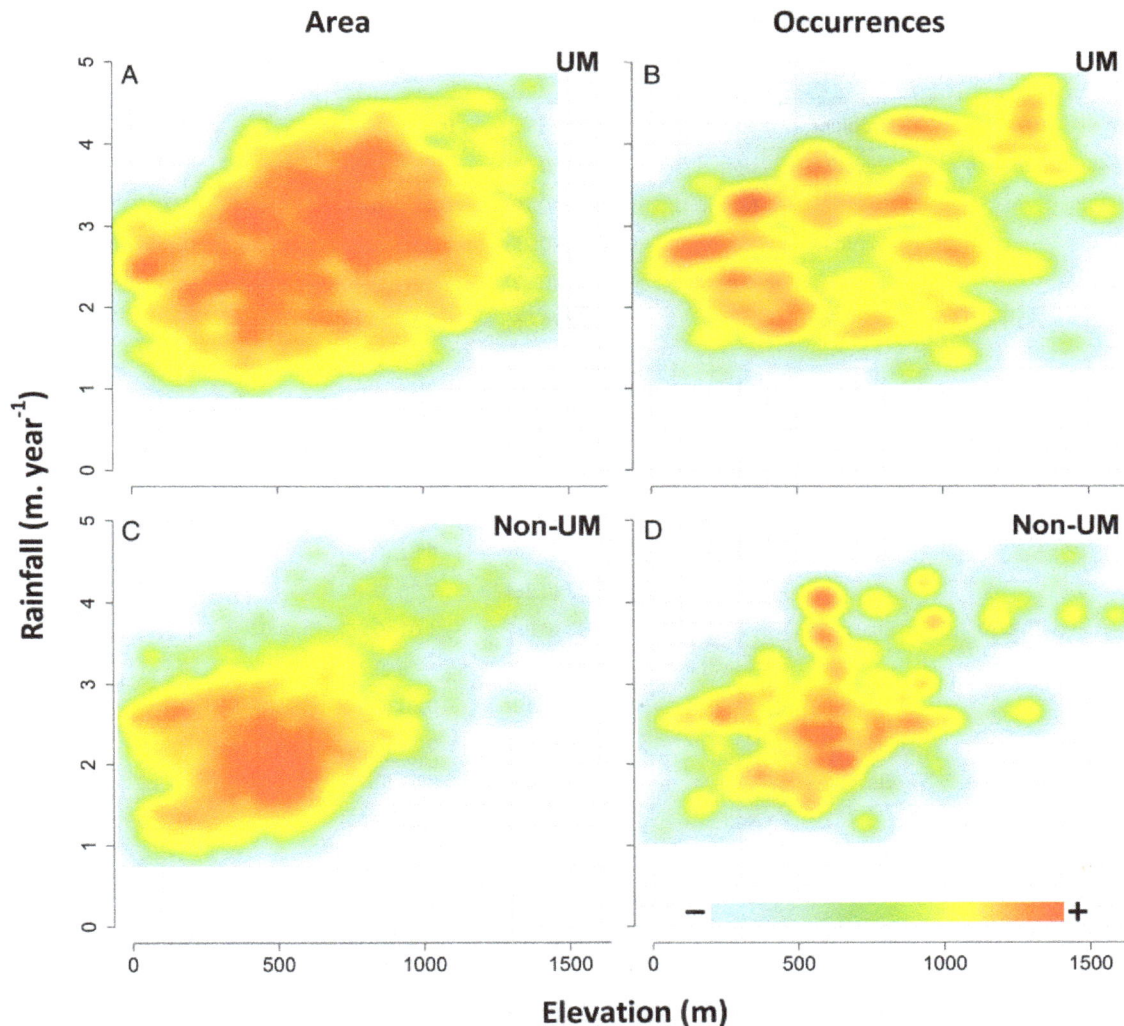

Occurrences

Figure 2. Relative density distributions of land area (A and C) and tree occurrences (B and D) on UM substrate (A and B) and non-UM substrates (C and D) along the rainfall and elevation gradients.

occurrences were dense, so the γ-diversity observed deviated from the available area.

However occurrences-based rarefaction curves clearly attest a lower rate of species accumulation at higher classes of elevation and rainfall on both substrates (Fig. 5). Such a decrease was more pronounced on UM than on non-UM substrates. Indeed, on UM substrates, Fisher's α decreased by ~40 % from mid to high elevation or rainfall, while on non-UM substrates it decreased by 25 and 30 % from low to high elevation and rainfall.

Geographical and environmental distributions of species

Although our list of tree species was drawn from occurrences in rainforest plots, tree species distribution extended to many other vegetation types (see Fig. 1). Indeed, ~5 % of them (38 species) occurred only in rainforests. Less

than 10% (66 species) occurred strictly on UM substrates and 13% (90 species) on non-UM. However, if we consider that a threshold of 90% of the occurrences reveals a species affiliation to a substrate, then 29% (205 species) were affiliated to UM and 23% (161 species) to non-UM. Finally ~80 % of the rainforest species (561 species) occurred once in 'maquis' through a total number of 18 503 occurrences.

With respect to other environmental features, half of the species exhibited an elevation tolerance (the difference between the minimum and maximum elevation where a species has been recorded) higher than 895 m (891 ± 332 m on average) and a rainfall tolerance higher than 2.4 m year^{-1} (2.2 ± 0.8 m year^{-1} on average), with no significant deviation according to affiliations to substrate (Kruskal–Wallis rank sum test, P-values >0.05). This range of tolerances represents more than half of

Table 2. Linear log–log correlation between the available rainforest area (AREA), the number of occurrences (N), the observed γ-diversity (R_{obs}) and the theoretical γ-diversity (R_{theo}) along the elevation gradient and the rainfall gradient and on UM and non-UM substrates separately ($*P < 0.05$, $**P < 0.01$, $***P < 0.001$).

Model	Gradient	Substrates	Slope (SE)	Intercept (SE)	R^2	F value	Df	P-value
$\log(N) = a \times \log(AREA) + b$	Elevation	UM	0.44 (0.08)	−1.52 (0.30)	0.71	34.65	14	***
		Non-UM	0.52 (0.10)	−1.39 (0.44)	0.65	25.78	14	***
	Rainfall	UM	0.85 (0.25)	−0.62 (0.86)	0.45	11.32	14	**
		Non-UM	0.87 (0.32)	−0.71 (1.16)	0.34	7.33	14	*
$\log(R_{obs}) = a \times \log(N) + b$	Elevation	UM	0.83 (0.06)	7.56 (0.22)	0.92	163.9	15	***
		Non-UM	0.81 (0.06)	7.56 (0.23)	0.92	175.3	15	***
	Rainfall	UM	0.64 (0.04)	7.03 (0.12)	0.96	334.1	13	***
		Non-UM	0.84 (0.06)	7.56 (0.22)	0.95	229.9	13	***
$\log(R_{obs}) = a \times \log(AREA) + b$	Elevation	UM	0.36 (0.06)	6.31 (0.22)	0.74	40.27	14	***
		Non-UM	0.44 (0.09)	6.49 (0.36)	0.65	25.88	14	***
	Rainfall	UM	0.49 (0.16)	6.52 (0.55)	0.40	8.84	13	*
		Non-UM	0.83 (0.26)	7.38 (0.94)	0.44	10.16	13	**
$\log(R_{theo}) = a \times \log(AREA) + b$	Elevation	UM	0.37 (0.06)	−1.01 (0.26)	0.7	32.9	14	***
		Non-UM	0.32 (0.07)	−1.16 (0.28)	0.63	24	14	***
	Rainfall	UM	0.57 (0.18)	−0.60 (0.60)	0.44	10.27	13	**
		Non-UM	0.78 (0.27)	−0.13 (0.96)	0.39	8.44	13	*
$\log(R_{theo}) = a \times \log(N) + b$	Elevation	UM	0.89 (0.06)	0.42 (0.20)	0.93	213.8	15	***
		Non-UM	0.65 (0.07)	−0.23 (0.27)	0.84	82.05	15	***
	Rainfall	UM	0.70 (0.05)	−0.13 (0.17)	0.94	214.3	13	***
		Non-UM	0.80 (0.07)	0.11 (0.31)	0.90	111.1	13	***
$\log(R_{obs}) = a \times \log(R_{theo}) + b$	Elevation	UM	0.93 (0.03)	7.18 (0.09)	0.98	734.8	15	***
		Non-UM	1.09 (0.13)	7.45 (0.34)	0.83	72.02	15	***
	Rainfall	UM	0.89 (0.04)	7.10 (0.10)	0.97	486.1	13	***
		Non-UM	1.00 (0.05)	7.33 (0.14)	0.97	480.3	13	***

the elevation and rainfall ranges available across the rainforest of *Grande Terre*. In contrast, the vast majority of species exhibited a spatial distribution highly aggregated at the scale of the *Grande Terre* (92 % of the species with $I_{mst} \geq 0.5$) as well as in plots (90 % of the species with $I_{mst} \geq 0.5$).

Discussion

Tree occurrences dataset

We compiled occurrences from herbarium specimens, plot inventories, photographs and observations to draw up a first assessment of tree distribution and diversity in New Caledonian rainforests. These occurrences cover almost a quarter of the main island, *Grande Terre*, providing a fairly comprehensive view of the actual distribution of rainforest tree species. Herbarium specimens have

long been used to study species and diversity distributions (Lavoie 2013) and the biases associated with a taxonomic approach are widely recognized (Ahrends *et al.* 2011). Species accumulation is more rapid when using herbarium specimens than when using plot inventories but their relative abundance is not reliable (e.g. Garcillán and Ezcurra 2011). To our knowledge, the use of field photographs is less common. The development of GPS and photographic technologies now generates a huge amount of high-quality georeferenced and retrospectively verifiable information. In our study, even though this data source was less substantial, we stress that in the future it is likely to become a critical source of reliable data, by involving parataxonomists, in particular, through collaborative networks (e.g. Basset *et al.* 2004). Even if observations may be more doubtful than from other datasets (i.e. there is no way to check data *a posteriori*), these rapid

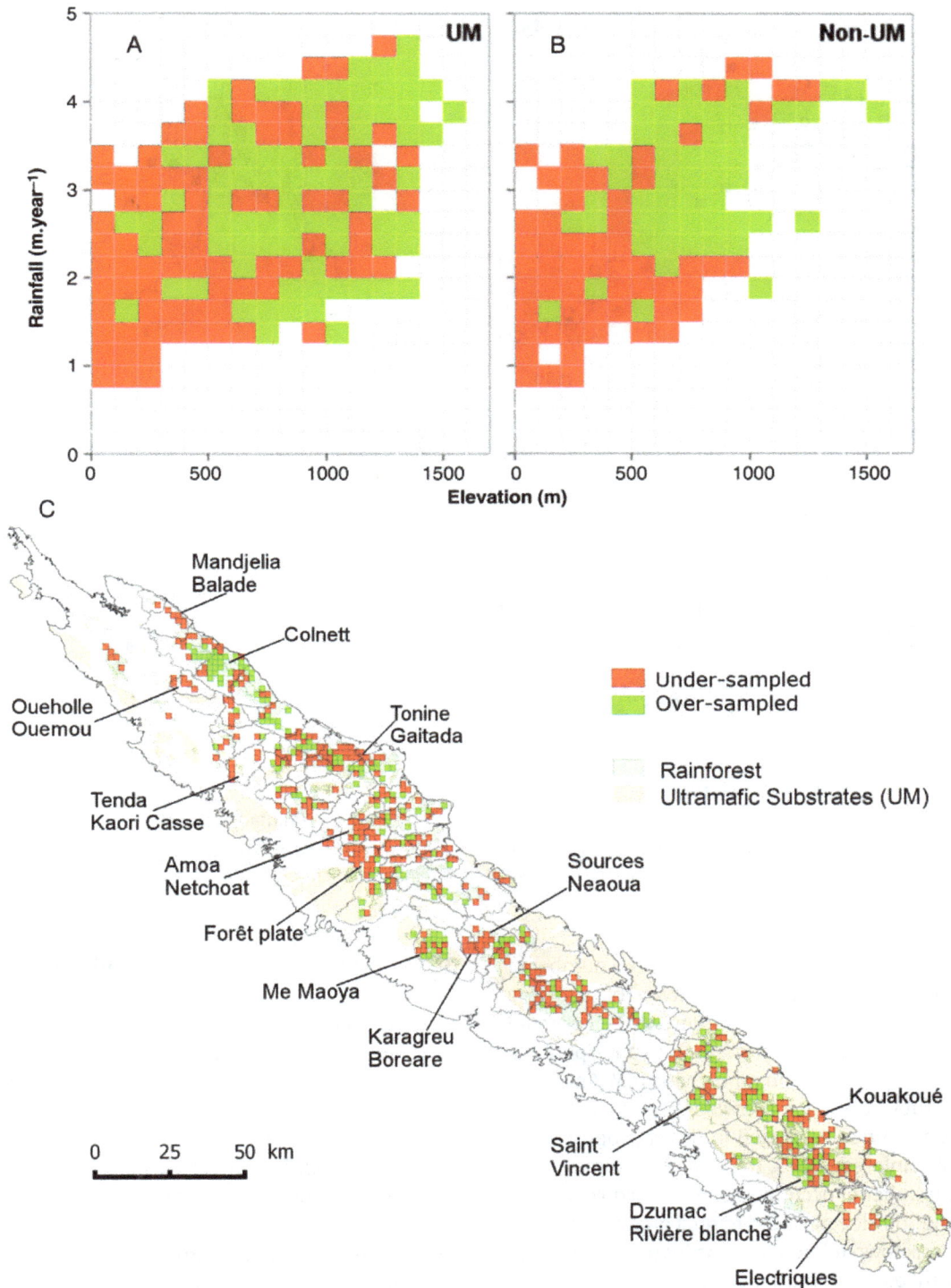

Figure 3. Environmental representativeness of occurrences on UM (A) and non-UM substrates (B) and geographical projection of no-data cells (C). Cells in red and green do not contain data. Cells in red are under-represented in our dataset and so are priorities for future botanical surveys to improve the knowledge of species distribution.

botanical surveys provide the very substantial quantity of data required to enhance our understanding of species distribution (Guitet *et al.* 2014). The complementarity of these datasets provides information on both the spatial and the ecological distribution of taxa and an assessment of γ-diversity.

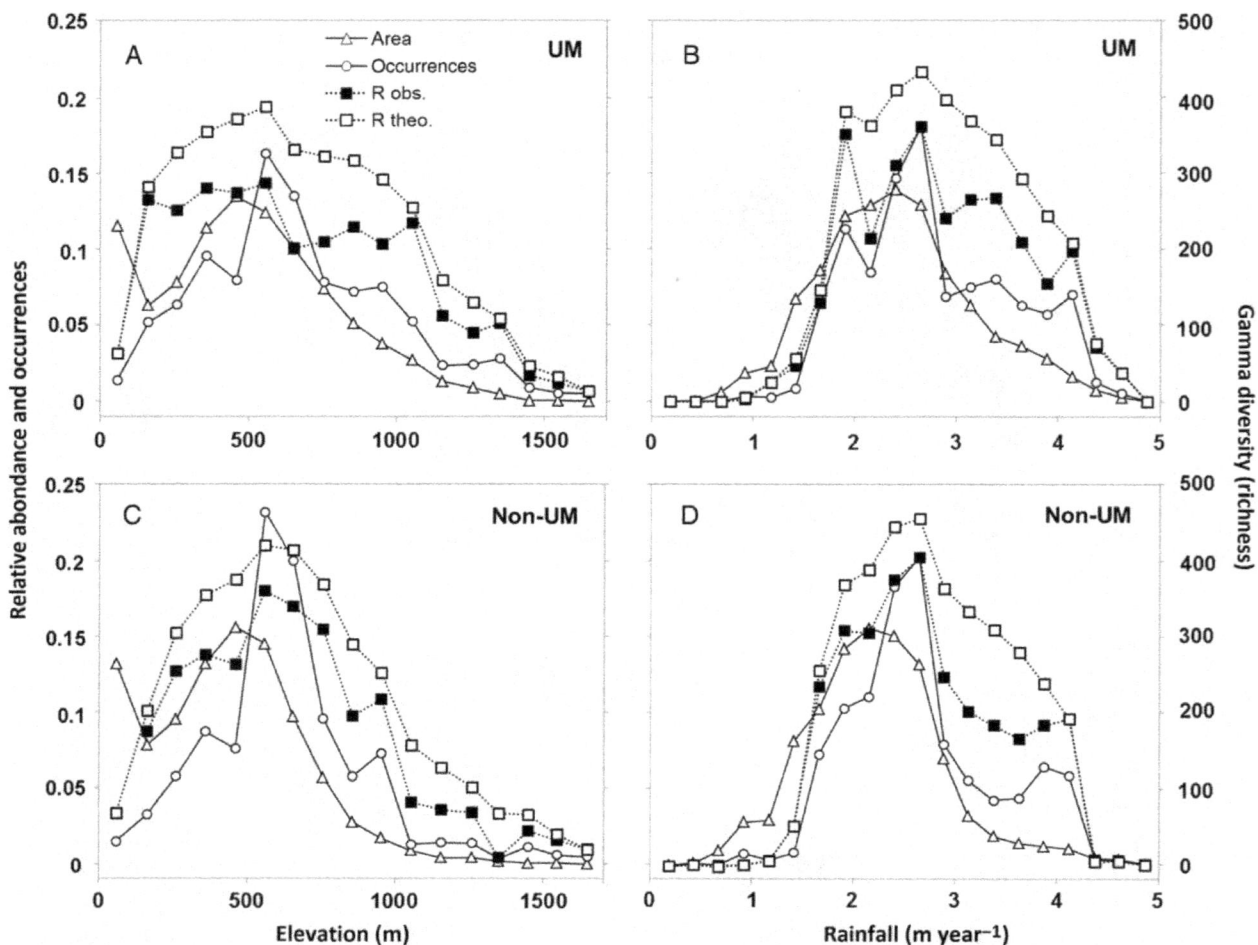

Figure 4. Distribution of observed and theoretical γ-diversities, the density of rainforest area and the density of tree occurrences on UM substrates (A and B) and non-UM substrates (C and D) along the elevation and rainfall gradients.

The dataset is unbalanced with regard to the number of occurrences found on UM substrates (two-thirds of the dataset). This pattern results from a huge dataset collected over decades on UM substrates to assess and anticipate the environmental impacts of mining activities (McCoy et al. 1999). In a cruel irony, our knowledge of tree species distribution on UM has increased in direct proportion to the decline in rainforest areas. In addition, the mismatch between the number of occurrences and available rainforest areas along the elevation and rainfall gradients reveals substantially unknown rainforest areas (notably at low elevation and with rainfall). These areas (see Fig. 3) should be sampled in priority to improve the representativeness of our occurrence data and subsequently to enhance our knowledge of New Caledonian tree species distribution. In the face of the high level of threat to New Caledonian rainforests, this information is critical to build reliable species distribution models under the current climate and also under putative future climatic conditions.

Geographical and environmental distributions of γ-diversity

The log–log correlations between observed or theoretical species γ-diversity and the available land area along the elevation gradient suggest that elevation impacts γ-diversity mainly through the so-called 'area effect' as a consequence of the basic species/area relationship (Sanders 2002). This effect was also found to drive plant γ-diversity on other high-elevation islands, including the vascular flora of Borneo (Grytnes et al. 2008), the palm flora of New Guinea (Bachman et al. 2004), the epiphytic flora of Taiwan (Hsu et al. 2014) and the fern flora of La Réunion Island (Karger et al. 2011). At the highest elevations and rainfalls, the species accumulation rates (α-Fisher) reveal a faster saturation of species richness. This may reflect a bias in our species selection method focussed on plots rarely distributed in such extremes values. However, few tree species are known to be high-elevation specialists in New Caledonia (Nasi et al. 2002).

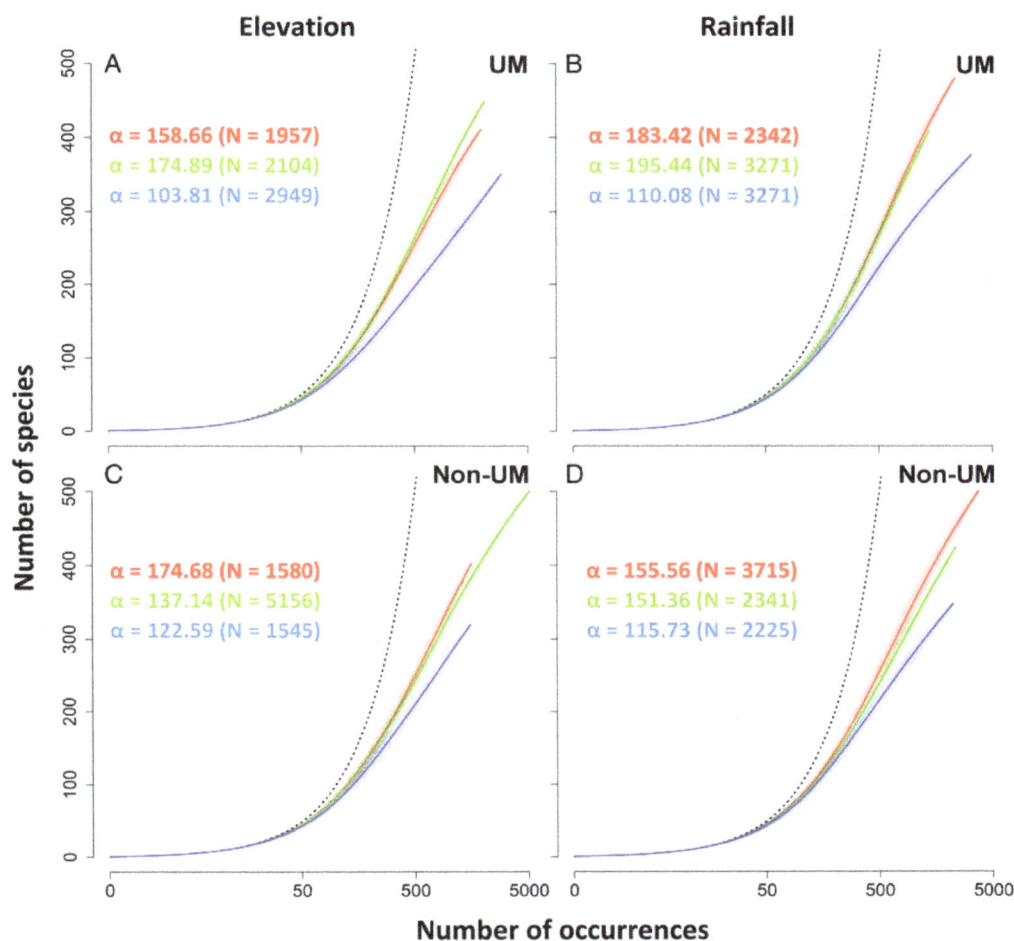

Figure 5. Species richness occurrences-based rarefaction curves compiled for low (red), mid (green) and high (blue) classes of elevation (≤400, 400–800 and >800 m, respectively) and rainfall (≤2.5, 2.5–3.0 and >3.0 m year^{-1}, respectively) on UM (A and B) and non-UM substrates (C and D), where α is Fisher's index and N is the total number of occurrences (the dotted line is the theoretical maximum rate of accumulation).

Furthermore, elevation in New Caledonia remains too low to record abiotic factors such as extreme low temperatures, freezing or extreme levels of solar radiation which can radically change a flora due to plant physiological limitations (Ghalambor et al. 2006; McCain and Grytnes 2010). Finally, the weaker correlation between γ-diversity and area along the rainfall gradient suggests that rainfall is likely to act as a stronger environmental driver of γ-diversity than elevation.

Geographical and environmental distributions of species

Most of the tree species selected from rainforest plots occurred beyond the rainforest boundaries (~60 %). This habitat transgression may be explained by a high tolerance of some species to open habitats such as 'maquis'. However, it could also be partially due to the scale of the digitalization of the vegetation map, likely to be inappropriate for exhaustively delineating rainforests which are highly fragmented from low to mid-elevation (Morat et al. 1999). This fragmentation results from a dramatic decrease in rainforest areas, which now occur on only 50 % of the pre-human surfaces (Jaffré et al. 1998). As a consequence, occurrences recorded in small relict patches were not included in the rainforest dataset but they did contribute in identifying the range of species tolerance with respect to substrate, elevation and rainfall.

Surprisingly, few species are substrate-specialists and our findings depart markedly from a balanced distribution in the three edaphic compartments (UM, non-UM and generalist). The adjective 'affiliated' should be used rather than 'specialist' since some edaphic transgressions could be explained by soil properties that could mitigate deficiencies and toxicity arising on UM substrates rather than by the substrate itself (Read et al. 2006; Ibanez et al. 2014). This tolerance to such contrasting habitats leads to an imbalance in species frequency; few were very abundant (124 species that account for 50 % of our dataset) while the vast majority remained rare in our dataset. Hyperdominance or oligarchy has previously been reported in other locations (Pitman *et al.* 2001, 2013; ter Steege *et al.* 2013), including in Pacific islands (Keppel *et al.* 2011), and is commonly related to low environmental heterogeneity, combined with a tendency towards species–habitat associations rather than obligate associations. In New Caledonia, tree oligarchic species always occur beyond the boundaries of substrates and rainforest habitat, even though they are clearly affiliated with UM substrate and rainforest. Thus, while constraints provided by UM substrates are often invoked to explain the originality and the diversification of the New Caledonian flora (Jaffré *et al.* 1987; Pintaud and Jaffré 2001; Pillon *et al.* 2010; Barrabé 2013), our findings nuance the specific relationship between UM substrate and trees distribution. They suggest that the large spatial variability in environmental conditions in UM rainforests could also have been involved in the diversification processes.

Finally, the large tolerance of species to UM substrates, elevation and rainfall contrasts with their spatial aggregation at the scale of *Grande Terre* as well as within plots network. This pattern supports the hypothesized low dispersibility among rainforest tree species (Seidler and Plotkin 2006). The loss of dispersibility in island plants is one of the 'insular syndromes' described by Carlquist (Carlquist 1974). Three hypotheses have been proposed by Carlquist to explain the loss of dispersibility in island plants: (i) the 'precinctiveness', which means that most seedlings germinate close to the parent plant because these habitats are more likely to be favourable than those further away; (ii) some island plants have shifted from a pioneering plant syndrome (*r*-strategists) to an exacerbated forest syndrome (*K*-strategists) and (iii) contact with the original dispersal vector may have been lost. The first hypothesis fits well with the observed aggregation of rainforest tree species while the third could be consistent with the low diversity of potential animal dispersal agents (Carpenter *et al.* 2003). However the second hypothesis runs counter to the very high tolerance of species to environmental gradients, including the observed transgressions beyond the habitat boundaries.

As a result, although the γ-diversity follows classical variations (e.g. hump-shaped) along environmental gradients, the distribution of individual species results more from stochastic processes rather than from deterministic ones (Leibold 1995; Chase and Myers 2011; Rosindell *et al.* 2011). This is in line with previous results from Ibanez *et al.* (2014) showing that the α-diversity is particularly low and the β-diversity particularly high in New Caledonian rainforests.

Conclusions

This study is the first to attempt to describe tree diversity in New Caledonian rainforests through extensive datasets collected over several decades. The collection effort is clearly critical if a realistic view of the observed γ-diversity is to be obtained, while the complementarity of several data collection methods provides both a comprehensive and dense coverage of tree species distribution. The wide ranges of tree species distribution with respect to substrates, elevation and rainfall contrast (i) with their spatial aggregation, (ii) with the small extent of rainforest core and (iii) with the lowest γ-diversity observed at high elevation or rainfall. Our results suggest a uniform γ-diversity from low to mid-elevations (<800 m) and rainfalls (<2.5 m year^{-1}), mostly dependent of occurrences and availability of rainforest areas. Above these thresholds, the decrease in the γ-diversity could be more related to the increase in rainfall through biological processes that we must further investigate in the future. Lastly, this study calls for new botanical data to be collected, mainly in non-UM rainforests under low elevation and low rainfall, to better estimate the total γ-diversity with respect to the large amplitude of ecological conditions available on *Grande Terre*.

Sources of Funding

Our work was partially funded by the DDEE-SIEC, 'La direction du développement économique et de l'environnement-Service impact environmental et conservation' of the Northern Province of New Caledonia.

Contributions by the Authors

P.B. conceived the study, managed the database, compiled the datasets, proceeded with the GIS-analysis and led the writing of this article. T.I. conceived the study, led the statistical analyses and actively contributed to the writing. R.P. actively contributed to the data compilation, the statistical analyses and the writing. H.V. was involved in the collection of many tree species occurrences in all the datasets used in this article. V.H. was also involved in the collection of many tree species occurrences in all the datasets used in this article. E.B. was fully involved in the setting up of 1 ha plots and reviewed the final manuscript. T.J. contributed to the overall coherence, robustness and balance of this article by providing the essential references and contributing from his experience of the ecology of New Caledonian forests, especially those on UM soils.

Acknowledgements

The authors wish to thank all the contributors who have

collected data over lengthy periods, without whose considerable efforts this study would not have been possible. This dedication includes botanists who have collected many samples as well as technicians and students who have participated in the overall effort of observation of New Caledonia trees. We also thank the two anonymous referees for their invaluable help in improving the manuscript. Finally, we express our sincere thanks to J.J. Cassan and the SIEC ('Service Impact environmental et conservation') of the Northern Province of New Caledonia for their unwavering commitment to our work.

Supporting Information

The following additional information is available in the online version of this article –

File S1. List of the 702 tree species according to their endemic (*E*) or indigenous (*I*) status, the total number of occurrences (*N*), the number of occurrences on ultramafic (UM) substrates, the number of occurrences in rainforest (Forest) and the ranges of elevation and rainfall calculated by the difference between the maximum and the minimum values where the species have been recorded.

File S2. Simplified list of reports related to inventory studies and surveys in New Caledonian forests that have most contributed to the 'observations' tree occurrences dataset.

Literature Cited

Ahrends A, Rahbek C, Bulling MT, Burgess ND, Platts PJ, Lovett JC, Kindemba VW, Owen N, Sallu AN, Marshall AR, Mhoro BE, Fanning E, Marchant R. 2011. Conservation and the botanist effect. *Biological Conservation* **144**:131–140.

Bachman S, Baker WJ, Brummitt N, Dransfield J, Moat J. 2004. Elevational gradients, area and tropical island diversity: an example from the palms of New Guinea. *Ecography* **27**:299–310.

Barrabé L. 2013. *Systématique et Evolution du genre Psychotria (Rubiaceae) en Nouvelle-Calédonie*. PhD Thesis, Université de la Nouvelle-Calédonie.

Barry RG. 2008. *Mountain weather and climate*. Cambridge, UK: Cambridge University Press.

Basset Y, Novotny V, Miller SE, Weiblen GD, Missa O, Stewart AJA. 2004. Conservation and biological monitoring of tropical forests: the role of parataxonomists. *Journal of Applied Ecology* **41**:163–174.

Bonvallot J, Gay J-Ch, Habert É. 2012. *Atlas de la Nouvelle-Calédonie*. Marseille-Nouméa: IRD-Congrès de la Nouvelle-Calédonie.

Carlquist S. 1974. *Island biology*. New York: Columbia University Press.

Carpenter RJ, Read J, Jaffré T. 2003. Reproductive traits of tropical rain-forest trees in New Caledonia. *Journal of Tropical Ecology* **19**:351–365.

Caujapé-Castells J, Tye A, Crawford DJ, Santos-Guerra A, Sakai A, Beaver K, Lobin W, Florens FBV, Moura M, Jardim R, Kueffer C.

2010. Conservation of oceanic island floras: present and future global challenges. *Perspectives in Plant Ecology, Evolution and Systematics* **12**:107–129.

Chase JM, Myers JA. 2011. Disentangling the importance of ecological niches from stochastic processes across scales. *Philosophical Transactions of the Royal Society of London. Series B, Biological Sciences* **366**:2351–2363.

Cluzel D, Chiron M-D. 1998. Upper Eocene unconformity and pre-obduction events in New Caledonia. *Comptes-Rendus de l'Académie des Sciences. Science de la terre et des planètes* **327**: 485–491.

DEM-DTSI. 2012. Modèle Numérique de Terrain au pas de 50 m—Service de la Géomatique et de la Télédétection. Gouvernement de la Nouvelle-Calédonie. http://www.geoportal.gouv.nc/geoportal/catalog/search/resource/details.page?uuid=%7BF3D33715-1A2E-4D2F-938F-A98AC53E94C6%7D (10 December 2014).

DIMENC/SGNC-BRGM. 2010. Les massifs de péridotites (massifs miniers, ultrabasiques, ultramafiques et nappes des ophiolites) au 1/1.000.000ème. http://www.geoportal.gouv.nc/geoportal/catalog/search/resource/details.page?uuid=%7B76E02B1B-D501-4B33-B13E-F5A711217CD4%7D (10 December 2014).

Espeland M, Murienne J. 2011. Diversity dynamics in New Caledonia: towards the end of the museum model? *BMC Evolutionary Biology* **11**:254.

Fritsch E. 2012. Les sols. In: Bonvallot J, Gay J-C, Habert É, eds. *Atlas de la Nouvelle-Calédonie*. Marseille-Nouméa: IRD-Congrès de la Nouvelle-Calédonie, 73–76.

Garcillán PP, Ezcurra E. 2011. Sampling procedures and species estimation: testing the effectiveness of herbarium data against vegetation sampling in an oceanic island. *Journal of Vegetation Science* **22**:273–280.

Ghalambor CK, Huey RB, Martin PR, Tewksbury JJ, Wang G. 2006. Are mountain passes higher in the tropics? Janzen's hypothesis revisited. *Integrative and Comparative Biology* **46**:5–17.

Gotelli NJ, Colwell RK. 2001. Quantifying biodiversity: procedures and pitfalls in the measurement and comparison of species richness. *Ecology Letters* **4**:379–391.

Grandcolas P, Murienne J, Robillard T, Desutter-Grandcolas L, Jourdan H, Guilbert E, Deharveng L. 2008. New Caledonia: a very old Darwinian island? *Philosophical Transactions of the Royal Society of London. Series B, Biological Sciences* **363**:3309–3317.

Grytnes J-A, Felde VA. 2014. Diversity patterns in a diversity hotspot. *Applied Vegetation Science* **17**:381–383.

Grytnes J-A, Beaman JH, Romdal TS, Rahbek C. 2008. The mid-domain effect matters: simulation analyses of range-size distribution data from Mount Kinabalu, Borneo. *Journal of Biogeography* **35**: 2138–2147.

Guitet S, Sabatier D, Brunaux O, Hérault B, Aubry-Kientz M, Molino J-F, Baraloto C. 2014. Estimating tropical tree diversity indices from forestry surveys: a method to integrate taxonomic uncertainty. *Forest Ecology and Management* **328**:270–281.

Herbert J. 2006. Distribution, habitat and red list status of the New Caledonian endemic tree Canacomyrica monticola (Myricaceae). *Biodiversity and Conservation* **15**:1459–1466.

Hillebrand H. 2004. On the generality of the latitudinal diversity gradient. *The American Naturalist* **163**:192–211.

Hortal J, Lobo JM, Jiménez-Valverde A. 2007. Limitations of biodiversity databases: case study on seed-plant diversity in Tenerife, Canary Islands. *Conservation Biology* **21**:853–863.

Hsu RC-C, Wolf JHD, Tamis WLM. 2014. Regional and elevational patterns in vascular epiphyte richness on an East Asian island. *Biotropica* **46**:549–555.

Ibanez T, Munzinger J, Dagostini G, Hequet V, Rigault F, Jaffré T, Birnbaum P. 2014. Structural and floristic diversity of mixed tropical rain forest in New Caledonia: new data from the New Caledonian Plant Inventory and Permanent Plot Network (NC-PIPPN). *Applied Vegetation Science* **17**:386–397.

Jaffré T. 1980. *Etude écologique du peuplement végétal des sols dérivés de roches ultrabasiques en Nouvelle-Calédonie.* Paris: ORSTOM.

Jaffré T. 1993. The relationship between ecological diversity and floristic diversity in New Caledonia. *Biodiversity Letters* **1**:82–87.

Jaffré T, Veillon J-M. 1990. Etude floristique et structurale de deux forêts denses humides sur roches ultrabasiques en Nouvelle-Calédonie. *Bulletin Museum National d'Histoire Naturelle, Paris, Série 4, Section B, Adansonia* **12**:243–273.

Jaffré T, Morat P, Veillon J-M, MacKee HS. 1987. Changements dans la végétation de la Nouvelle-Calédonie au cours du tertiaire : la végétation et la flore des roches ultrabasiques. *Bulletin Museum National d'Histoire Naturelle, Paris, Série 4, Section B, Adansonia* **9**: 365–391.

Jaffré T, Bouchet P, Veillon J-M. 1998. Threatened plants of New Caledonia: is the system of protected areas adequate? *Biodiversity and Conservation* **7**:109–135.

Jaffré T, Munzinger J, Lowry PP. 2010. Threats to the conifer species found on New Caledonia's ultramafic massifs and proposals for urgently needed measures to improve their protection. *Biodiversity and Conservation* **19**:1485–1502.

Jaffré T, Rigault F, Munzinger J. 2012. La végétation. In: Bonvallot J, Gay J-Ch, Habert É, eds. *Atlas de la Nouvelle-Calédonie.* Marseille-Nouméa: IRD-Congrès de la Nouvelle-Calédonie, 77–80.

Karger DN, Kluge J, Krömer T, Hemp A, Lehnert M, Kessler M. 2011. The effect of area on local and regional elevational patterns of species richness. *Journal of Biogeography* **38**:1177–1185.

Keppel G, Tuiwawa MV, Naikatini A, Rounds IA. 2011. Microhabitat specialization of tropical rain-forest canopy trees in the Sovi Basin, Viti Levu, Fiji Islands. *Journal of Tropical Ecology* **27**:491–501.

Kessler M, Kluge J, Hemp A, Ohlemüller R. 2011. A global comparative analysis of elevational species richness patterns of ferns. *Global Ecology and Biogeography* **20**:868–880.

Kumar S, Stohlgren TJ. 2009. Maxent modeling for predicting suitable habitat for threatened and endangered tree *Canacomyrica monticola* in New Caledonia. *Journal of Ecology and Natural Environment* **1**:94–98.

Lavoie C. 2013. Biological collections in an ever changing world: Herbaria as tools for biogeographical and environmental studies. *Perspectives in Plant Ecology, Evolution and Systematics* **15**: 68–76.

Leibold MA. 1995. The niche concept revisited: mechanistic models and community context. *Ecology* **76**:1371–1382.

Lomolino MV. 2001. Elevation gradients of species–density: historical and prospective views. *Global Ecology and Biogeography* **10**:3–13.

McCain CM. 2007. Area and mammalian elevational diversity. *Ecology* **88**:76–86.

McCain CM, Grytnes J-A. 2010. Elevational gradients in species richness. In: Jonsson R, ed. *Encyclopedia of life sciences—ecology.*

Chichester, UK: John Wiley & Sons, Ltd., 10p. doi: 10.1002/9780470015902.a0022548.

McCoy S, Jaffré T, Rigault F, Ash JE. 1999. Fire and succession in the ultramafic maquis of New Caledonia. *Journal of Biogeography* **26**: 579–594.

Météo-France. 2007. *Atlas climatique de la Nouvelle-Calédonie.* Noumea: Météo France en Nouvelle-Calédonie.

Morat P. 1993. Our knowledge of the flora of New Caledonia: endemism and diversity in relation to vegetation types and substrates. *Biodiversity Letters* **1**:72–81.

Morat P, Jaffré T, Veillon J-M. 1999. Menaces sur les taxons rares de la Nouvelle-Calédonie. Actes du Colloque sur les espèces végétales menacées de France. *Bulletin de la Société Botanique du Sud-Ouest* **19**:129–144.

Morat P, Jaffré T, Tronchet F, Munzinger J, Pillon Y, Veillon J-M, Chalopin M, Birnbaum P, Rigault F, Dagostini G, Tinel J, Lowry PP. 2012. Le référentiel taxonomique Florical et les caractéristiques de la flore vasculaire indigène de la Nouvelle-Calédonie. *Adansonia* **34**:179–221.

Murienne J. 2009. Testing biodiversity hypotheses in New Caledonia using phylogenetics. *Journal of Biogeography* **36**:1433–1434.

Murienne J, Grandcolas P, Piulachs MD, Bellés X, D'Haese C, Legendre F, Pellens R, Guilbert E. 2005. Evolution on a shaky piece of Gondwana: is local endemism recent in New Caledonia? *Cladistics* **21**:2–7.

Nasi R, Jaffré T, Sarrailh J-M. 2002. Les forêts de montagnes de Nouvelle-Calédonie. *Bois et Forêts des Tropiques* **274**:5–17.

Neall VE, Trewick SA. 2008. The age and origin of the Pacific islands: a geological overview. *Philosophical Transactions of the Royal Society B: Biological Sciences* **363**:3293–3308.

Nogués-Bravo D, Araújo MB, Romdal T, Rahbek C. 2008. Scale effects and human impact on the elevational species richness gradients. *Nature* **453**:216–219.

Oksanen J, Blanchet FG, Kindt R, Legendre P, Minchin PR, O'Hara RB, Simpson GL, Solymos P, Stevens MHH, Wagner H. 2013. Vegan: community ecology package. R package version 2.2 (http://vegan.r-forge.r-project.org, 10 December 2014).

Pillon Y. 2012. Time and tempo of diversification in the flora of New Caledonia. *Botanical Journal of the Linnean Society* **170**:288–298.

Pillon Y, Munzinger J, Amir H, Lebrun M. 2010. Ultramafic soils and species sorting in the flora of New Caledonia. *Journal of Ecology* **98**:1108–1116.

Pintaud J-C, Jaffré T. 2001. Pattern of diversification of Palms on ultramafic rocks in New Caledonia. *South African Journal of Sciences* **97**:548–550.

Pintaud J-C, Jaffré T, Veillon J-M. 1999. Conservation status of New Caledonia palms. *Pacific Conservation Biology* **5**:9–15.

Pitman NCA, Terborgh JW, Silman MR, Núñez VP, Neill DA, Cerón CE, Palacios WA, Aulestia M. 2001. Dominance and distribution of tree species in upper Amazonian terra firme forests. *Ecology* **82**:2101–2117.

Pitman NCA, Silman MR, Terborgh JW. 2013. Oligarchies in Amazonian tree communities: a ten-year review. *Ecography* **36**:114–123.

Poncet V, Munoz F, Munzinger J, Pillon Y, Gomez C, Couderc M, Tranchant-Dubreuil C, Hamon S, de Kochko A. 2013. Phylogeography and niche modelling of the relict plant *Amborella trichopoda* (Amborellaceae) reveal multiple Pleistocene refugia in New Caledonia. *Molecular Ecology* **22**:6163–6178.

Quantum GIS Development Team. 2014. QGIS Geographic Information System. Open Source Geospatial Foundation. http://qgis.osgeo.org.

Rahbek C. 2005. The role of spatial scale and the perception of large-scale species-richness patterns. *Ecology Letters* **8**: 224–239.

Read J, Jaffré T, Godrie E, Hope GS, Veillon J-M. 2000. Structural and floristic characteristics of some monodominant and adjacent mixed rainforests in New Caledonia. *Journal of Biogeography* **27**:233–250.

Read J, Jaffré T, Ferris JM, McCoy S, Hope GS. 2006. Does soil determine the boundaries of monodominant rain forest with adjacent mixed rain forest and maquis on ultramafic soils in New Caledonia? *Journal of Biogeography* **33**:1055–1065.

Rosindell J, Hubbell SP, Etienne RS. 2011. The unified neutral theory of biodiversity and biogeography at age ten. *Trends in Ecology and Evolution* **26**:340–348.

Sanders NJ. 2002. Elevational gradients in ant species richness: area, geometry, and Rapoport's rule. *Ecography* **25**:25–32.

Schmidt-Lebuhn AN, Knerr NJ, González-Orozco CE. 2012. Distorted perception of the spatial distribution of plant diversity through uneven collecting efforts: the example of Asteraceae in Australia. *Journal of Biogeography* **39**:2072–2080.

Seidler TG, Plotkin JB. 2006. Seed dispersal and spatial pattern in tropical trees. *PLoS Biology* **4**:e344.

ter Steege H, Pitman NCA, Sabatier D, Baraloto C, Salomão RP, Guevara JE, Phillips OL, Castilho CV, Magnusson WE, Molino J-F, Monteagudo A, Núñez Vargas P, Montero JC, Feldpausch TR, Coronado ENH, Killeen TJ, Mostacedo B, Vasquez R, Assis RL, Terborgh J, Wittmann F, Andrade A, Laurance WF, Laurance SGW, Marimon BS, Marimon B-H Jr, Guimarães Vieira IC, Amaral IL, Brienen R, Castellanos H, Cárdenas López D, Duivenvoorden JF, Mogollón HF, de Almeida Matos FD, Dávila N, García-Villacorta R, Stevenson Diaz PR, Costa F, Emilio T, Levis C, Schietti J, Souza P, Alonso A, Dallmeier F, Montoya AJD, Fernandez Piedade MT, Araujo-Murakami A, Arroyo L, Gribel R, Fine PVA, Peres CA, Toledo M, Aymard CGA, Baker TR, Cerón C, Engel J, Henkel TW, Maas P, Petronelli P, Stropp J, Zartman CE, Daly D, Neill D, Silveira M, Paredes MR, Chave J, de Andrade Lima Filho D, Jørgensen PM, Fuentes A, Schöngart J, Cornejo Valverde F, Di Fiore A, Jimenez EM, Peñuela Mora MC, Phillips JF, Rivas G, van Andel TR, von Hildebrand P, Hoffman B, Zent EL, Malhi Y, Prieto A, Rudas A, Ruschell AR, Silva N, Vos V, Zent S, Oliveira AA, Schutz AC, Gonzales T, Trindade Nascimento M, Ramirez-Angulo H, Sierra R, Tirado M, Umaña Medina MN, van der Heijden G, Vela CIA, Vilanova Torre E, Vriesendorp C, Wang O, Young KR, Baider C, Balslev H, Ferreira C, Mesones I, Torres-Lezama A, Urrego Giraldo LE, Zagt R, Alexiades MN, Hernandez L, Huamantupa-Chuquimaco I, Milliken W, Palacios Cuenca W, Pauletto D, Valderrama Sandoval E, Valenzuela Gamarra L, Dexter KG, Feeley K, Lopez-Gonzalez G, Silman MR. 2013. Hyperdominance in the Amazonian Tree Flora. *Science* **342**:1243092.

Terry JP, Kostaschuk RA, Wotling G. 2008. Features of tropical cyclone-induced flood peaks on Grande Terre, New Caledonia. *Water and Environment Journal* **22**:177–183.

Whittaker RJ, Araújo MB, Jepson P, Ladle RJ, Watson JEM, Willis KJ. 2005. Conservation biogeography: assessment and prospect. *Diversity and Distributions* **11**:3–23.

Wulff AS, Hollingsworth PM, Ahrends A, Jaffré T, Veillon J-M, L'Huillier L, Fogliani B. 2013. Conservation priorities in a biodiversity hotspot: analysis of narrow endemic plant species in New Caledonia. *PLoS ONE* **8**:e73371.

10

Low redundancy in seed dispersal within an island frugivore community

Kim R. McConkey[1,3]* and Donald R. Drake[2]

[1] School of Biological Sciences, Victoria University of Wellington, PO Box 600, Wellington, New Zealand
[2] Department of Botany, University of Hawai'i at Manoa, 3190 Maile Way, Honolulu, HI 96822, USA
[3] Present address: School of Natural Sciences and Engineering, National Institute of Advanced Studies, Indian Institute of Science Campus, Bangalore, India

Associate Editor: Anna Traveset

Abstract. The low species diversity that often characterizes island ecosystems could result in low functional redundancy within communities. Flying foxes (large fruit bats) are important seed dispersers of large-seeded species, but their redundancy within island communities has never been explicitly tested. In a Pacific archipelago, we found that flying foxes were the sole effective disperser of 57 % of the plant species whose fruits they consume. They were essential for the dispersal of these species either because they handled >90 % of consumed fruit, or were the only animal depositing seeds away from the parent canopy, or both. Flying foxes were especially important for larger-seeded fruit (>13 mm wide), with 76 % of consumed species dependent on them for dispersal, compared with 31 % of small-seeded species. As flying foxes decrease in abundance, they cease to function as dispersers long before they become rare. We compared the seed dispersal effectiveness (measured as the proportion of diaspores dispersed beyond parent crowns) of all frugivores for four plant species in sites where flying foxes were, and were not, functionally extinct. At both low and high abundance, flying foxes consumed most available fruit of these species, but the proportion of handled diaspores dispersed away from parent crowns (quality) was significantly reduced at low abundance. Since alternative consumers (birds, rodents and land crabs) were unable to compensate as dispersers when flying foxes were functionally extinct, we conclude that there is almost no redundancy in the seed dispersal function of flying foxes in this island system, and potentially on other islands where they occur. Given that oceanic island communities are often simpler than continental communities, evaluating the extent of redundancy across different ecological functions on islands is extremely important.

Keywords: Ecological redundancy; flying foxes; frugivore; fruit bats; functional extinction; Pacific islands; *Pteropus*; seed dispersal.

* Corresponding author's e-mail address: kimmcconkey@gmail.com

Introduction

Resilience to disturbance is greatest in ecosystems that have high species diversity because of the buffering effect diversity can have on function (Mayfield et al. 2010; Dalerum et al. 2012; Reich et al. 2012). When multiple species perform a given ecosystem function, there is redundancy within the system, and the function may be fully or partially maintained following perturbations in species populations (Dalerum et al. 2012). As ecosystems lose species, however, associated declines in functional redundancy increase the vulnerability of these ecosystems to further change (Reich et al. 2012). Islands are characterized by inherently low species diversity compared with continents (MacArthur 1965; Whittaker and Fernández-Palacios 2007), and they have been disproportionately further depleted by human-mediated extinctions (e.g. Olson and James 1982; Steadman et al. 1991; Steadman 2006). Hence, current island ecosystems might exhibit especially low functional redundancy, which makes the ongoing human-mediated disturbances to them (Brooks et al. 2002; Whittaker and Fernández-Palacios 2007) a serious threat to their stability (Cox et al. 1991; Traveset et al. 2012). An alternative view is that island systems may be somewhat buffered against low functional redundancy because island species are often generalists, or even super-generalists, in their diet and habitat use (Banack 1998; Olesen et al. 2002). Hence, understanding the vulnerability of island species to a lack of functional redundancy is complicated, but important, to ensure that functional ecosystems are maintained.

Fruit bats in the family Pteropodidae are effective seed dispersers throughout the Old World tropics (Richards 1990; Rainey et al. 1995; Banack 1998; Hodgkison et al. 2003; Bollen et al. 2004; Nyhagen et al. 2005). Flying foxes (Pteropus spp.) are predominantly found on islands, with a distribution stretching from the coast of East Africa, through tropical Asia, to Polynesia. Simplified frugivore communities exist on many of these islands, with especially low diversity in the tropical Pacific (Steadman 2006). Here, flying foxes, many species of which declined following human discovery of the islands (Steadman 2006), have generalist diets (Banack 1998) and are often regarded as 'keystone' seed dispersers, particularly for large-seeded plants, because of a relative lack of other large frugivores (Cox et al. 1991; Rainey et al. 1995; McConkey and Drake 2002). In many places, the only extant alternative dispersers of large-seeded species are pigeons, whose role is limited by their gape size (Meehan et al. 2002; McConkey et al. 2004a), and non-volant animals (e.g. rats and crabs) that may sometimes disperse seeds (Lee 1985, 1988; O'Dowd and Lake 1991; McConkey et al. 2003; Pérez et al. 2008; Shiels and Drake 2011) but are unlikely to be functionally similar to flying foxes. Hence, this community might be expected to show very low redundancy in seed dispersal, especially for large-seeded fruit (Meehan et al. 2002). With ~80 % of plant species displaying zoochorous dispersal mechanisms on some islands (Fall et al. 2007), reductions to flying fox populations could have widespread effects on the ecosystem.

The abundance of flying foxes relative to the number of food-bearing trees has a direct, non-linear relationship with their function as seed dispersers (McConkey and Drake 2006). When flying fox abundance is low, the animals remain within the fruiting plant to feed, dropping all unswallowed seeds (flying foxes cannot swallow seeds >4 mm) directly underneath. As fruiting plants fill with animals, feeding territories become fully occupied, forcing any additional flying foxes to 'raid' the occupied trees for fruit, which they take to another tree to consume. Only 'raiders' are likely to disperse large seeds beyond parent crowns (Richards 1990); consequently, flying foxes become functionally extinct as seed dispersers once their abundance drops below a habitat-specific threshold at which 'raiding' begins (Richards 1990; McConkey and Drake 2006). If the frugivore community has low functional redundancy, then declines in flying fox populations resulting from habitat loss, introduced predators, hunting (Wiles et al. 1997; Brooke 2001; Jenkins et al. 2007; Palmeirin et al. 2007) and cyclones (Pierson et al. 1996; McConkey et al. 2004b) will have large consequences for the island ecosystems in which they occur. Given that >70 % of island flying fox species are threatened, near threatened or lacking sufficient information for assessment (IUCN 2014), testing their functional redundancy as dispersers on islands has become urgent.

Our aim was to assess the functional redundancy of an island population of flying foxes in the seed dispersal of their main food plants. We addressed the following hypotheses: (i) relative to other dispersers, flying foxes disperse a disproportionately high proportion of seeds of large-seeded species (defined here as dispersal quantity); (ii) flying foxes disperse a higher proportion of handled seeds away from parent crowns than other dispersers, and mean dispersal distances are greater (defined here as dispersal quality); (iii) flying foxes have greater 'seed dispersal effectiveness' (SDE = quantity × quality, Schupp et al. 2010) for large diaspores than other dispersers and (iv) total SDE of plants is reduced at sites where flying foxes are functionally extinct.

Methods

Research was conducted in the Vava'u archipelago of Tonga in Western Polynesia between June 1999 and

June 2001. The archipelago includes 64 islands (total land area 143.3 km^2; range: <1–96 km^2) spread over ~750 km^2 of ocean. The vegetation was mainly mature rain forest 20–25 m tall (described in Franklin *et al.* 1999). The frugivore community consisted of the insular flying fox *Pteropus tonganus* (body mass averages 428 g; this and following measurements are taken from Gibbons and Watkins 1982; Meehan *et al.* 2002; del Hoyo *et al.* 2013), Pacific pigeon *Ducula pacifica* (395 g), three small dove species (*Ptilinopus perousii* (90 g), *P. porphyraceus* (110 g) and *Alopecoenas stairi* (syn. *Gallicolumba stairi* (Gray, 1856), 171 g)), four even smaller passerines that are at least partially frugivorous (*Pycnonotus cafer* (34 g), *Aplonis tabuensis* (60 g), *Foulehaio carunculatus* (31 g) and *Lalage maculosa* (30 g)), three rat species (*Rattus norvegicus* (215 g), *R. rattus* (140 g) and *R. exulans* (92 g)), crabs (*Coenobita* spp.; measurements not available) and a rarely observed iguana (*Brachylophus fasciatus*; 160 g).

Identifying alternative dispersers for flying fox-dispersed seeds

Over a 2-year sampling period, we identified consumers of different plant species through direct observations and feeding signs left on fruit and seeds. We conducted systematic and opportunistic searches for handled fruit and dispersed seeds within the forest and under daytime bat roosts, as part of seed dispersal studies on flying foxes (McConkey and Drake 2006) and Pacific pigeons (McConkey *et al.* 2004a; Meehan *et al.* 2005). Flying foxes can disperse seeds by swallowing and defecating (seeds <4 mm in diameter) or by carrying the fruit away from the parent plant to eat elsewhere and subsequently dropping the unconsumed seeds (Richards 1990). Diaspores (single- or multi-seeded dispersal units) handled by flying foxes were identified by distinctive impressions left in the pulp (triangular teeth marks in the pulp adhering to the seed), by corresponding wads of spat out pulp or by recovery from bat faeces (Fig. 1). Since the searches were conducted under the mature rain forest canopy, there might be a bias to finding species with larger diaspores. However, smaller-seeded species were deposited in faeces under roosts or fruiting trees, and the easily recognizable wads of fruit pulp also enabled their identification. Hence, we believe the bias to be small.

We classified species with fruits eaten by flying foxes as: (i) 'commonly-eaten' species, having multiple records of flying fox foraging and dispersal; (ii) 'rarely-eaten' species, which were plants that were found often but had only one or two records of feeding by flying foxes; these bore bat-teeth marks but most of the pulp remained and (iii) 'damaged' species, having fruits whose seeds

Figure 1. Fruit and diaspores of *Pleiogynium timoriense* showing signs of flying fox feeding. (A) Fruit handled by flying fox. Triangular tooth mark is visible; (B) entire unhandled fruit; (C) day-old and (D) fresh endocarp with most flesh chewed off; (E) day-old and (F) fresh spat out wads of pulp.

were eaten by flying foxes. These might still be dispersed by flying foxes if carried away from the parent plant for consumption but dropped before the seed was destroyed; this was observed during the study.

Other frugivores also leave unique impressions on diaspores: pigeon-dispersed diaspores are defecated in identifiable scats or regurgitated (single clean diaspores), crabs leave linear claw marks in the pulp and rodents leave incisor marks in the pulp or diaspores (we were unable to distinguish rat species and discuss them collectively). During the study period, we recorded all diaspores regurgitated (n = 10 samples) by pigeons or found in their scats (n = 67; McConkey *et al.* 2004a), in rat husking stations (n = 13 720 diaspores; McConkey *et al.* 2003) or handled by crabs (n = 140 diaspores).

For small seeds (defined below), the feeding sign of doves and passerines (collectively referred to as 'small birds' hereafter) is probably not always distinguishable from pigeons. Our list of dispersed species for these birds was generated from both direct observations and pigeon-dispersed species that had diaspores small enough (<14 mm wide) to be swallowed by doves, which were the largest birds within this category (Meehan *et al.* 2002).

To investigate the different contributions of frugivores according to diaspore size, we classified diaspores into four size categories based on the ability of birds and flying foxes to swallow them (cf. Meehan *et al.* 2002): *small* (S <4 mm), which can be swallowed by all dispersers, except crabs; *medium* (M = 4–13 mm), which can be swallowed by small birds and pigeons; *large* (L = 14–27 mm), which can be swallowed only by pigeons and *extra large* (XL >27 mm), which cannot be swallowed by any dispersers.

Measuring quantity, quality and SDE

We conducted intensive seed dispersal studies on a selection of the most common species in the diet of flying foxes ($n = 83$ plants of 14 species; McConkey and Drake 2006). All had ripe fruit containing diaspores too large for flying foxes to swallow and were easily detected on the forest floor. Two species had diaspores that were small enough to be swallowed by some small birds (M), seven could be swallowed only by pigeons (L) and five exceeded the gape of any frugivore (XL) (Table 1).

Studies were conducted on eight islands over 15 independent visits (5–9 trees per visit). From each plant's trunk to 45 m beyond the crown edge, four 2-m-wide transects were checked for freshly dispersed (handled) seeds and fallen fruits every morning for 3–5 consecutive days. If a fruiting conspecific plant was found along this transect or near to the edge, the transect extended halfway between the trees. We ensured that each tree had at least two transects extending the full 45 m, and the majority had three or four. We identified the animal responsible for the handled seeds by their feeding sign. Seed densities (seeds m^{-2} day^{-1}) were calculated for seeds dispersed under the plant crown and at 1-m increments from the crown edge. We used the seed density to calculate the total number of diaspores dispersed by each frugivore at each 5-m distance increment; total seed fall was the product of the recorded seed density at that distance category and the total area around the tree at that distance (calculated using the canopy radii and the formula of an ellipse).

Table 1. Plant species consumed by flying foxes in the Vava'u Islands of Tonga during the study period. Species are divided into categories of increasing seed width: small (S < 4 mm), which can be swallowed by flying foxes and all birds; medium (M = 4–13 mm), which can be swallowed by small birds and pigeons; large (L = 14–27 mm), which can be swallowed only by pigeons; and extra large (XL > 27 mm), which cannot be swallowed by anything (cf. Meehan et al. 2002). Alternative dispersers are taken from our observations as well as previously published records and seed size (Meehan et al. 2002): P, pigeon; S, small birds; R, rats; C, crabs; when an animal acts as a predator, it is in small letters, so Rr means rats act as both dispersers and predators. A question mark means possible disperser. Abbreviations after the species name indicate species used for more intensive studies. [1]Species is not commonly consumed by flying foxes; [2]syn. *Planchonella grayana*; [3]species not in italics have seeds that are eaten by flying foxes. In the case of *Maniltoa grandiflora*, we recorded a single seed that had minimal damage and had been dispersed away from the canopy. Hence, these species might still be effectively dispersed, but both were uncommon in the diet and could not be fully assessed.

Seed width (mm)	Have other dispersers		No other dispersers	
	Species	Eaten by	Species	Eaten by
Small (<4)	Ficus 4 spp.,	P S R?	Melodinus vitiensis	
			Passiflora aurantia[1]	R?
Medium (4–13)	Micromelum minutum	P S r	Pouteria grayana[2] PG	r c
	Morinda citrifolia	P		
	Diospyros elliptica DE	P S r c		
	Podocarpus pallidus[1,3]	P		
	Vavaea amicorum[1]	P S		
	Jasminum didyum[1]	P S		
Large (14–27)	Syzygium clusiifolium SC	P	Maniltoa grandiflora[1,3]	r
	Termialia litoralis TL	P		
	Chionanthus vitiensis CV	P Rr		
	Pleiogynium timoriense PTi	P Rr C		
	Syzygium dealatum SD	P		
	Faradaya amicorum FA	P		
	Guettarda speciosa GS	P r		
	Ochrosia vitiensis[1]	P		
	Canarium harveyi[1]	P r		
	Hernandia nymphaeifolia[1]	P		
Extra large (>27)	Mangifera indica MI	r C	Burckella richii BR	r
	Pandanus tectorius PTe	Rr C	Terminalia catappa TC	r
	Neisosperma oppositifolium	Rr C		
	Inocarpus fagifer IF	r Cc		

Seed dispersal effectiveness (Schupp *et al.* 2010) integrates the quantity and quality components of seed dispersal. We defined 'quantity' as the proportion of all seeds that had been handled and dispersed by a particular disperser species and 'quality' as the proportion of these seeds that were dispersed at least 5 m from the parent canopy by the same species. The 5-m cut-off was chosen because whole, unhandled fruits frequently bounced or rolled ≤5 m from the edge of the parent crown, suggesting that handled fruits found in the same range could have been dropped from within the fruiting crown. Our measure of dispersal quality (and therefore SDE) is incomplete since we do not know the recruitment potential of the seeds dispersed at different distances. Here, we base quality on the minimum assumption that seeds dispersed away from parent crowns are more likely to establish than seeds dispersed under the crown, and this Janzen–Connell effect has shown to be a common scenario for most tropical trees (Swamy and Terborgh 2010). We calculated SDE for plant species for which we had data for three or more individual plants.

Functional extinction of flying foxes and SDE

The effectiveness of flying foxes as seed dispersers is non-linearly related to their abundance since they become functionally extinct once their abundance drops below a habitat-specific threshold (McConkey and Drake 2006). Flying foxes were abundant in the island group during our study, but local abundance at any site varies temporally and spatially as individuals track fruit supplies around the islands (McConkey and Drake 2007). We used this natural variability to compare SDE values for flying foxes in sites where they were at low abundance relative to food availability (i.e. functionally extinct) with sites where their relative abundance was high. We also used the SDE landscape (Schupp *et al.* 2010) to investigate the relative contributions of other frugivores at these sites and determine whether they compensated for flying fox loss. Site categories used followed McConkey and Drake (2006).

Statistical analysis

To evaluate whether the SDE of flying foxes was significantly altered by their abundance, we compared SDE values of all species from low- and high-abundance sites using a *t*-test. *t*-Tests were also used to compare the quantity and quality values in low- and high-abundance sites for one species, *Pleiogynium timoriense*, for which we had >10 studied trees in each category. We used the conservative approach of identifying non-overlapping confidence intervals to evaluate differences in mean dispersal distances among species. *Z*-tests were used to compare the proportion of diaspores dispersed under conspecific crowns by flying foxes, with other frugivores. We only used trees studied at sites where flying fox abundance was high (i.e. where they were not functionally extinct) to determine the proportions dispersed by flying foxes. Statistical analyses were done using Sigmastat 3.5.

Results

Alternative dispersers for flying fox-dispersed seeds

Fruits of 30 plant species being handled by flying foxes at the study site were recorded (22 commonly-eaten species, 8 rarely-eaten species, of which 2 had the seeds partially eaten) and 6 of these had no alternative dispersers. Six of the consumed plant species had small diaspores and four of these had several alternative dispersers (Table 1, Fig. 2). One of the seven species with medium diaspores (*Planchonella grayana* H.St.John) was dispersed only by flying foxes; its seeds were encased in large, hard fruits that prevented bird feeding (Fig. 2). Most species that had large diaspores ($n = 11$) and were fed on by flying foxes were also dispersed by pigeons, while species with extra large diaspores ($n = 6$) were occasionally dispersed by rats or crabs ($n = 4$). Interestingly, crab-dispersed species were recorded only in this largest category (Fig. 2). Rats often consumed the same species as flying foxes, but for the majority of these, the rats destroyed the diaspores (Fig. 2). Overall, pigeons dispersed most of the species dispersed by flying foxes, but they could not handle the largest seeds, for which terrestrial animals were the only alternative dispersers (Fig. 2).

Quantitative contributions of frugivores

Although most plant species had multiple potential dispersers, flying foxes were the predominant dispersers for 14 species they commonly consumed, being responsible for >90 % of the dispersed diaspores for 12 species and >70 % for the remaining 2 ($n = 83$ individual plants in total) (Fig. 3). We recorded pigeon dispersal for six of these species, but rates approached 10 % for only two. Crabs and rats were each recorded two times, and rats dispersed a relatively high proportion of *Pandanus tectorius* diaspores.

Distance distribution of dispersed seeds

Flying foxes and rats deposited the lowest proportion of handled diaspores under conspecific canopies (31 % ($n = 45$ plants) and 29 % ($n = 37$), respectively), and there was no statistical difference between them ($Z = -1.05$, $P = 0.15$) (Fig. 4A). Both pigeons ($n = 8$) and crabs ($n = 10$) were significantly more likely to deposit handled diaspores under conspecific canopies than flying foxes were (65 and 89 %, respectively) (pigeons: $Z = -2.64$, $P = 0.004$; crabs: $Z = -2.81$, $P = 0.002$). If all handled

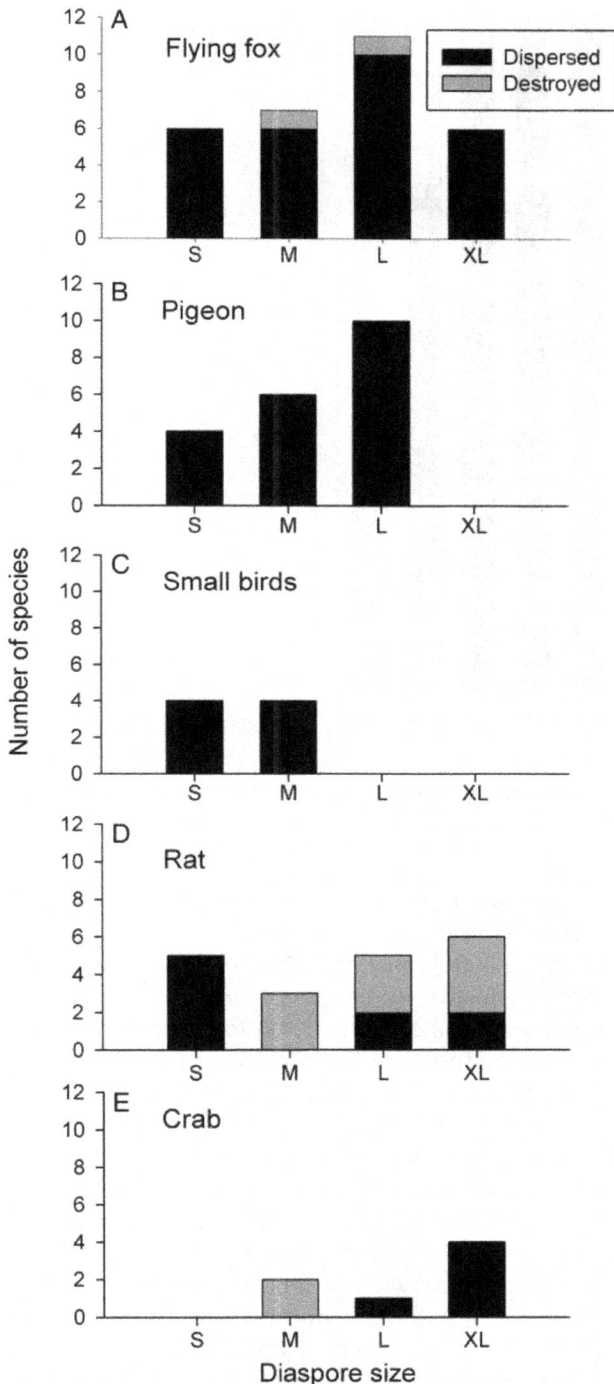

Figure 2. Overlap in seed dispersal services provided by flying foxes and other fruit-eating animals in Tonga. The figure shows the number of plant species that are dispersed (black) or destroyed (grey) by (A) flying foxes and how many of these species are also consumed by other animals (B–E). Dispersed species are arranged into four categories of diaspore size: small (S < 4 mm), which is the gape limit of flying foxes; medium (M = 4–13 mm), which is the gape limit of small birds; large (L = 14–27 mm), which is the gape limit of pigeons and extra large (XL > 27 mm) cannot be swallowed by anything (cf. Meehan et al. 2002).

diaspores are included (with those deposited under conspecific canopies), flying foxes dispersed diaspores a mean distance of 12.7 m from the canopy edge, which is significantly further than all other dispersers (the 95 % confidence intervals do not overlap) (Fig. 4B). Crabs deposited diaspores very close to conspecific crowns on average, while mean dispersal distances for pigeons and rats were intermediate (Fig. 4B). If only seeds dispersed away from conspecific canopies are considered, flying foxes still dispersed seeds significantly further than all other dispersers, followed by pigeons (Fig. 4C).

Seed dispersal effectiveness and functional extinction

Flying foxes were the only effective seed disperser for seven of the eight species for which we had data for three or more individual plants (Fig. 5). For these plants, flying foxes displayed a greater-than-average effectiveness (their SDE value is above the isocline representing average SDE) and all other animals that handled fruit had SDE values at or near zero. The exception was *P. tectorius*, which was also dispersed by introduced rats.

The SDE of flying foxes was greater in sites where they were abundant than in sites where they were not ($t = 3.13$, df $= 4$, $P = 0.0096$; high abundance, mean $\pm$ 1 SD, 0.25 ± 0.13; low abundance, 0.05 ± 0.09). When flying foxes were abundant (relative to food plant availability), they generally produced high-quantity and high-quality components of seed dispersal (i.e. they dispersed many seeds and dispersed them away from the parent canopy). However, they were low-quality dispersers of the extra large, non-native *Mangifera indica* (Fig. 5). At low abundance, flying foxes still consumed relatively large amounts of fruit (high quantity) but dispersed little away from the crown (low quality). This pattern was confirmed statistically for one species (which had sufficient individuals to test); for *P. timoriense*, there was no difference in the quantity of seeds handled by flying foxes in low- and high-abundance sites ($t = 0.06$, df $= 33$, $P = 0.47$), but there was a difference in the quality of seed dispersal (i.e. proportion of seeds dispersed away from conspecific crowns; $t = -2.38$, df $= 33$, $P = 0.011$). Alternative frugivores did not compensate for this reduced dispersal, probably partly due to the fact that more fruit were not generally available for consumption in the absence of flying foxes.

Flying foxes are potentially critical for at least 57 % of their 30 food species (Fig. 6). We considered them not critical for 33 % of species because alternative dispersers existed that could handle at least 10 % of the available seeds, some of which would be dispersed beyond the parent crown. This category also includes species with fruit

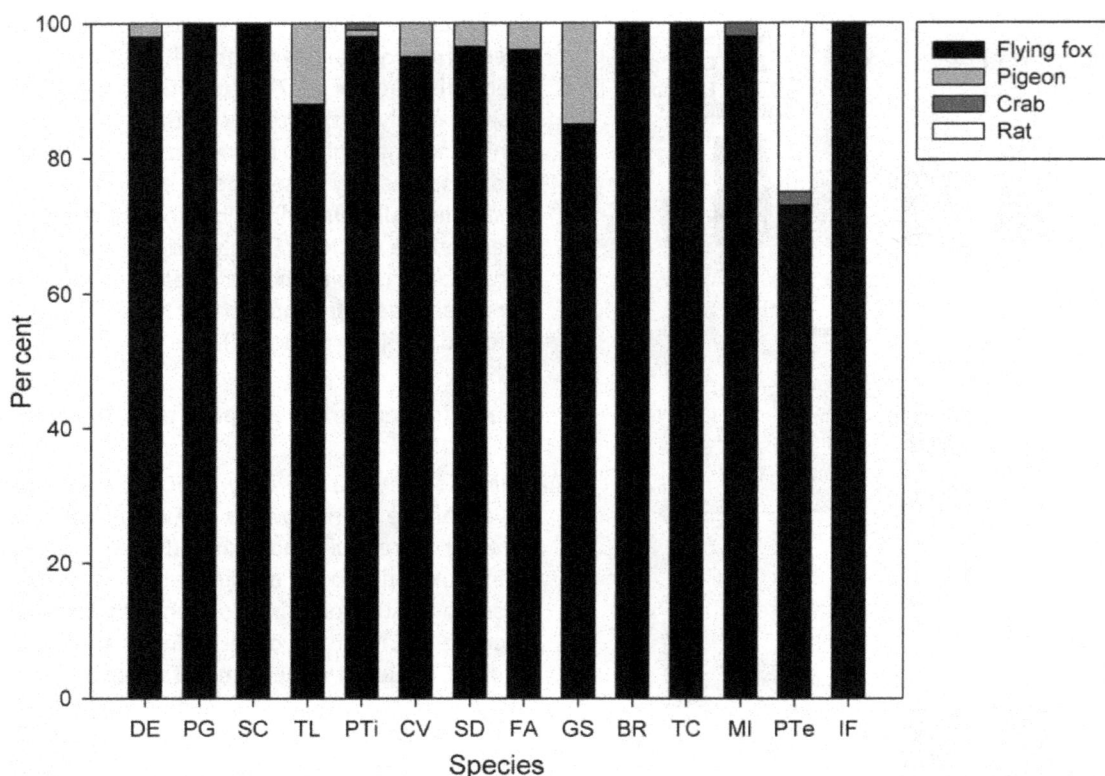

Figure 3. Percentage of seeds dispersed by different animal species for 14 plant species commonly consumed by flying foxes (see Table 1 for species identity). Plant species are arranged according to increasing diaspore size.

characteristics suitable for bird dispersal. We lacked sufficient information to evaluate three consumed species. Flying foxes were potentially critical for 76 % of the species with L or XL diaspores (>14 mm) and 31 % of species with M or S diaspores.

Discussion

Flying foxes are essential components of the frugivore community in Tonga, and probably in many other island ecosystems, because they fulfil a non-redundant role in seed dispersal, especially for large-seeded plants. Although both flying foxes and Pacific pigeons (McConkey *et al.* 2004*a*) ate a diverse range of plant species, suggesting that they might be diet generalists, this was not reflected in a significant overlap in their roles as seed dispersers. Pacific pigeons are the only other major disperser of large seeds (Meehan *et al.* 2002; McConkey *et al.* 2004*a*), but they consume <40 % of the species dispersed by flying foxes and disperse very few seeds of these species. Elsewhere in the tropical Pacific, pigeon species are perhaps equally important as flying foxes where they overlap, but for a different subset of the available species, especially those with smaller seeds (Meehan *et al.* 2002, 2005; Fall *et al.* 2007). Crabs and rats are

primarily considered seed predators (O'Dowd and Lake 1991; Green *et al.* 1997; McConkey *et al.* 2003; Lindquist and Carroll 2004; Pérez *et al.* 2008) but were capable of providing effective seed dispersal for some of the species dispersed by flying foxes. Flying foxes have been proposed as 'keystone' species in the tropical Pacific because of their potential importance as seed dispersers and pollinators (Cox *et al.* 1991; McConkey and Drake 2002; Scanlon *et al.* 2014). Our results confirm this status based solely on their seed dispersal function, with 57 % of all species consumed by flying foxes reliant on them for seed dispersal (76 % of larger-seeded species).

In island systems with low functional redundancy, seed predators may assume important roles as seed dispersers. In our study system, flying foxes damaged the seeds of two species for which we could identify no functional seed disperser, and granivorous rats and crabs were often the only alternative consumer of some species. Rats can disperse very small seeds internally (Williams *et al.* 2000; Shiels and Drake 2011), but frequently carry larger-seeded fruit to husking stations for processing, where some seeds may be abandoned and germinate (McConkey *et al.* 2003; Shiels and Drake 2011). Crabs were more likely than rats to disperse seeds (Lee 1985; Krishna and Somanathan 2014), but

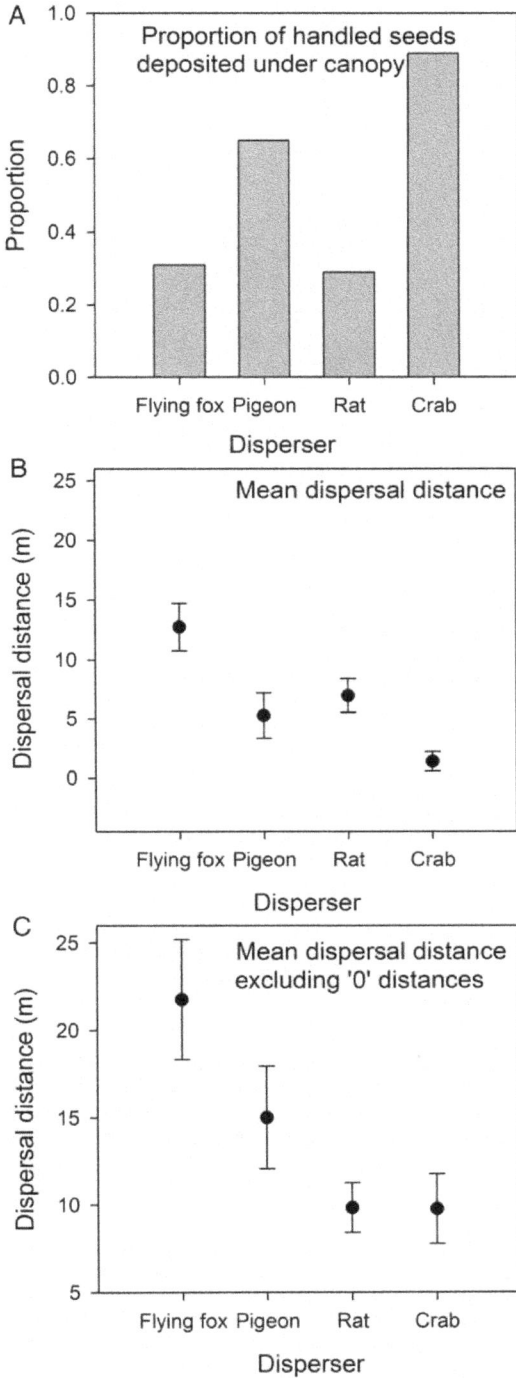

Figure 4. Summary of seed dispersal distances achieved by different dispersers. (A) Proportion of handled seeds deposited under the canopy of parent plants by each disperser. (B) Mean seed dispersal distances with 95 % confidence intervals. All handled seeds are included in this calculation, including those dispersed under parent crowns. (C) Mean distances (with 95 % confidence intervals) across all handled seeds that were dispersed away from the parent crowns (i.e. seeds dispersed 0 m are excluded). For both (B) and (C), values with non-overlapping confidence intervals are significantly different.

still destroyed seeds of three of the seven species we identified as crab-consumed. However, partial consumption of a seed's cotyledons does not always result in seed death. Provided the seed retains an intact embryonic axis, seed germination is possible and germination speed can be enhanced (Dalling *et al.* 1997; Takakura 2002; Pérez *et al.* 2008). Tolerance to cotyledon damage increases with seed size (Mack 1998). This was the only potential mode of seed dispersal for one common plant species (*Maniltoa grandiflora*) whose large seeds were consumed by both flying foxes and rats. Three plant species dispersed by flying foxes had multi-seeded diaspores that could be efficiently dispersed by seed predators. Rats removed the diaspores from the vicinity of the parent crown, but frequently destroyed only some seeds, leaving the remainder viable (D. R. Drake and K. R. McConkey unpubl. data).

An important difference in the seed dispersal capabilities of flying foxes compared with the more sedentary rats and crabs is their respective abilities to disperse seeds over long distances. The large-seeded species that might rely on any of these animals for dispersal are not carried passively in the gut of the animal, but rather actively in the mouth (or pincers). Flying foxes are highly mobile and may carry seeds as far as 10 km (Shapcott 1998), although shorter distances are more common. Flying foxes are the only means by which some of the large-seeded (particularly, XL) plant species may regularly reach another island; we recorded a bat-handled *Terminalia litoralis* seed that had no conspecific tree on the island and must have been carried at least 1.8 km. Without flying foxes, these long-distance dispersal events will not occur, except for coastal species that have buoyant seeds or can float by 'rafting' (Fall *et al.* 2007). Even local dispersal events are dominated by flying foxes in our study system. Crabs move fruits away from the source to avoid competition with other crabs (Lee 1985), and whereas distances are <10 m from parent crowns, they may leave seeds in burrows where the seed is protected from rodent predation (Lee 1985; Krishna and Somanathan 2014). Similarly, rats carry seeds to areas nearby where they can be sheltered from predators while feeding (McConkey *et al.* 2003), usually resulting in short dispersal distances (but sometimes up to 20 m). Given the low plant species diversity on these islands (Franklin *et al.* 2006) and the often close spacing of conspecifics, these distances may be adequate for escaping distance-dependent mortality (Chimera and Drake 2011; Comita *et al.* 2014) but may not be as effective in reaching gaps and enhancing gene flow as the more scattered dispersal patterns achieved by flying foxes are.

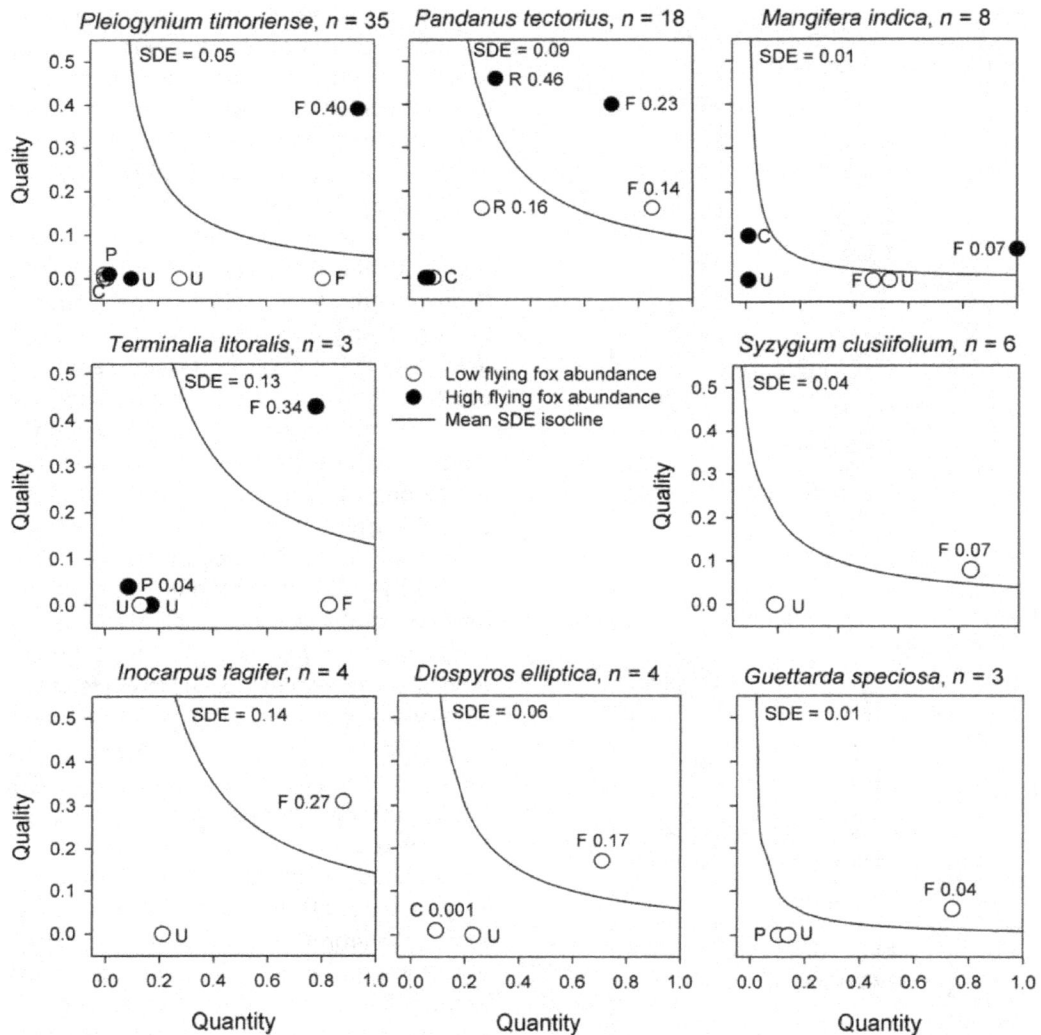

Figure 5. Seed dispersal effectiveness (SDE) of flying foxes and other seed dispersers for eight species commonly consumed by flying foxes. The figure shows the quantity (x-axis, proportion of seeds dispersed) and quality (y-axis; proportion of seeds moved at least 5 m from the canopy edge) components of SDE. The average SDE is indicated by the isocline (and value is noted). Animals above the isocline have greater-than-average effectiveness. Seed dispersal effectiveness values for independent animal species (F, flying fox; P, pigeon; C, crab; R, rat) and uneaten fruit (U) are written next to their label. Animals with SDE not shown had SDE = 0, since no seeds were dispersed away from the crown.

Contemporary ecosystems that persist on the islands in the tropical Pacific have a mélange of fruit-eating animals and fruits with varied origins and novel interactions (Steadman 1993; Shiels and Drake 2011; Spotswood et al. 2012). Losses in functional redundancy associated with disperser extinction or extirpation may have been partially supplemented by animal introductions (Schlaepher et al. 2011). Archaeological evidence indicates that the pre-human frugivore assemblage in Tonga was more diverse than today's, with two flying fox species and three large pigeon species (Steadman 1993, 2006; Meehan et al. 2002). Associated with the loss of some species has been the establishment of possibly one pigeon species (*D. pacifica*, Steadman 1993; Koopman and Steadman

1995) and three rat species. Although the rats may be functioning primarily as seed predators, they also provide a potentially important backup dispersal system for some species—particularly those with multi-seeded diaspores or with seeds that can germinate after partial damage (Pérez et al. 2008).

Island ecosystems that are dependent on flying foxes require not merely enough animals to maintain a viable population, but sufficient numbers for them to continue to disperse seeds and sustain the forests they ultimately depend on. Flying foxes become functionally or ecologically extinct as dispersers before their numbers are low enough to be considered 'rare' (McConkey and Drake 2006). At sites where they were functionally extinct, flying

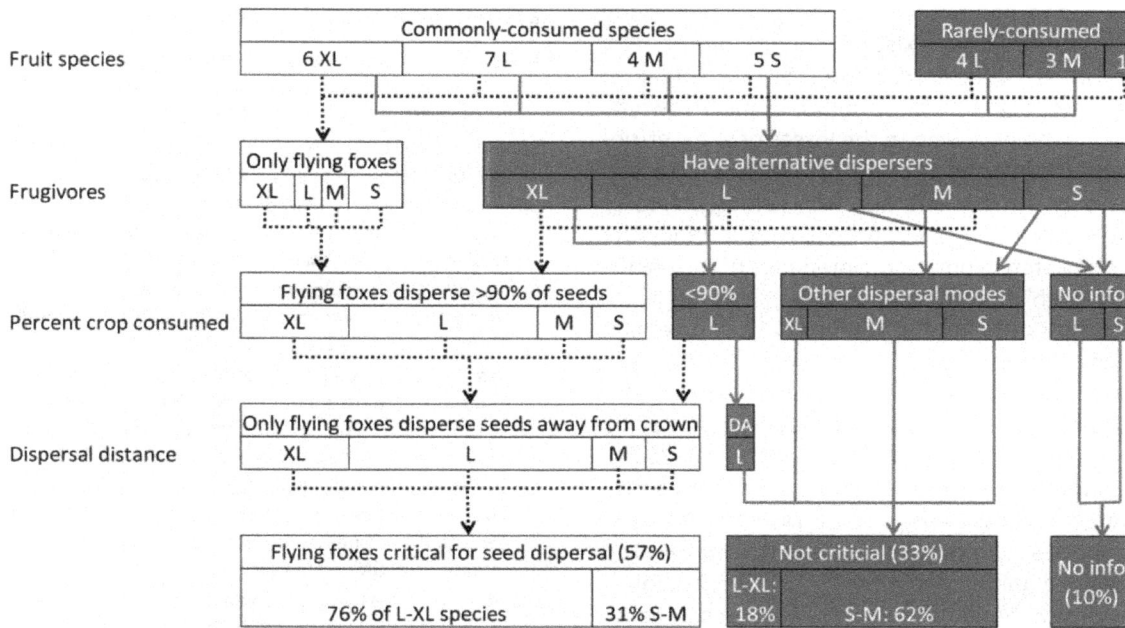

Figure 6. Percentage of fruit species consumed by flying foxes that are potentially dependent on them for seed dispersal. Species with small (S), medium (M), large (L) and extra large (XL) diaspores are indicated separately. Width of bars is proportional to the number of species represented within it. 'Fruit species' shows the number of species that are common or rare in the flying fox diet. 'Frugivores' indicates the proportion of species that have or lack an alternative disperser. 'Percent crop consumed' distinguishes those species for which alternative frugivores make almost no contribution (bats disperse >90 % of seeds), those species studied for which flying foxes dispersed <90 % of diaspores and those species that were not studied but which show features more consistent with bird or water dispersal. 'Dispersal distance' indicates whether flying foxes are the only disperser moving seeds away from the canopy or whether other species also contribute (DA, dispersal away).

foxes continued to consume significant quantities of fruit, but dropped all—or almost all—seeds under the parent crown (McConkey and Drake 2006) where they may suffer higher mortality (Chimera and Drake 2011; Comita et al. 2014). Alternative frugivores did not compensate for this reduced seed dispersal role of flying foxes; while this may have been partly due to the fact that flying foxes continued to consume many fruit (making them unavailable to other consumers), the presence of fallen, unconsumed fruit under canopies suggest that these plant species are not heavily fed on by other frugivores regardless of flying fox density. This confirms the lack of redundancy in the seed dispersal network in our study system.

The Tongan flying fox that was the focus of our study is considered to be declining on some Pacific Islands (e.g. Cook Islands, Cousins and Compton 2005), and maintaining stable populations on others (e.g. Fiji, Scanlon et al. 2014), while its status remains unclear in most regions (Hamilton and Helgen 2008). A population decline of 80 % was caused by a cyclone that occurred after our study (McConkey et al. 2004b); it is not known to what extent the population has recovered, although it should be fairly robust to these periodic disturbances, provided hunting is not significant. Ongoing population monitoring

is essential to ensure that this flying fox species, and others, can continue to perform their keystone roles in seed dispersal and possibly other ecological functions.

Conclusions

In many simple island communities, bats are a dominant provider of ecosystem services. Two flying fox species disperse or pollinate nearly 80 % of canopy-trees in Samoa (Banack 1998), and bat species in Fiji serviced 42 % of plant species that were important to local communities (Scanlon et al. 2014). A single species of flying fox, *P. tonganus*, may disperse at least 50 % of the overstorey tree species in Vava'u (Fall et al. 2007). The same study found that 77 % of the plants were adapted for bird dispersal, but with eight extant fruit-eating birds in this archipelago, this guild probably has more functional redundancy. Although overlap exists in the diets of flying foxes, birds and other dispersers, our study shows that this rarely translates into redundancy in the dispersal service provided by flying foxes, and more than half of the fruit species they consume can depend on them for dispersal. Given that the functional role of flying foxes in seed dispersal (and possibly pollination) can be severely affected by population decline (McConkey and Drake 2006), and that many island populations of *Pteropus* are already threatened, maintaining existing populations is very important. This is a difficult task given the often negative perceptions local agricultural communities have towards flying fox populations, owing to the losses of crops attributed to them (Scanlon *et al.* 2014). Promoting increased awareness of the important, and very vulnerable, roles of flying foxes in maintaining forests is potentially the most important step to ensure maintenance of the unique ecosystems in which they occur.

Island communities are inherently low in diversity, and the lack of redundancy we found in this simple island system may be typical of islands. The loss of a range of animal species, from reptiles (Hansen *et al.* 2008; Traveset *et al.* 2012) to birds (Caves *et al.* 2013), has been shown to have profound consequences for seed dispersal processes on islands, whereas the roles of similar species within mainland habitats have often gone unnoticed (Moura *et al.* 2015). Low ecological redundancy may characterize many island ecosystems, and this is likely to extend to other interactions as well, such as predation (Rogers *et al.* 2012), pollination and herbivory. In fact, pollination has been shown to be even more vulnerable in some island systems than seed dispersal (Kelly *et al.* 2010; Anderson *et al.* 2011). Consequently, identifying the critical species within island ecosystems across a range of ecological interactions is imperative for conservation.

Sources of Funding

Our work was funded by the Wildlife Conservation Society (USA), Victoria University of Wellington (New Zealand), Percy Sladen Memorial Trust (UK) and Polynesian Airlines (Samoa).

Contributions by the Authors

Both authors conducted fieldwork and were involved in the writing of the manuscript.

Acknowledgements

We thank Nola Parsons, Leigh Bull, Rachel McClellan, Tom Morrell, Nic Gorman, Defini Tau'alupe and Hayley Meehan for their help in collecting data. Logistical support was provided by Pat and Keith McKee, Aleiteisi Tangi, Leonati, Filipe Tonga, Tavake Tonga, Tevita Mose, Malini Moa and Liongi Po'oi. We would like to thank the Tongan Government for permission to carry out the research.

Literature Cited

Anderson SH, Kelly D, Ladley JJ, Molloy S, Terry J. 2011. Cascading effects of bird functional extinction reduce pollination and plant density. *Science* **331**:1068–1071.

Banack SA. 1998. Diet selection and resource use by flying foxes (genus *Pteropus*). *Ecology* **79**:1949–1967.

Bollen A, Van Elsacker L, Ganzhorn JU. 2004. Relations between fruits and disperser assemblages in a Malagasy littoral forest: a community-level approach. *Journal of Tropical Ecology* **20**:599–612.

Brooke AP. 2001. Population status and behaviours of the Samoan flying fox (*Pteropus samoensis*) on Tutuila Island, American Samoa. *Journal of Zoology* **254**:309–319.

Brooks TM, Mittermeier RA, Mittermeier CG, da Fonesca GAB, Rylands AB, Konstant WR, Flick P, Pilgrim J, Oldfield S, Magin G, Hilton-Taylor C. 2002. Habitat loss and extinction in the hotspots of biodiversity. *Conservation Biology* **16**:909–923.

Caves EM, Jennings SB, Hillerislambers J, Tewksbury JJ, Rogers HS. 2013. Natural experiment demonstrates that bird loss leads to cessation of dispersal of native seeds from intact to degraded forests. *PLoS ONE* **8**:e65618.

Chimera CG, Drake DR. 2011. Could poor seed dispersal contribute to predation by introduced rodents in a Hawaiian dry forest? *Biological Invasions* **13**:1029–1042.

Comita LS, Queenborough SA, Murphy SJ, Eck JL, Xu K, Krishnadas M, Beckman N, Zhu Y. 2014. Testing predictions of the Janzen–Connell hypothesis: a meta-analysis of experimental evidence for distance- and density-dependent seed and seedling survival. *Journal of Ecology* **102**:845–856.

Cousins JA, Compton SG. 2005. The Tongan flying fox *Pteropus tonganus*: status, public attitudes and conservation in the Cook Islands. *Oryx* **39**:196–203.

Cox PA, Elmqvist T, Pierson ED, Rainey WE. 1991. Flying foxes as strong interactors in South Pacific Island ecosystems: a conservation hypothesis. *Conservation Biology* **5**:448–454.

Dalerum F, Cameron EZ, Kunkel K, Somers MJ. 2012. Interactive effects of species richness and species traits on functional diversity and redundancy. *Theoretical Ecology* **5**:129–139.

Dalling JW, Harms KE, Aizprúa R. 1997. Seed damage tolerance and seedling resprouting ability of *Prioria copaifera* in Panamá. *Journal of Tropical Ecology* **13**:481–490.

del Hoyo J, Elliott A, Sargatal J, Christie DA, de Juana E. 2013. *Handbook of the birds of the world alive*. Barcelona: Lynx Editions. http://www.hbw.com (11 June 2015).

Fall PL, Drezner TD, Franklin J. 2007. Dispersal ecology of the lowland rain forest in the Vava'u island group, Kingdom of Tonga. *New Zealand Journal of Botany* **45**:393–417.

Franklin J, Drake DR, Bolick LA, Smith DS, Motley TJ. 1999. Rain forest composition and patterns of secondary succession in the Vava'u Island Group, Tonga. *Journal of Vegetation Science* **10**:51–64.

Franklin J, Wiser SK, Drake DR, Burrows LE, Sykes WR. 2006. Environment, disturbance history and rain forest composition across the islands of Tonga, Western Polynesia. *Journal of Vegetation Science* **17**:233–244.

Gibbons JRH, Watkins IF. 1982. Behavior, ecology, and conservation of South Pacific banded iguanas, *Brachylophus*, including a newly discovered species. In: Burghardt GM, Rand AS, eds. *Iguanas of the world*. New Jersey: Noyes Publication, 418–441.

Green PT, O'Dowd DJ, Lake PS. 1997. Control of seedling recruitment by land crabs in rain forest on a remote oceanic island. *Ecology* **78**:2474–2486.

Hamilton S, Helgen K. 2008. *Pteropus tonganus. The IUCN red list of threatened species.* Version 2014.3. www.iucnredlist.org (21 April 2015).

Hansen DM, Kaiser CN, Müller CB. 2008. Seed dispersal and establishment of endangered plants on oceanic islands: the Janzen–Connell model, and the use of ecological analogues. *PLoS ONE* **3**:e2111.

Hodgkison R, Balding ST, Zubaid A, Kunz TH. 2003. Fruit bats (Chiroptera: Pteropodidae) as seed dispersers and pollinators in a lowland Malaysian rain forest. *Biotropica* **35**:491–502.

IUCN 2014. The IUCN red list of threatened species. Version 2014. http://www.iucnredlist.org.

Jenkins RKB, Andriafidison D, Razafimanahaka HJ, Rabearivelo A, Razafindrakoto Z, Ratsimandresy Z, Andrianandrasana RH, Razafimahatratra E, Racey PA. 2007. Not rare, but threatened: the endemic Madagascar flying fox *Pteropus rufus* in a fragmented landscape. *Oryx* **41**:263–271.

Kelly D, Ladley JJ, Robertson AW, Anderson SH, Wotton DM, Wiser SK. 2010. Mutualisms with the wreckage of an avifauna: the status of bird pollination and fruit dispersal in New Zealand. *New Zealand Journal of Ecology* **34**:66–85.

Koopman KF, Steadman DW. 1995. Extinction and biogeography of bats on 'Eua, Kingdom of Tonga. *American Museum Novitates* **3125**:1–13.

Krishna S, Somanathan H. 2014. Secondary removal of *Myristica fatua* (Myristicaceae) seeds by crabs in *Myristica* swamp forests in India. *Journal of Tropical Ecology* **30**:259–263.

Lee MAB. 1985. The dispersal of *Pandanus tectorius* by the land crab *Cardisoma carnifex. Oikos* **45**:169–173.

Lee MAB. 1988. Food preferences and feeding behaviour of the land crab *Cardisoma carnifex. Micronesica* **21**:275–279.

Lindquist ES, Carroll CR. 2004. Differential seed and seedling predation by crabs: impacts on tropical coastal forest composition. *Oecologia* **141**:661–671.

MacArthur RH. 1965. Patterns of species diversity. *Biological Reviews* **40**:510–533.

Mack AL. 1998. An advantage of large seed size: tolerating rather than succumbing to seed predators. *Biotropica* **30**:604–608.

Mayfield MM, Bonser SP, Morgan JW, McNamara S, Vesk PA. 2010. What does species richness tell us about functional trait diversity to land-use change. *Global Ecology and Biogeography* **19**:423–431.

McConkey KR, Drake DR. 2002. Extinct pigeons and declining bat populations: are large seeds still being dispersed in the tropical Pacific? In: Levey D, Silva W, Galetti M, eds. *Frugivory and seed dispersal: ecological, evolutionary and conservation.* Wallingford: CAB International, 381–396.

McConkey KR, Drake DR. 2006. Flying foxes cease to function as seed dispersers long before they become rare. *Ecology* **87**: 271–276.

McConkey KR, Drake DR. 2007. Indirect evidence that flying foxes track food resources among islands in a Pacific archipelago. *Biotropica* **39**:436–440.

McConkey KR, Drake DR, Meehan HJ, Parsons N. 2003. Husking stations provide evidence of seed predation by introduced rodents in Tongan rain forests. *Biological Conservation* **109**:221–225.

McConkey KR, Meehan HJ, Drake DR. 2004a. Seed dispersal by Pacific pigeons (*Ducula pacifica*) in Tonga, Western Polynesia. *EMU* **104**: 369–376.

McConkey KR, Drake DR, Franklin J, Tonga F. 2004b. Effects of Cyclone Waka on flying foxes (*Pteropus tonganus*) in the Vava'u Islands of Tonga. *Journal of Tropical Ecology* **20**:555–561.

Meehan HJ, McConkey KR, Drake DR. 2002. Potential disruptions to seed dispersal mutualisms in Tonga, Western Polynesia. *Journal of Biogeography* **29**:695–712.

Meehan HJ, McConkey KR, Drake DR. 2005. Early fate of *Myristica hypargyraea* seeds dispersed by *Ducula pacifica* in Tonga, Western Polynesia. *Austral Ecology* **30**:374–382.

Moura ACDA, Cavalcanti L, Leite-Filho E, Mesquita DO, McConkey KR. 2015. Can green iguanas compensate for vanishing seed dispersers in the Atlantic forest fragments of north-east Brazil? *Journal of Zoology* **295**:189–196.

Nyhagen DF, Turnbull SD, Olesen JM, Jones CG. 2005. An investigation into the role of the Mauritan flying fox, *Pteropus niger*, in forest regeneration. *Biological Conservation* **122**:491–497.

O'Dowd DJ, Lake PS. 1991. Red crabs in rain forest, Christmas Island: removal and fate of fruits and seeds. *Journal of Tropical Ecology* **7**: 113–122.

Olesen JM, Eskildsen LI, Venkatasamy S. 2002. Invasion of pollination networks on oceanic islands: importance of invader complexes and endemic super generalists. *Diversity and Distributions* **8**: 181–192.

Olson SL, James HF. 1982. Fossil birds from the Hawaiian Islands: evidence for wholesale extinction by man before western contact. *Science* **4560**:633–635.

Palmeirin JM, Champion A, Naikatini A, Niukula J, Tuiwawa M, Fisher M, Yabaki-Gounder M, Thorsteinsdóttir S, Qalovaki S, Dunn T. 2007. Distribution, status and conservation of the bats of the Fiji Islands. *Oryx* **41**:509–519.

Pérez HE, Shiels AB, Zaleski HM, Drake DR. 2008. Germination after simulated rat damage in seeds of two endemic Hawaiian palm species. *Journal of Tropical Ecology* **24**:555–558.

Pierson ED, Elmqvist T, Rainey WE, Cox PA. 1996. Effects of tropical cyclonic storms on flying fox populations on the South Pacific islands of Samoa. *Conservation Biology* **10**:438–451.

Rainey WE, Pierson ED, Elmqvist T, Cox PA. 1995. The role of flying foxes (Pteropodidae) in oceanic island ecosystems of the Pacific. *Symposia of the Zoological Society of London* **67**:79–96.

Reich PB, Tilman D, Isbell F, Mueller K, Hobbie SE, Flynn DFB, Eisenhauer N. 2012. Impacts of biodiversity loss escalate through time as redundancy fades. *Science* **336**:589–592.

Richards GC. 1990. The spectacled flying-fox, *Pteropus conspicillatus* (Chiroptera: Pteropodidae), in North Queensland. 2. Diet, seed dispersal and feeding ecology. *Australian Mammalogy* **13**:25–31.

Rogers H, Hille Ris Lambers J, Miller R, Tewksbury JJ. 2012. 'Natural experiment' demonstrates top-down control of spiders by birds on a landscape level. *PLoS One* **7**:e43446.

Scanlon A, Petit S, Bottroff G. 2014. The conservation status of bats in Fiji. *Oryx* **48**:451–459.

Scanlon AT, Petit S, Tuiwawa M, Naikatini A. 2014. High similarity between a bat-serviced plant assemblage and that used by humans. *Biological Conservation* **174**:111–119.

Schlaepfer MA, Sax DF, Olden JD. 2011. The potential conservation value of non-native species. *Conservation Biology* **25**:428–437.

Schupp EW, Jordano P, Gómez JM. 2010. Seed dispersal effectiveness revisited: a conceptual review. *New Phytologist* **188**:333–353.

Shapcott A. 1998. The patterns of genetic diversity in *Carpentaria acuminata* (Arecaceae), and rainforest history in northern Australia. *Molecular Ecology* **7**:833–847.

Shiels AB, Drake DR. 2011. Are introduced rats (*Rattus rattus*) both seed predators and dispersers in Hawaii? *Biological Invasions* **13**:883–894.

Spotswood EN, Meyer J-Y, Bartolome JW. 2012. An invasive tree alters the structure of seed dispersal networks between birds and plants in French Polynesia. *Journal of Biogeography* **39**:2007–2020.

Steadman CW. 2006. *Extinction and biogeography of tropical Pacific birds*. Chicago: University of Chicago Press.

Steadman DW. 1993. Biogeography of Tongan birds before and after human impact. *Proceedings of the National Academy of Sciences of the USA* **90**:818–822.

Steadman DW, Stafford TW, Donahue DJ, Jull AJT. 1991. Chronology of Holocene vertebrate extinction in the Galápagos islands. *Quaternary Research* **36**:126–133.

Swamy V, Terborgh JW. 2010. Distance-responsive natural enemies strongly influence seedling establishment patterns of multiple species in an Amazonian rain forest. *Journal of Ecology* **98**:1096–1107.

Takakura K. 2002. The specialist seed predator *Bruchidius dorsalis* (Coleoptera: Bruchidae) plays a crucial role in the seed germination of its host plant, *Gleditsia japonica* (Leguminosae). *Functional Ecology* **16**:252–257.

Traveset A, González-Varo JP, Valido A. 2012. Long-term demographic consequences of a seed dispersal disruption. *Proceedings of the Royal Society B: Biological Sciences* **279**:3298–3303.

Whittaker RJ, Fernández-Palacios JM. 2007. *Island biogeography: ecology, evolution, and conservation*, 2nd edn. New York: Oxford University Press.

Wiles GJ, Engbring J, Otobed D. 1997. Abundance, biology, and human exploitation of bats in the Palau Islands. *Journal of Zoology* **241**:203–227.

Williams PA, Karl BJ, Bannister P, Lee WG. 2000. Small mammals as potential seed dispersers in New Zealand. *Austral Ecology* **25**:523–532.

11

Patterns in floral traits and plant breeding systems on Southern Ocean Islands

Janice M. Lord*

Botany Department, University of Otago, PO Box 56, Dunedin 9054, New Zealand

Guest Editor: Christoph Kueffer

Abstract. The harsh climatic conditions and paucity of potential pollinators on Southern Ocean Islands (SOIs; latitude 46°S–55°S) lead to the expectation that anemophily or self-fertilization are the dominant modes of plant sexual reproduction. However, at least some species have showy inflorescences suggesting biotic pollination or dimorphic breeding systems necessitating cross-pollination. This study investigates whether anemophily and self-compatibility are common on SOIs, whether species or genera with these traits are more widespread or frequent at higher latitudes, and whether gender dimorphy is correlated with anemophily, as might occur if reliance on pollinators was a disadvantage. Of the 321 flowering plant species in the SOI region, 34.3 % possessed floral traits consistent with anemophily. Compatibility information was located for 94 potentially self-fertilizing species, of which 92.6 % were recorded as partially or fully self-compatible. Dioecy occurred in 7.1 % of species overall and up to 10.2 % of island floras, but has not clearly arisen *in situ*. Gynodioecy occurred in 3.4 % of species. The frequency of anemophily and gender dimorphy did not differ between the SOI flora and southern hemisphere temperate reference floras. At the species level, gender dimorphy was positively associated with fleshy fruit, but at the genus level it was associated with occurrence in New Zealand and a reduced regional distribution. Anemophily was more prevalent in genera occurring on subantarctic islands and the proportion of species with floral traits suggestive of biotic pollination was significantly higher on climatically milder, cool temperate islands. These results support the contention that reliance on biotic pollinators has constrained the distribution of species on SOIs; however, it is also clear that the reproductive biology of few SOI species has been studied *in situ* and many species likely employ a mixed mating strategy combining biotic pollination with self-fertilization.

Keywords: Cool temperate; dioecy; gynodioecy; self-incompatibility; southern hemisphere; subantarctic; wind-pollination.

Introduction

Isolated islands provide an unparalleled opportunity to study evolutionary processes in plants (Carlquist 1977; Lloyd 1985; Sakai *et al.* 1995; Grant 1998; Losos and Ricklefs 2009). In particular, the study of island floras has made a significant contribution to our understanding of the evolution of plant reproductive strategies, as islands often lack guilds of pollinators common on larger landmasses and new colonists can be further disadvantaged by small population sizes, increasing the risk of inbreeding depression

* Corresponding author's e-mail address: janice.lord@otago.ac.nz

(Carlquist 1977; Grant 1998). Compared with mainland relatives, island plants tend to have smaller, less brightly coloured floral displays, lower flower visitation rates and a greater incidence of anemophily (wind pollination) (Carlquist 1977; Inoue *et al.* 1995; Barrett 1998). Self-compatible hermaphrodites, capable of producing seed via self-fertilization, should have an advantage in establishing on isolated islands (Baker 1955). Selection should also act strongly in favour of the evolution of self-compatibility in biotically pollinated, out-crossing taxa if pollinators are scarce (Barrett 1998). However, islands are also associated with a high incidence of dioecy (Carlquist 1977; Bawa 1982; Sakai *et al.* 1995; Barrett 1998; Webb *et al.* 1999; see Table 1 for breeding system terminologies), possibly due to selection for mechanisms that reduce the likelihood of inbreeding depression, as well as the well-known correlation between dioecy and fleshy fruits, which could make long-distance dispersal more likely (Bawa 1982; Sakai *et al.* 1995; Sakai and Weller 1999; Webb *et al.* 1999). Baker and Cox (1984) further linked the frequency of dioecy in island floras to moist tropical climates and the probable source flora, but highlighted the paucity of the available data. Despite the importance of islands in understanding these evolutionary processes in plants, plant reproductive traits have still been investigated in remarkably few island floras and complete data are still lacking even for well-studied islands (Bernardello *et al.* 2001; Newstrom and Robertson 2005; Chamorro *et al.* 2012).

The Southern Ocean is home to numerous island groups which have received relatively little attention from island biologists. This region is here defined following Greve *et al.* (2005) as including the southern waters influenced by westerly winds as well as the subantarctic region influenced by the Antarctic Convergence. Southern Ocean Islands (SOIs) between latitude 46°S and 55°S have in common their relative isolation, cold oceanic climate and a dominance of herbaceous and graminoid vascular plants. Islands in the region vary in size from <3 to >3000 km² and vascular plant species richness relates strongly to island area as well as sea surface temperature (Chown *et al.* 1998). The floras of SOIs have derived substantially from survivors of a native subantarctic flora, rather than post-glacial maximum long-distance dispersal colonists (Wagstaff and Hennion 2007; Van der Putten *et al.* 2010; Wagstaff *et al.* 2011). However, long-distance dispersal from larger landmasses to the north such as Australia, New Zealand and South America, as well as within and among SOI groups, has produced strong patterns of nestedness and regionalization (Chown *et al.* 1998; Greve *et al.* 2005; Van der Putten *et al.* 2010).

With the exception of the Falklands Islands (Islas Malvinas), the islands of the Southern Ocean lack butterflies and bees (Gressitt 1964; Lloyd 1985; Donovan 2007; Schermann-Legionnet *et al.* 2007; Convey *et al.* 2010). Furthermore, the climate is characterized by high winds and low temperatures which would appear unsuitable

Table 1. An overview of flowering plant breeding system terminologies and the frequency of selected breeding systems worldwide (from Richards 1997; Renner 2014).

System	Description	Frequency
Self-compatible	Pollination of a receptive stigma by pollen from the same plant leads to viable seed.	~61 % of species
Self-incompatible	Pollination of a receptive stigma by pollen from the same plant (or another plant carrying the same genetic recognition factors) does not lead to viable seed.	~39 % of species
Hermaphrodite	Plants monomorphic. All flowers on an individual plant are functionally male and female (i.e. cosexual). Plants can potentially self-pollinate by pollen transfer to a receptive stigma within or between flowers.	~72 % of species
Monoecy	Monomorphic. Flowers on an individual plant are either male or female. Plants can potentially self-pollinate by pollen transfer from male flowers to female flowers.	5–6 % of species
Gynomonoecy	Monomorphic. Flowers on an individual plant are either cosexual or female. Plants can potentially self-pollinate either by pollen transfer within cosexual flowers or transfer of pollen from cosexual to female flowers.	~2.8 % of species, common in Asteraceae
Andromonoecy	Monomorphic. Flowers on an individual plant are either cosexual or male. Plants can potentially self-pollinate by pollen transfer within cosexual flowers.	~1.5 % of species
Dioecy	Dimorphic. Plants are either entirely female or entirely male. Plants cannot self-pollinate.	5–6 % of species
Gynodioecy	Dimorphic. Plants are either entirely female or entirely hermaphrodite. Females cannot self-pollinate. Hermaphrodites can potentially self-pollinate.	~7 % of species

for flying insects; in fact flightlessness has evolved in many SOI insect groups (Gressitt 1964; Chown *et al.* 1998). This has led to the suggestion that plants on SOIs mainly rely on anemophily (pollen transfer via wind) or self-fertilization, rather than biotic pollination, in order to reproduce sexually (Smith 1984; Bergstrom *et al.* 1997; Schermann-Legionnet *et al.* 2007; Convey *et al.* 2010). On some SOIs, however, dioecious species, as well as species with showy floral displays, are a conspicuous component of the flora. If the depauperate insect fauna and harsh climatic conditions on SOIs have selected against reliance on biotic pollination, it might be that species with showy flowers, while conspicuous, are atypical and, further, that dimorphic breeding systems are successful only in conjunction with anemophily. Experimental studies of SOI plant breeding systems and pollination modes have produced a range of results. Bergstrom *et al.* (1997) demonstrated autogamous self-compatibility or suspected facultative cleistogamy in seven species on Macquarie Island and Schermann-Legionnet *et al.* (2007) described effective self-pollination in *Pringlea antiscorbutica* (Brassicaceae) on Îles Kerguelen. However, Nicholls (2000) and Lord *et al.* (2013) found varying degrees of reliance on biotic pollination on Campbell Island, including self-incompatibility in two species.

The aim of this study was to examine the frequency and distribution of reproductive traits among native flowering plants on SOIs. In particular, I tested whether self-compatibility and anemophily are common in the SOI region compared with mainland floras at similar latitudes, whether anemophilous and self-compatible species occur on more SOIs, as might be expected if such species encountered fewer barriers to establishment and whether species with floral traits suggestive of biotic pollination are restricted to islands with milder climates. I also tested whether gender dimorphic breeding systems (dioecy and gynodioecy) are particularly uncommon, associated with anemophily, or show taxonomic or biogeographic associations with the New Zealand flora, which has a high incidence of gender dimorphism (Webb *et al.* 1999).

Methods

Islands included

The 11 SOI groups used in this study include all of the islands mentioned in Van der Putten *et al.* (2010) and all of the island groups south of latitude 46°S used by Chown *et al.* (1998) and Greve *et al.* (2005) with the exception of the Bounty Islands. The Bounty Islands are not included as only one flowering plant species has been described from the island group (Amey *et al.* 2007; de Lange *et al.* 2013), and it has not been relocated recently (J. Hiscock, New Zealand Department of Conservation,

pers. comm.). Following Smith (1984), the climate zone of the Snares, Antipodes, Auckland, Campbell and Falklands Islands was classed as cool temperate and that of Prince Edward & Marion Islands, Îles Crozet, Îles Kerguelen, Heard and MacDonald, South Georgia and Macquarie Islands was classed as subantarctic. Data on total island group land area, median latitude and longitude and phytogeographic province were obtained from Chown *et al.* (1998) and Van der Putten *et al.* (2010).

Reference floras

Reference floras from New Zealand, Stewart Island, Tierra del Fuego and alpine Patagonia were used for statistical analysis of SOI flora composition. These areas are relevant to the SOI flora as they extend beyond latitude 45°S, include cool temperate forest and alpine vegetation and encompass the South Pacific and South Atlantic phytogeographic provinces of Van der Putten *et al.* (2010). No mainland reference flora was available for the SOI South Indian province identified in Van der Putten *et al.* (2010) as no large landmass extends below 35°S in the Indian Ocean. Stewart Island (~47°S, 168°E; 1720 km²) lies 27 km south of mainland New Zealand, to which it was connected during the last glacial maximum. A list of native flowering plant species was obtained from Wilson (1982), and floral traits and breeding systems were extracted from Allan (1961), Moore and Edgar (1970), Webb *et al.* (1988) and Edgar and Connor (2000). The frequency of self-compatibility in native New Zealand flowering plants was obtained from Newstrom and Robertson (2005). Breeding system, pollination mode and self-compatibility information for the native flowering plant flora of alpine Chilean Patagonia (~50°S, 73°W) were obtained from Arroyo and Squeo (1990). The flora of Tierra del Fuego (~54°S, 69°W; Moore 1983) was used as an additional reference to determine whether particular families were over represented in the SOI flora.

The SOI flora

Data on SOI plant species occurrences and reproductive traits were obtained from Greene (1964), Moore (1968), Huntly (1971), Johnson and Campbell (1975), Meurk (1975), Godley (1989), Chown *et al.* (1998), Hay *et al.* (2004), Broughton and McAdam (2005), Upson (2012) and the Flora of New Zealand series (Allan 1961; Moore and Edgar 1970; Webb *et al.* 1988; Edgar and Connor 2000), which includes all SOIs in the Pacific province (Table 2). The complete SOI flora used in this study consisted of 321 flowering plant species in 51 families and 150 genera, of which 81 species (25.2 %) were endemic to the region. The majority of species (62.5 %) were eudicots and the majority of genera (57.7 %) were represented by a single species; only 20 genera (13.3 %)

Table 2. Features of Southern Ocean Islands and their flowering plant floras. Province follows Van der Putten *et al.* (2010). SC, fully or partially self-compatible species; Comp. known, species for which information on the ability to self-fertilize was located.

Island group	Median Lat. (°S)	Median Long.	Area (km²)	Province	Total species	Dioecious/ gynodioecious	SC species/ comp. known
Îles Crozet	46.24	51.22°E	356	Indian	17	0/1	9/9
Prince Edward & Marion Islands	46.77	37.35°E	334	Indian	15	0/1	8/8
The Snares	48.12	166.6°E	3	Pacific	13	1/0	1/1
Îles Kerguelen	49.37	69.5°E	7200	Indian	22	0/1	11/11
Antipodes Islands	49.7	178.8°E	21	Pacific	49	5/0	11/12
Auckland Islands	50.83	166.0°E	626	Pacific	138	11/1	18/21
Falklands Islands	51.5	59.50°W	8500	Atlantic	151	9/9	63/67
Campbell Island	52.5	169.2°E	113	Pacific	99	10/1	18/21
Heard & MacDonald Islands	53.07	73.05°E	371	Indian	11	0/1	4/4
South Georgia	54.25	37.0°W	3755	Atlantic	16	0/1	9/9
Macquarie Island	54.62	158.9°E	128	Pacific	36	2/1	10/11

contained four or more species. Single island occurrences constituted the majority of records (64.7 %); only 15 species occurred on five or more islands. The largest genera in the SOI region are *Carex* (17 species), *Poa* (15) and *Ranunculus* (11), and the largest families are Poaceae (47 species), Asteraceae (46), Cyperaceae (35), Apiaceae (21, including *Stilbocarpa* following Mitchell *et al.* 1999) and Orchidaceae (18). The relative contributions to the SOI flora of these families, and the fifth category, 'others', were similar to the Stewart Island flora ($\chi^2 = 4.46$, $P > 0.4$, df = 5), but differed from both the alpine flora of Patagonia and the flora of Tierra del Fuego ($\chi^2 = 22.69$, $P < 0.001$, df = 5; $\chi^2 = 15.43$, $P < 0.01$, df = 5, respectively). In both of the latter cases, the SOI flora had significantly more orchid species (cell χ^2 values >5), but did not differ in the relative contributions of the other five categories.

Sources of species trait information

All species names were cross-checked for synonymy, taxonomical validity and authority using the online taxonomic resource The Plant List (2013). Further information on flora composition, floral traits and breeding systems was obtained via keyword searches on Web of Science, Google Scholar and internet search engines for journal articles, reports and dissertations, using genus, species and island names as search terms. British Antarctic Survey Bulletins and publications were also searched for relevant information. For some non-endemic species, information could only be obtained from species descriptions in the floras of Tierra del Fuego (Moore 1983) and Patagonia (Arroyo and Squeo 1990; Arroyo *et al.* 1992). Taxonomic treatments of SOI genera were also examined for statements concerning breeding system characteristics of species

and subgenera, as well as statements concerning the uniformity of breeding systems within genera.

Data analysis

Species were assigned one of four floral types based on descriptions of reproductive structures in the literature. Floral type 'A' (FTa) lacked sterile display structures and possessed long-exserted anthers and stigmas, consistent with anemophily. Floral type 'B' (FTb) showed investment in conspicuous, usually coloured petals or other sterile display structures, potentially indicating signalling to animal pollinators. Species with inconspicuous petals but lacking exserted sex organs were classed as ambiguous (FTab). Species with very small (<2 mm diameter), solitary flowers lacking exserted sex organs may rely on self-fertilization, but as direct evidence of this was generally lacking, such species were classified as 'minute' (FTm). Where variability in breeding system was mentioned for a species, the most commonly occurring breeding system was used in analyses. For SOI endemic species, the information obtained generally did not mention variation among island populations. All species traits and sources of information are listed in **Supporting Information— Table S1**.

χ^2 Tests of independence were used to determine taxonomic bias in the SOI flora, and bias in the frequency of gender dimorphic breeding systems, self-compatibility and wind pollination compared with the reference floras. All χ^2 analyses were performed using Statistix v.9 (Analytical Software).

Generalized linear models (GLMs) with a binary error distribution and a logit link function were used to test explanatory relationships between self-compatibility, floral type, gender dimorphy, island occurrence and taxonomic affinity with the New Zealand flora. The low level of endemicity and lack of species-rich lineages suggest that most SOI species have dispersed to, or among, SOIs independently. Thus individual species can be treated as separate tests of the ecological association between a trait and the ability to establish on an island (Westoby et al. 1995). From the limited experimental data available even closely related endemic species can differ in reproductive traits (e.g. Pleurophyllum, Lord et al. 2013). Initial analyses used all 321 species and treated floral type, dimorphic breeding system and self-compatibility as binary response variables (1 for presence, 0 for absence). In order to test the leverage of larger genera in which trait correlations may be shared due to niche conservatism (Lord et al. 1995), analyses were repeated at the level of genus. The prevalence of traits among island floras was also analysed in relation to the island attributes latitude, climate zone and phytogeographic province, with an additional factor allowing for the possibility that increased representation of uniformly anemophilous taxa such as Poaceae and Cyperaceae may bias the model. Analyses involving genera and islands treated the number of species within a genus or on an island as 'trials' and the number of species exhibiting a particular floral type or a gender dimorphic breeding system as 'events'. Optimal models were determined using backwards stepwise regression, with the inclusion of a predictor variable dependent on the significance of Type III MS and Wald χ^2 values. All regressions were performed in SPSS Statistics V. 22 (IBM Corporation).

Results

Self-compatibility

Compatibility information was located for 31.5 % (94 species) of the 298 SOI species that were potentially capable of at least partial self-fertilization (dioecious species excluded) and included 89 monomorphic species and the hermaphroditic morphs of five gynodioecious species. Full or partial self-compatibility has been reported for 83 of these 94 species (92.6 %), including the hermaphroditic morphs of four gynodioecious species. Compatibility information derived directly from SOIs was located for 71 of these species (23.8 % of potentially self-fertilizing species). For the remaining 23 species (all non-SOI endemics), information was extracted from species descriptions or studies from mainland areas. Of the 71 species for which SOI-specific information was available, 95.8 % were fully or partially self-compatible. The GLM found no relationship between self-compatibility and species distribution among islands, climate zones and biogeographic provinces (Fig. 1, Tables 2 and 3).

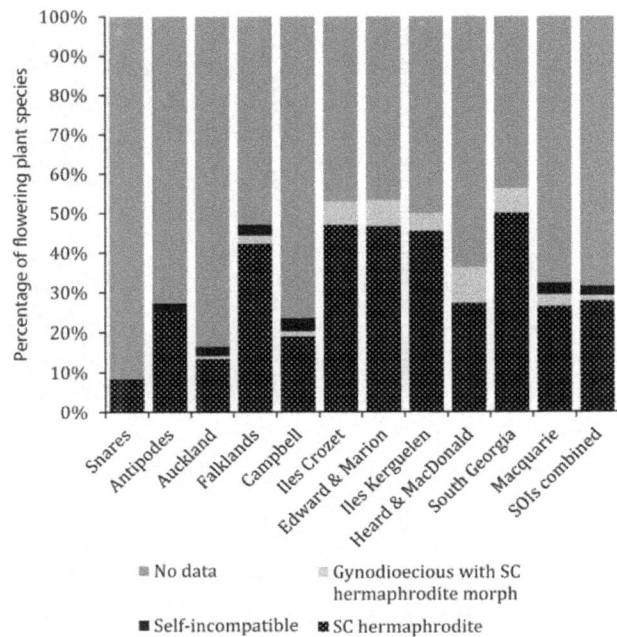

Figure 1. The frequency of known obligate outcrossing (self-incompatible or dioecious) species, fully or partially self-compatible species and species lacking compatibility information for flowering plants native to 11 Southern Ocean Islands. SC, self-compatible. The five island groups clustered on the left side of the horizontal axis are cool temperate. The six island groups clustered to the right of the horizontal axis are subantarctic. Island groups within each cluster are ordered by latitude.

Floral types

For the SOI flora as a whole, FTa species, possessing floral traits consistent with anemophily, were less common (34.3 %) than FTb species, possessing floral traits suggestive of biotic pollination (49.5 %). Floral type 'A' species, as opposed to other floral types combined, were no more common in the SOI flora than in the floras of Stewart Island ($\chi^2 = 0.28$, $P > 0.5$, df = 1) or alpine Patagonia ($\chi^2 = 0.88$, $P > 0.5$, df = 1). Generalized linear models found that FTa species were more common in genera that also occurred in New Zealand, and were more common in genera that occurred in the Pacific Province and on subantarctic islands (Table 3). Neither latitude, climate zone nor the prevalence of Poaceae and Cyperaceae explained the prevalence of FTa species on islands, but FTb species were more common on cool temperate as opposed to subantarctic islands (Fig. 2, Table 3).

Breeding systems

Hermaphroditism was the most common breeding system overall (218 species, 67.9 %), followed by monoecy (38, 11.8 %), gynomonoecy (28, 8.7 %), dioecy (23, 7.1 %), gynodioecy (11, 3.4 %) and andromonoecy (3, 0.9 %). The number of species with hermaphroditic vs. other

Table 3. Regression models for relationships among gender dimorphy, floral traits, distribution and taxonomic affinities of species, genera and islands. CZIS, binary, island in (1) subantarctic climate zone, (0) cool temperate climate zone; CZOCC, categoric; 0, only in cool temperate climate zone; 1, in both climate zones; 2, only in subantarctic climate zone; FTα, binary; 1, anemophilous floral traits; 0, other types; FTb, binary; 1, floral traits suggesting biotic pollination; 0, other types; FF, binary; 1, fleshy fruit; 0, dry fruit; LAT, island latitude in degrees; NZG, binary; 1, genus present in New Zealand; 0, absent; ALT/IND/PAC: binary, 1, species or genus in Atlantic, Indian or Pacific provinces as defined by Van der Putten et al. (2010); 0, absent; PROV, categoric, province; TOTIS, number of islands on which a species or genus occurs; TOTPROV, number of provinces in which a species or genus occurs; df, degrees of freedom. Model χ^2 = Omnibus χ^2 test of model significance, df = 1.

Dataset	Response variable	Predictors tested	df	Significant predictors	Coefficient	Wald χ^2 (P value)	Model χ^2 (P value)
All species (N = 321)	FTα (0,1)	ATL	1				
		CZOCC	2				
		IND	1				
		NZG	1	NZG	2.324	14.601 (0.000)	26.372 (0.000)
		PAC	1				
		TOTIS	1				
		TOTPROV	1				
All species (N = 321)	Dimorphic (0,1)	ATL	1				
		CZOCC	2				
		FTα	1				
		FF	1	FF	2.775	36.957 (0.000)	34.778 (0.000)
		IND	1				
		NZG	1				
		PAC	1				
		TOTIS	1				
		TOTPROV	1				
Comp. known (N = 94)	Self-compatible (0,1)	ATL	1	None			
		CZOCC	2				
		IND	1				
		NZG	1				
		PAC	1				
		TOTIS	1				
		TOTPROV	1				

Group	Variable	Factor	df	Level	Coef	Statistic (p)	Overall (p)
149 Genera	Number of species with FTa = 1	ATL	1				
		CZOCC	2	CZOCC-1	0.908	11.679 (0.001)	
				CZOCC-2	1.358	1.124 (0.289)	34.349 (0.000)
		IND	1				
		NZG	1				
		PAC	1	PAC	1.144	7.242 (0.007)	
		TOTIS	1				
		TOTPROV	1				
149 Genera	Number of gender dimorphic species	ATL	1				
		CZOCC	2				
		IND	1				
		NZG	1	NZG	6.187	30.136 (0.000)	
		PAC	1	PAC	2.235	4.221 (0.040)	107.221 (0.000)
		TOTIS	1				
		TOTPROV	1	TOTPROV	-1.536	12.754 (0.000)	
11 Islands	Number of gender dimorphic species	LAT	1				
		PROV	1	None			
		CZIS	1				
11 Islands	Number of species with FTa = 1	LAT	1				
		PROV	1	None			
		CZIS	1				
11 Islands	Number of species with FTb = 1	LAT	1				
		PROV	1				
		CZIS	1	CZIS	-0.931	16.433 (0.000)	17.842 (0.000)

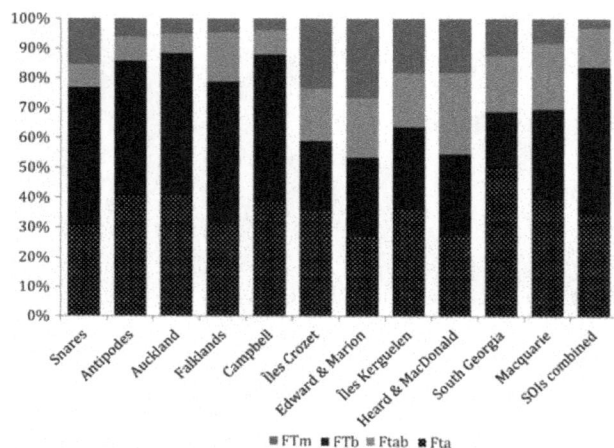

Figure 2. The frequency of floral types among flowering plants native to 11 Southern Ocean Island groups. Arrangement of island groups along the horizontal axis follows Figure 1. FTa, floral traits consistent with anemophily; FTb, flora traits suggestive of biotic pollination; FTab, ambiguous anemophilous or biotically pollinated flowers; FTm, minute flowers.

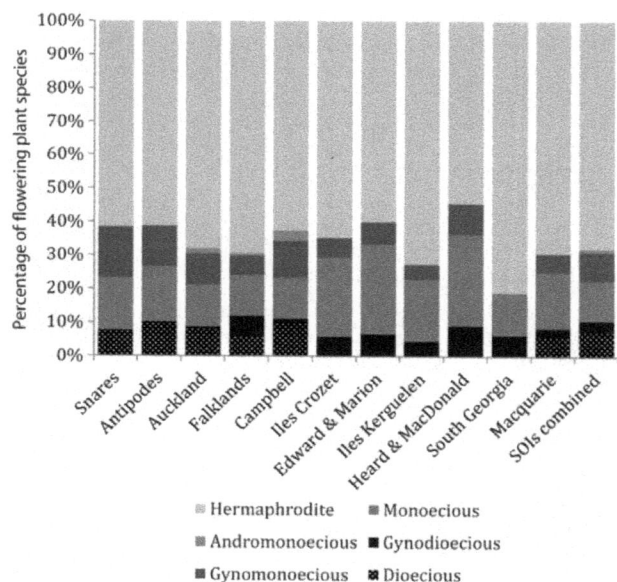

Figure 3. The frequency of breeding system classes among flowering plants native to 11 Southern Ocean Island groups. Arrangement of island groups along the horizontal axis follows Figure 1. See Table 1 for breeding system definitions.

breeding systems did not differ significantly among the 11 island groups ($\chi^2 = 5.31$, $P > 0.05$, df = 10; Fig. 3). The proportion of species with gender dimorphic vs. gender monomorphic breeding systems did not differ significantly between the SOI flora and either the Stewart Island flora or the alpine flora of Patagonia ($\chi^2 = 3.50$, $P > 0.05$, df = 1; $\chi^2 = 1.90$, $P > 0.05$, df = 1 respectively). Dioecy only featured in the floras of the Falklands Islands and

island groups in the south Pacific Province, where it occurred in up to 10.2 % of species (Table 2). The only gender dimorphic species found on other islands was *Acaena magellanica*.

Generalized linear models of species traits found that gender dimorphy was not associated with anemophily but was positively associated with fleshy fruit (Table 3). Genera shared with New Zealand or occurring in the Pacific province tended to have a greater proportion of gender dimorphic SOI species than other genera; however, gender dimorphism was negatively associated with the total number of biogeographic provinces occupied. Among islands, the number of species with dimorphic breeding systems was not explained by island latitude, province or climate zone.

Discussion

Given the paucity of typical pollinating groups and the relentlessly cold and windy climatic conditions on SOIs, flowering plants could be expected to rely on wind pollination or self-fertilization for sexual reproduction (Smith 1984; Bergstrom *et al.* 1997; Schermann-Legionnet *et al.* 2007; Convey *et al.* 2010). However, the disadvantages of self-fertilization in small populations could favour the evolution of gender dimorphic breeding systems on SOIs as has been suggested for other island groups (Carlquist 1977; Bawa 1982; Sakai *et al.* 1995; Webb *et al.* 1999). This study has brought together the available data on SOI plants to test predictions concerning self-compatibility, anemophily and gender dimorphy. The SOI flora has a very high incidence of self-compatibility, 95.8 % of species for which information specifically from SOIs was available are described as at least partially self-compatible. Self-compatibility is thus considerably more common on SOIs than in New Zealand (63.9 %, Newstrom and Robertson 2005) or in the alpine flora of Patagonia (69.7 %, Arroyo and Squeo 1990), and also exceeds high values reported for other island groups, e.g. 80 % of native Galapagos Islands species ($N = 55$, Chamorro *et al.* 2012) and 85 % of native Juan Fernandez Islands species ($N = 18$, Bernardello *et al.* 2001). While information on a greater proportion of the SOI flora was obtained in this study compared with those of Chamorro *et al.* (2012) and Bernardello *et al.* (2001), the quality of the data is unknown and mostly stems from simple statements rather than experimentation, thus the level of self-compatibility might be overstated. However, even if further experimental work modifies this value, it still represents an extreme on a global scale.

Unlike self-compatibility, floral traits consistent with anemophily were no more common overall in the SOI region compared with southern New Zealand and southern South America. Anemophily did, however, show a relationship with climate zone; genera represented on subantarctic islands possessed a higher proportion of anemophilous species than genera restricted to cool temperate islands, and cool temperate island floras showed a higher incidence of species with floral traits suggestive of biotic pollination. The fact that no relationship was observed between anemophily and latitude reflects the influence of the Southern Ocean on the terrestrial flora of SOIs. The Antarctic Polar Frontal Zone, which is responsible for significantly cooler sea surface temperatures around subantarctic, as opposed to cool temperate, islands (Chown et al. 1998), varies in latitude from ~48°S to 63°S depending on longitude and is remarkably stable apart from seasonal movements (Moore et al. 1999). Thus climates on the subantarctic islands of this study (south of the Polar Front) are considerably colder (~4 °C mean drop in sea surface temperature, data from Chown et al. 1998) than cool temperate islands to the north of the front. This boundary also marks a dramatic drop in vascular plant species richness which is strongly linked to insect species richness (Chown et al. 1998).

Despite climatic constraints apparently favouring anemophily on harsher subantarctic islands, the fact that the largest class of floral types consisted of petaloid, often coloured, floral displays, indicates that biotic pollination cannot be ruled out as a means of sexual reproduction among SOI plants. Notes of insects observed visiting flowers of SOI species were found throughout the literature researched for this study; however, observations were seldom recorded systematically. It is highly likely that many plant species employ a mixed mating strategy combining opportunistic biotic pollination with the reproductive assurance of self-fertilization. For example, on Auckland and Campbell Islands, flower-visiting moths and flies are easily overlooked as they are only active during rare periods of sunshine, and nocturnal insects, including a native Orthopteran, are likely important pollinators (Lord et al. 2013; Lord unpubl. data). Such findings and studies of other island systems (e.g. Olesen and Valido 2003; Olesen et al. 2012) highlight the opportunity for novel plant–pollinator relationships to evolve on isolated islands. A further indication that biotic pollination is still important on SOIs is the finding that gender dimorphy was not associated with anemophily, so many gender dimorphic species must rely on biotic pollinators for sexual reproduction, as has been found by Lord et al. (2013) for two dioecious species on Campbell Island. However, the absence of dioecious species from all subantarctic islands apart from Macquarie Island suggests a reliance on cross-pollination might reduce the success of a species establishing and/or persisting on harsher subantarctic islands.

The frequency of gender dimorphy among SOI species was substantially higher than values reported for other high-latitude islands (e.g. Iceland and British Isles both 3 %; Baker and Cox 1984). While gender dimorphy at the species level was strongly related to fleshy fruits, suggesting advantages associated with long-distance dispersal (Bawa 1982; Sakai et al. 1995; Webb et al. 1999), at the genus-level gender dimorphy showed a strong effect of taxonomic affinity with the New Zealand flora, which has a high incidence of gender dimorphy (Webb et al. 1999). This supports the contention of Baker and Cox (1984) that a potential source flora with a high prevalence of dimorphy is a major factor in explaining patterns of breeding systems on islands. However, unlike the classic pacific examples of the evolution of gender dimorphy (Hawaiian Islands, Sakai et al. 1995; New Zealand, Webb et al. 1999), gender dimorphic breeding systems were not a feature of species-rich lineages on SOIs (which were generally lacking). Furthermore, no gynodioecious species were SOI endemics and all SOI endemic dioecious species were in genera with dioecious or dimorphic species elsewhere. So while a dimorphic breeding system is clearly a viable reproductive strategy in the Southern Ocean region, until more data are available on phylogenetic affinities among SOI plants and their mainland relatives, there is no clear evidence that gender dimorphic breeding systems have evolved in situ on these isolated islands.

Conclusions

Very little experimental data exist concerning the reproductive ecology of SOI plants in situ. This lack of data is surprising given the relative simplicity of these floras and the long history of botanical study on many of these islands. Many SOI species show floral features consistent with biotic pollination and a number have been shown to be reliant on floral visitors for pollen transfer. Furthermore, species capable of autonomous self-fertilization can still benefit from out-crossing; for example, in a mainland study of gynodioecious Fuchsia excorticata, which occurs on Auckland Islands, fewer than 10 % of progeny derived from self-pollination in hermaphrodites survived and none had flowered after 11 years (Robertson et al. 2011). Virtually nothing is known about floral visitors to petaloid SOI species, but from the little information available it is clear that insect–plant interactions on SOIs require more research. Detailed studies such as those of Bergstrom et al. (1997), Schermann-Legionnet et al. (2007) and Lord et al. (2013) are required to definitively determine breeding systems and reliance on biotic pollinators, and the results of Lord et al. (2013) suggest that even nocturnal or flightless invertebrates may be capable of providing pollination services, so warrant closer study. Southern Ocean Island plants are being subjected to continued and increasing pressure from introduced species, human impacts and the likely effects of climate change (Meurk 1982; Smith and Steenkamp 1990; Copson and Whinam 2001; Chapuis *et al.* 2004; Le Roux *et al.* 2005; Shaw *et al.* 2005; McGlone *et al.* 2007; Convey *et al.* 2010). Management to assist recovery from disturbance and promote regeneration following restoration efforts requires an understanding of fundamental plant ecology including reproductive strategies.

Sources of Funding

Funding for this project was provided by a University of Otago Research and Study Leave Grant and a Fulbright New Zealand Travel Award.

Acknowledgements

Thanks to the New Zealand Department of Conservation, the New Zealand Navy and the Sir Peter Blake Trust for continued support for research on Southern Ocean Islands. This project arose from conversations with Dr A. Sakai and Prof. S. Weller. Barbara Anderson provided statistical advice, and together with Bronwyn Lowe and two anonymous referees, provided valuable feedback on earlier drafts.

Supporting Information

The following additional information is available in the online version of this article –

Table S1. List of all 321 flowering plant species used in analyses, their occurrence on 11 Southern Ocean Islands, breeding system, self-compatibility information (where available), floral traits and sources of information.

Literature Cited

Allan HH. 1961. *Flora of New Zealand, Vol. I*. Wellington, New Zealand: Government Printer.

Amey J, Lord JM, de Lange P. 2007. First record of a vascular plant from the Bounty Islands: *Lepidium oleraceum* (nau, Cook's scurvy grass) (Brassicaceae). *New Zealand Journal of Botany* **45**:87–90.

Arroyo MTK, Squeo F. 1990. Relationships between plant breeding systems and pollination. In: Kawano S, ed. *Biological approaches and evolutionary trends in plants*. London: Academic Press, 205–227.

Arroyo MTK, von Bohlen CP, Cavieres L, Marticorena C. 1992. Survey of the alpine flora of Torres del Paine National Park, Chile. *Gayana Botanica* **49**:47–70.

Australian Biological Resources Study. 1993. *Flora of Australia*. Vol. 50. Canberra: Australian Governmental Publishing Service.

Baker HG. 1955. Self-compatibility and establishment after 'long-distance' dispersal. *Evolution* **9**:347–349.

Baker HG, Cox PA. 1984. Further thoughts on dioecism and islands. *Annals of the Missouri Botanical Garden* **71**:244–253.

Barrett SCH. 1998. The reproductive biology and genetics of island plants. In: Grant PR, ed. *Evolution on Islands*. Oxford: Oxford University Press, 18–34.

Bawa KS. 1982. Outcrossing and the incidence of dioecism in island floras. *The American Naturalist* **119**:866–871.

Bergstrom DM, Selkirk PM, Keenan HM, Wilson ME. 1997. Reproductive behaviour of ten flowering plant species on subantarctic Macquarie Island. *Opera Botanica* **132**:109–120.

Bernardello G, Anderson GJ, Stuessy TF, Crawford DJ. 2001. A survey of floral traits, breeding systems, floral visitors, and pollination systems of the angiosperms of the Juan Fernández Islands (Chile). *The Botanical Review* **67**:255–308.

Broughton DA, McAdam JH. 2005. A checklist of the native vascular flora of the Falkland Islands (Islas Malvinas): new information on the species present, their ecology, status and distribution. *The Journal of the Torrey Botanical Society* **132**:115–148.

Carlquist S. 1977. *Island biology*. New York: Columbia University Press.

Chamorro S, Heleno R, Olesen JM, McMullen CK, Traveset A. 2012. Pollination patterns and plant breeding systems in the Galápagos: a review. *Annals of Botany* **110**:1489–1501.

Chapuis J-L, Frenot Y, Lebouvier M. 2004. Recovery of native plant communities after eradication of rabbits from the subantarctic Kerguelen Islands, and influence of climate change. *Biological Conservation* **117**:167–179.

Chown SL, Gremmen NJM, Gaston KJ. 1998. Ecological biogeography of Southern Ocean Islands: species-area relationships, human impacts, and conservation. *The American Naturalist* **152**:562–575.

Convey P, Key RS, Key RJD. 2010. The establishment of a new ecological guild of pollinating insects on sub-Antarctic South Georgia. *Antarctic Science* **22**:508–512.

Copson G, Whinam J. 2001. Review of ecological restoration programme on subantarctic Macquarie Island: pest management progress and future directions. *Ecological Management and Restoration* **2**:129–138.

De Lange PJ, Heenan PB, Houliston GJ, Rolfe JR, Mitchell AD. 2013. New *Lepidium* (Brassicaceae) from New Zealand. *PhytoKeys* **24**:1–147.

Donovan BJ. 2007. Apoidea (Insecta: Hymenoptera). *Fauna of New Zealand* **57**:1–295.

Edgar E, Connor HE. 2000. *Flora of New Zealand, Vol. 5. Grasses*. Lincoln, New Zealand: Manaaki Whenua Press.

Godley EF. 1989. Flora of Antipodes Island. *New Zealand Journal of Botany* **27**:531–564.

Grant PR. 1998. *Evolution on Islands*. Oxford, UK: Oxford University Press.

Greene SW. 1964. *The vascular flora of South Georgia*. British Antarctic Survey Scientific Reports No. 45. London, UK: British Antarctic Survey.

Gressitt JL. 1964. *Insects of Campbell Island*. Pacific Insects Monograph 7. Hawaii: Bishop Museum.

Greve M, Gremmen NJM, Gaston KJ, Chown SL. 2005. Nestedness of Southern Ocean island biotas: ecological perspectives on a biogeographical conundrum. *Journal of Biogeography* **32**:155–168.

Hay CH, Warham J, Fineran BA. 2004. The vegetation of The Snares, islands south of New Zealand, mapped and discussed. *New Zealand Journal of Botany* **42**:861–872.

Huntly BJ. 1971. Vegetation. In: Van Zinderen EM, Winterbottom JM, Dyer RA, eds. *Marion and Prince Edward Islands. Report on the South African Biological and Geological Expedition 1965–1966*. Cape Town: AA Balkema.

Inoue K, Maki M, Masuda M. 1995. Evolution of *Campanula* flowers in relation to insect pollinators on islands. In: Lloyd DG, Barrett SCH, eds. *Floral biology*. New York: Chapman and Hall, 377–400.

Johnson PN, Campbell DJ. 1975. Vascular plants of the Auckland Islands. *New Zealand Journal of Botany* **13**:665–720.

Le Roux PC, McGeoch MA, Nyakatya MJ, Chown SL. 2005. Effects of a short-term climate change experiment on a sub-Antarctic keystone plant species. *Global Change Biology* **11**:1628–1639.

Lloyd DG. 1985. Progress in understanding the natural history of New Zealand plants. *New Zealand Journal of Botany* **23**:707–722.

Lord JM, Westoby M, Leishman M. 1995. Seed size and phylogeny in six temperate floras: constraints, niche conservatism, and adaptation. *The American Naturalist* **146**:349–364.

Lord JM, Huggins L, Little LM, Tomlinson VR. 2013. Floral biology and flower visitors on subantarctic Campbell Island. *New Zealand Journal of Botany* **51**:168–180.

Losos JB, Ricklefs RE. 2009. Adaptation and diversification on islands. *Nature* **457**:830–836.

McGlone M, Wilmshurst J, Meurk C. 2007. Climate, fire, farming and the recent vegetation history of subantarctic Campbell Island. *Earth and Environmental Science Transactions of the Royal Society of Edinburgh* **98**:71–84.

Meurk CD. 1975. Contributions to the flora and plant ecology of Campbell Island. *New Zealand Journal of Botany* **13**:721–742.

Meurk CD. 1982. Regeneration of subantarctic plants on Campbell Island following exclusion of sheep. *New Zealand Journal of Ecology* **5**:51–58.

Mitchell AD, Meurk CD, Wagstaff SJ. 1999. Evolution of *Stilbocarpa*, a megaherb from New Zealand's sub-antarctic islands. *New Zealand Journal of Botany* **37**:205–211.

Moore DM. 1968. *The vascular flora of the Falklands Islands*. British Antarctic Survey No. 60. London, UK: British Antarctic Survey.

Moore DM. 1983. *Flora of Tierra del Fuego*. Oswestry: Anthony Nelson.

Moore JK, Abbott MR, Richman JG. 1999. Location and dynamics of the Antarctic Polar Front from satellite sea surface temperature data. *Journal of Geophysical Research* **104**:3059–3073.

Moore LB, Edgar E. 1970. *Flora of New Zealand, Vol. 2*. Wellington, New Zealand: Government Printer.

Newstrom L, Robertson A. 2005. Progress in understanding pollination systems in New Zealand. *New Zealand Journal of Botany* **43**:1–59.

Nicholls V. 2000. Ecology and ecophysiology of subantarctic Campbell Island megaherbs. MSc Thesis. Massey University, New Zealand.

Olesen JM, Valido A. 2003. Lizards as pollinators and seed dispersers: an island phenomenon. *Trends in Ecology and Evolution* **18**:177–181.

Olesen JM, Alarcón M, Ehlers BK, Aldasoro JJ, Roquet C. 2012. Pollination, biogeography and phylogeny of oceanic island bellflowers (Campanulaceae). *Perspectives in Plant Ecology, Evolution and Systematics* **14**:169–182.

Renner SS. 2014. The relative and absolute frequencies of angiosperm sexual systems: dioecy, monoecy, gynodioecy, and an updated online database. *American Journal of Botany* **101**:1588–1596.

Richards AJ. 1997. *Plant Breeding Systems*, 2nd edn. London: Chapman & Hall.

Robertson AW, Kelly D, Ladley JJ. 2011. Futile selfing in the trees *Fuchsia excorticata* (Onagraceae) and *Sophora microphylla* (Fabaceae): inbreeding depression over 11 years. *International Journal of Plant Sciences* **172**:191–198.

Sakai AK, Weller SG. 1999. Gender and sexual dimorphism in flowering plants: a review of terminology, biogeographic patterns, ecological correlates, and phylogenetic approaches. In: Geber MA, Dawson TE, Delph LF, eds. *Gender and sexual dimorphism in flowering plants*. Berlin: Springer, 1–32.

Sakai AK, Wagner WL, Ferguson DM, Herbst DR. 1995. Biogeographical and ecological correlates of dioecy in the Hawaiian flora. *Ecology* **76**:2530–2543.

Schermann-Legionnet A, Hennion F, Vernon P, Atlan A. 2007. Breeding system of the subantarctic plant species *Pringlea antiscorbutica* R. Br. and search for potential pollinators in the Kerguelen Islands. *Polar Biology* **30**:1183–1193.

Shaw JD, Hovenden MJ, Bergstrom DM. 2005. The impact of introduced ship rats (*Rattus rattus*) on seedling recruitment and distribution of a subantarctic megaherb (*Pleurophyllum hookeri*). *Austral Ecology* **30**:118–125.

Smith RIL. 1984. Terrestrial plant biology of the sub-Antarctic and Antarctic. In: Laws RM, ed. *Antarctic ecology Vol. 1*. London: Academic Press, 61–162.

Smith VR, Steenkamp M. 1990. Climatic change and its ecological implications at a subantarctic island. *Oecologia* **85**:14–24.

The Plant List. 2013. Version 1.1. http://www.theplantlist.org/ (1 June).

Upson R. 2012. *Important plant areas of the Falklands Islands*. Unpublished report, Falklands Conservation.

Van der Putten N, Verbruggen C, Ochyra R, Verleyen E, Frenot Y. 2010. Subantarctic flowering plants: pre-glacial survivors or postglacial immigrants? *Journal of Biogeography* **37**:582–592.

Wagstaff SJ, Hennion F. 2007. Evolution and biogeography of *Lyallia* and *Hectorella* (Portulacaceae), geographically isolated sisters from the Southern Hemisphere. *Antarctic Science* **19**:417–426.

Wagstaff SJ, Breitwieser I, Ito M. 2011. Evolution and biogeography of *Pleurophyllum* (Astereae, Asteraceae), a small genus of megaherbs endemic to the Subantarctic Islands. *American Journal of Botany* **98**:62–75.

Walton DWH. 1982. Floral phenology in the South Georgian vascular flora. *British Antarctic Survey Bulletin* **55**:11–25.

Webb CJ, Sykes W, Garnock-Jones P. 1988. *Flora of New Zealand, Vol. 4*. Christchurch: Botany Division, Department of Scientific and Industrial Research.

Webb CJ, Lloyd DG, Delph LF. 1999. Gender dimorphism in indigenous New Zealand seed plants. *New Zealand Journal of Botany* **37**:119–130.

Westoby M, Leishman MR, Lord JM. 1995. On misinterpreting the 'phylogenetic correction'. *Journal of Ecology* **83**:531–534.

Wilson HD. 1982. *Field guide Stewart Island plants*. Christchurch, New Zealand: Field Guide Publications.

12

Genetic consequences of cladogenetic vs. anagenetic speciation in endemic plants of oceanic islands

Koji Takayama[1], Patricio López-Sepúlveda[2], Josef Greimler[3], Daniel J. Crawford[4], Patricio Peñailillo[5], Marcelo Baeza[2], Eduardo Ruiz[2], Gudrun Kohl[3], Karin Tremetsberger[6], Alejandro Gatica[7], Luis Letelier[8], Patricio Novoa[9], Johannes Novak[10] and Tod F. Stuessy[3,11]*

[1] Museum of Natural and Environmental History, Shizuoka, Oya 5762, Suruga-ku, Shizuoka-shi, Shizuoka 422-8017, Japan
[2] Departamento de Botánica, Universidad de Concepción, Casilla 160-C, Concepción, Chile
[3] Department of Botany and Biodiversity Research, University of Vienna, Rennweg 14, A-1030 Vienna, Austria
[4] Department of Ecology and Evolutionary Biology and the Biodiversity Institute, University of Kansas, Lawrence, KS 60045, USA
[5] Instituto de Ciencias Biológicas, Universidad de Talca, 2 Norte 685, Talca, Chile
[6] Institute of Botany, Department of Integrative Biology and Biodiversity Research, University of Natural Resources and Life Sciences, Gregor Mendel Straße 33, A-1180 Vienna, Austria
[7] Bioma Consultores S.A., Mariano Sanchez Fontecilla No. 396, Las Condes, Santiago, Chile
[8] Universidad Bernardo O'Higgins, Centro de Investigaciones en Recursos Naturales y Sustentabilidad, General Gana 1702, Santiago, Chile
[9] Jardín Botánico de Viña del Mar, Corporación Nacional Forestal, Camino El Olivar 305, Viña del Mar, Chile
[10] Institute for Applied Botany and Pharmacognosy, University of Veterinary Medicine, Veterinärplatz 1, A-1210 Vienna, Austria
[11] Herbarium, Department of Evolution, Ecology, and Organismal Biology, The Ohio State University, 1315 Kinnear Road, Columbus, OH 43212, USA

Guest Editor: Clifford Morden

Abstract. Adaptive radiation is a common mode of speciation among plants endemic to oceanic islands. This pattern is one of cladogenesis, or splitting of the founder population, into diverse lineages in divergent habitats. In contrast, endemic species have also evolved primarily by simple transformations from progenitors in source regions. This is anagenesis, whereby the founding population changes genetically and morphologically over time primarily through mutation and recombination. Gene flow among populations is maintained in a homogeneous environment with no splitting events. Genetic consequences of these modes of speciation have been examined in the Juan Fernández Archipelago, which contains two principal islands of differing geological ages. This article summarizes population genetic results (nearly 4000 analyses) from examination of 15 endemic species, involving 1716 and 1870 individuals in 162 and 163 populations (with amplified fragment length polymorphisms and simple sequence repeats, respectively) in the following genera: *Drimys* (Winteraceae), *Myrceugenia* (Myrtaceae), *Rhaphithamnus* (Verbenaceae), *Robinsonia* (Asteraceae, Senecioneae) and *Erigeron* (Asteraceae, Astereae). The results indicate that species originating

* Corresponding author's e-mail address: stuessy.1@osu.edu

anagenetically show high levels of genetic variation within the island population and no geographic genetic partitioning. This contrasts with cladogenetic species that show less genetic diversity within and among populations. Species that have been derived anagenetically on the younger island (1–2 Ma) contain less genetic variation than those that have anagenetically speciated on the older island (4 Ma). Genetic distinctness among cladogenetically derived species on the older island is greater than among similarly derived species on the younger island. An important point is that the total genetic variation within each genus analysed is comparable, regardless of whether adaptive divergence occurs.

Keywords: Adaptive radiation; anagenesis; cladogenesis; genetic diversity; phyletic speciation; Robinson Crusoe Islands.

Introduction

Oceanic islands have long stimulated biologists to investigate patterns and processes of evolution (e.g. Darwin 1842; Wallace 1881; Whittaker and Fernández-Palacios 2007; Bramwell and Caujapé-Castells 2011). These isolated land masses, far from continental source areas, offer opportunities for determining origins of immigrants and their evolutionary history after establishment. The low probability of long-distance dispersal and successful colonization, the reduction of genetic variation in founding populations and the challenges of adaptation to new environments are all features that combine to affect processes of evolution in island archipelagos, particularly speciation.

One dimension of speciation in island plants that has received considerable attention is adaptive radiation (Carlquist 1974; Whittaker and Fernández-Palacios 2007; Rundell and Price 2009). This is a process that begins with dispersal from the original immigrant population into different habitats on the same or neighbouring island. This isolation leads to divergence of the new segregate populations, each becoming rapidly adapted to divergent habitats (Schluter 2001), such that eventually new species are recognized taxonomically. This general process of speciation is usually diagrammed (Fig. 1) as splitting events or cladogenesis (Rensch 1959). A number of dramatic species complexes have developed in oceanic islands through adaptive radiation, such as illustrated by the lobelioids (Givnish et al. 2009) and silverswords (Carlquist et al. 2003) in Hawaii, Aeonium (Liu 1989; Jorgensen and Olesen 2001) and Echium (Böhle et al. 1996) in the Canary Islands and Scalesia (Eliasson 1974) in the Gálapagos archipelago.

In addition to speciation via adaptive radiation (involving cladogenesis), another process, anagenesis (Fig. 1), has recently been emphasized (Stuessy et al. 1990, 2006; Whittaker et al. 2008). Some immigrant populations, especially when arriving on an island with limited ecological opportunity, proliferate in size and accumulate genetic diversity mainly through mutation and recombination. After many generations (perhaps over a million or more years), genetic changes result in different morphology that may be treated as a distinct species. This process has been labelled anagenetic speciation (Stuessy et al. 2006), being one type of progenitor-derivative speciation (Crawford 2010). It has been estimated that at least one-quarter of all endemic plant species of oceanic islands have originated via anagenesis (Stuessy et al. 2006).

Some studies have been published on the genetic consequences of cladogenesis in endemic plants of different archipelagos. Böhle et al. (1996) examined chloroplast sequence variation among endemic species of Echium (Boraginaceae) of the Canary Islands, showing very little nucleotide divergence even though the morphological variation is striking. Likewise, Baldwin (2003) examined internal transcribed spacer regions of nuclear ribosomal DNA (ITS) variation among species of the Hawaiian silverswords (Asteraceae) and again, limited sequence variation was seen. The general result from these, and other studies, is that during cladogenesis, the immigrant population becomes fragmented, with each segment containing a limited range of genetic variation in comparison with the continental progenitor population (Baldwin et al. 1998). Maximum morphological divergence occurs but with low levels of observable genetic diversity (Frankham 1997). There is some evidence (Perugganan et al. 2003) that the genetic changes responsible for the morphological adaptations involve alterations in regulatory rather than structural genes.

Results so far with anagenesis show a strikingly different pattern. Most of the investigations have been done on endemic species of Ullung Island, in which at least 88 % of the endemic species have originated anagenetically (Stuessy et al. 2006). The island is young (1.8 Ma; Kim 1985), of low elevation (<1000 m) and relatively ecologically uniform (Yim et al. 1981). Pfosser et al. (2005), using amplified fragment length polymorphisms (AFLPs), examined island and Japanese populations of Dystaenia takesimana and D. ibukiensis, respectively, and the results showed high levels of genetic variation within D. takesimana in comparison with D. ibukiensis. Similar results have been obtained in assessing the origin of Acer takesimensis and A. okomotoanum (Takayama

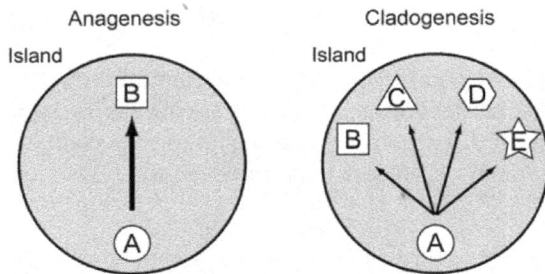

Figure 1. Diagram of the two principal modes of speciation in oceanic islands.

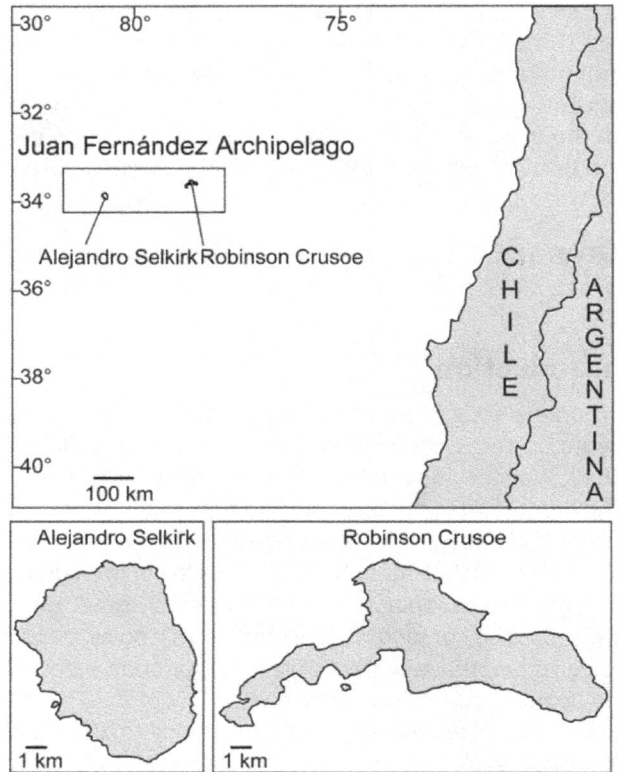

Figure 2. Location of the Juan Fernández Archipelago and its two major islands, Alejandro Selkirk (= Masafuera) and Robinson Crusoe (= Masatierra).

et al. 2012, 2013*a*). Because there is no partitioning of the immigrant population, it survives and proliferates, during which time it accumulates genetic variation through mutation and recombination. Eventually, the level of genetic diversity may even equal (or surpass) that observed in parental source populations (Stuessy 2007).

Because the above studies have been done on different genera in different island archipelagos, it would be useful to compare the genetic consequences of both types of speciation within groups of the same island system, preferably within the same island. In this fashion, more direct comparisons can be made because the general environment is the same. Important, obviously, is to locate plant groups that have originated via both anagenesis and cladogenesis within the same archipelago. A good choice for examining the genetic consequences of anagenesis and cladogenesis in endemic plants of oceanic islands is the Juan Fernández Archipelago, Chile. Approximately 64 % of the species have originated by cladogenesis and 36 % by anagenesis (Stuessy *et al.* 2006). From another perspective, it is estimated that 70 % of the *colonists* to the islands have diverged anagenetically, in contrast to only 30 % that have diverged via adaptive radiation (Stuessy *et al.* 1990).

The Juan Fernández Archipelago consists of two major islands (Fig. 2): Robinson Crusoe (= Masatierra), located 667 km west of continental Chile at 33°S latitude, and Alejandro Selkirk (= Masafuera) situated 181 km further westward into the Pacific Ocean. The former is known to be ~4 million years old and the latter 1–2 million years old (Stuessy *et al.* 1984). At present, these two islands are approximately the same size of 50 km^2 (Stuessy 1995). The flora is small, containing 78 native and 135 endemic vascular plant species (Danton *et al.* 2006). From a biogeographic standpoint, this setting is particularly favourable for generating initial hypotheses, because the near island (Robinson Crusoe) is also the older one, making it highly probable as the initial site for colonization of most groups. Furthermore, the older island is hypothesized to have been much larger when formed (Stuessy *et al.* 1998), making it a bigger target for dispersal from the mainland.

Numerous molecular markers now exist for assessing genetic variation within and among populations (Lowe *et al.* 2004). Amplified fragment length polymorphisms (Vos *et al.* 1995) have been used effectively to provide an overall evaluation of population genetic diversity (Tremetsberger *et al.* 2003; López-Sepúlveda *et al.* 2013*a*). These are treated as dominant markers and hence cannot be employed to determine allelic frequencies. An appropriate co-dominant and polymorphic marker that does allow allelic calculations are nuclear microsatellites or simple sequence repeats (SSRs). The challenge with this marker is to develop primers for locating sequences within the genome for comparison. Next-generation sequencing (NGS) methods are now available that allow this to be done much more easily and at reasonable cost (Takayama *et al.* 2011, 2013*b*). Numerous successful applications of SSRs have shown their efficacy to reveal genetic variation at the population level (Gleiser *et al.* 2008; Kikuchi *et al.* 2009; López-Sepúlveda *et al.* 2013*b*).

Studies using AFLPs and SSRs have already been published on a number of endemic taxa of the Juan Fernández Archipelago, representing groups that have undergone speciation via cladogenesis and anagenesis. The largest (and endemic) genus that has been investigated is *Robinsonia*

(Asteraceae; Takayama *et al.* 2015), which has seven species on Robinson Crusoe Island that have originated cladogenetically and one on Alejandro Selkirk Island that has evolved anagenetically. The genus *Erigeron* (Asteraceae; López-Sepúlveda *et al.* 2015) has six species that evolved cladogenetically on the younger island, Alejandro Selkirk. These two genera were selected because *Robinsonia* has speciated primarily via cladogenesis on the older island, and *Erigeron* has done so on the younger island. Regarding anagenesis, studies have been completed on *Drimys confertifolia* (Winteraceae; López-Sepúlveda *et al.* 2014) and *Rhaphithamnus venustus* (Verbenaceae; P. López-Sepúlveda, K. Takayama, D. J. Crawford, J. Greimler, P. Peñailillo, M. Baeza, E. Ruiz, G. Kohl, K. Tremetsberger, A. Gatica, L. Letelier, P. Novoa, J. Novak, T. F. Stuessy, submitted for publication), which occur on both islands of the archipelago. Investigations have also been completed on *Myrceugenia* (Myrtaceae; López-Sepúlveda *et al.* 2013b), which contains one endemic species on each of the islands. The available genetic data to date, therefore, come from 15 endemic species, plus 4 close continental relatives, summing to 1870 individuals in 163 populations.

The purposes of this article are to (i) summarize published data from AFLP and SSR investigations on endemic species of the genera *Drimys*, *Myrceugenia*, *Rhaphithamnus*, *Robinsonia* and *Erigeron*; (ii) compare and contrast differences in genetic diversity in groups that have undergone anagenetic or cladogenetic speciation and (iii) discuss the importance of considering modes of speciation for understanding levels of genetic diversity within endemic species of oceanic archipelagos.

Methods

The data summarized here (Table 1) provide the first comprehensive genetic comparisons (with AFLPs and SSRs) in the Juan Fernández Archipelago of species that have evolved by anagenesis and cladogenesis, based on consistent samplings, laboratory methods and modes of analysis. A number of earlier studies utilizing isozymes and DNA sequences have examined genetic variation in endemic species of these islands (e.g. Crawford *et al.* 1998, 2001a), but these investigations were not focussed on comparing modes of speciation. Genera in the present studies were selected for their representation of anagenesis and cladogenesis and for their occurrence on the two islands of different geological ages. The samples were collected during expeditions in February 2010 and 2011 from 1870 individuals in 163 populations in 15 endemic species, hence representing 14 % of the endemic angiosperms in the archipelago. The samples provide very good geographic coverage of populations over the landscape in both islands. The term population, as used here in the sense of sampling, refers to groups of individuals that were clearly delimited spatially in the field. The number of individuals analysed per population ranged from 1 to 31. The voucher data for these samples and details of data gathering and analysis are given in the respective publications.

Briefly, the following approaches were used for AFLPs. Four or six selective primer combinations were chosen. Numerous (24–85) primer trials were run with each genus to determine the best combination of primers for good resolution of individuals and populations. Data were obtained on an automated DNA sequencer (ABI 3130xl, Applied Biosystems, Waltham, MA, USA). Scoring was done using GeneMarker ver. 1.85 (SoftGenetics, State College, PA, USA). For analysis of AFLP data, the programs Arlequin 3.5.1.2 (Excoffier *et al.* 2005), FAMD ver. 1.25 (Schlüter and Harris 2006), R-Script AFLPdat (Ehrich 2006) and SPSS ver. 15.0 (SPSS; IBM, Armonk, NY, USA) were employed to determine total number of fragments (TNB), percentage of polymorphic fragments (PPB), Shannon Diversity Index (SDI), average gene diversity over loci (AGDOL) and rarity index (RI).

For SSRs, NGS methods (Takayama *et al.* 2011) were used to generate 6–12 loci, selected for their repeatability and scoring convenience. Polymerase chain reaction-amplified fragments were also run on the same automated sequencer and scored with GeneMarker ver. 1.85. Data analysis involved using GENEPOP 4.0 (Raymond and Rousset 1995), Micro-Checker 2.2.3 (van Oosterhout *et al.* 2004), FSTAT 2.9.3.2 and GENALEX 6 (Peakall and Smouse 2006). These allow analyses for observed proportion of heterozygotes (H_O), expected proportion of heterozygotes (H_E), number of alleles per locus (N_A), inbreeding coefficient (F_{IS}) and allelic richness standardized by five individuals (A_{R5}).

The overall pattern of higher genetic diversities in anagenetically derived species in comparison with cladogenetically derived ones was examined by a Student's *t*-test (average TNB, PPB, SDI, AGDOL and RI in AFLPs, and H_O, H_E, N_A and A_{R5} in SSRs) and shown in Table 2. To improve normality of H_O and H_E, a square-root transformation was applied. The overall patterns of higher genetic diversities in Robinson Crusoe Island (old) than Alejandro Selkirk Island (new) were also examined in the same way. The effects of two factors (speciation mode and island) and their interaction were analysed in a two-way ANOVA in R version 3.0.0 (R Core Team 2013) and shown in Table 3.

Data from both AFLPs and microsatellites were further analysed by assessing genetic distance (Nei *et al.* 1983) with the NeighborNet algorithm (Bryant and Moulton 2004) implemented by SplitsTree4 ver. 4.10 (Huson and Bryant 2006) and Population 1.2.30 (Langella 1999), respectively.

Table 1. Summary of measures of genetic diversity in endemic species of the Juan Fernández Archipelago that have originated by anagenesis or cladogenesis. All average values. Data from López-Sepúlveda et al. (2013a, b, 2014), Takayama et al. (2015) and P. López-Sepúlveda, K. Takayama, D. J. Crawford, J. Greimler, P. Peñailillo, M. Baeza, E. Ruiz, G. Kohl, K. Tremetsberger, A. Gatica, L. Letelier, P. Novoa, J. Novak, T. F. Stuessy, submitted for publication. TNB, total number of bands (fragments); PPB, percentage of polymorphic bands; SDI, Shannon Diversity Index; AGDOL, average gene diversity over loci; RI, rarity index; H_O, observed proportion of heterozygotes; H_E, expected proportion of heterozygotes; N_A, number of alleles per locus; F_{IS}, inbreeding coefficient; A_{RS}, allelic richness standardized by five individuals; RC, Robinson Crusoe Island; AS, Alejandro Selkirk Island.

Species	AFLPs							Microsatellites (SSRs)						
	No. of pops.	No. of plants	TNB	PPB	SDI	AGDOL	RI	No. of pops.	No. of plants	H_O	H_E	N_A	F_{IS}	A_{RS}
Anagenesis														
D. confertifolia (RC)	16	183	557	96.5	125.3	0.26	1.96	16	181	0.48	0.68	9.00	0.29	4.12
D. confertifolia (AS)	15	96	538	96.5	114.3	0.23	2.26	15	80	0.35	0.51	6.38	0.26	3.24
D. confertifolia (combined RC and AS)	31	279	576	100	134.7	0.28	2.06	31	261	0.44	0.68	9.88	0.33	4.13
M. fernandeziana (RC)	18	211	371	100	74.6	0.23	1.76	18	231	0.38	0.49	10.08	0.19	3.38
M. schulzei (AS)	13	129	417	100	96.2	0.28	3.39	13	155	0.39	0.61	10.33	0.35	3.79
R. venustus (RC)	20	143	440	99.3	96.4	0.25	2.80	20	140	0.17	0.23	4.22	0.31	1.83
R. venustus (AS)	4	18	271	57.3	60.8	0.18	2.34	4	11	0.30	0.34	2.33	0.13	2.12
R. venustus (combined RC and AS)	24	161	443	100	98.7	0.26	2.75	24	151	0.18	0.28	4.56	0.40	2.04
R. masafuerae (AS)	5	9	344	41.4	84.1	0.15	2.90	5	7	0.36	0.43	3.50	0.17	3.08
Cladogenesis														
Robinsonia gayana (RC)	10	123	592	77.2	111.0	0.16	2.39	10	134	0.34	0.42	6.30	0.28	3.04
R. gracilis (RC)	5	75	515	63.2	97.3	0.15	2.68	5	87	0.28	0.39	3.50	0.24	2.26
R. evenia (RC)	6	73	586	73.4	112.0	0.17	3.18	6	86	0.21	0.26	2.80	0.21	1.87
R. saxatilis (RC)	1	5	267	29.0	67.0	0.14	1.99	1	5	0.30	0.26	2.10	−0.22	2.10
Robinsonia (combined all RC species)	22	276	765	100	183.7	0.26	2.77	22	312	0.28	0.66	8.40	0.61	3.97
Robinsonia (combined all species)	27	285	766	100	265.0	0.26	2.68	27	319	0.29	0.67	8.70	0.61	4.02
E. fernandezianus (RC)	13	240	403	90.3	70.7	0.20	0.58	13	271	0.21	0.29	4.20	0.31	2.17
E. fernandezianus (AS)	19	172	426	95.3	81.1	0.23	0.81	19	200	0.17	0.50	7.50	0.72	3.27
E. fernandezianus (combined RC and AS)	32	412	433	97.5	81.7	0.23	0.68	32	471	0.20	0.40	8.00	0.64	2.86
E. ingae (AS)	2	21	315	61.3	62.0	0.18	0.62	2	25	0.20	0.34	2.90	0.55	2.04
E. luteoviridis (AS)	2	25	334	61.5	60.2	0.18	0.99	2	25	0.05	0.31	3.10	0.72	2.19
E. rupicola (AS)	9	175	377	81.8	69.5	0.20	0.67	9	211	0.17	0.36	4.40	0.57	2.43
E. turricola (AS)	3	10	269	49.3	57.6	0.19	0.50	3	10	0.24	0.53	3.40	0.57	2.94
E. stuessyi (AS)	1	8	306	66.7	82.4	0.28	0.81	2	11	0.20	0.25	2.10	0.53	1.89

Continued

Table 1. Continued

Species	AFLPs							Microsatellites (SSRs)						
	No. of pops.	No. of plants	TNB	PPB	SDI	AGDOL	RI	No. of pops.	No. of plants	H_O	H_E	N_A	F_{IS}	A_{RS}
Erigeron (combined all AS species)	36	411	443	100	95.1	0.26	0.74	37	482	0.17	0.62	9.20	0.76	2.85
Erigeron (combined all species)	49	651	444	100	94.2	0.26	0.68	50	753	0.18	0.56	9.50	0.73	3.46
Total and averages														
Anagenesis	91	789	419.7	84.4	93.1	0.23	2.49	91	805	0.35	0.47	6.55	0.24	3.08
Cladogenesis	71	927	399.1	68.1	79.2	0.19	1.38	72	1065	0.2	0.4	3.8	0.41	2.38
Robinson Crusoe	89	1053	466.4	78.6	94.3	0.19	2.17	89	1135	0.30	0.38	5.28	0.20	2.60
Alejandro Serkirk	73	663	359.7	71.1	76.8	0.21	1.53	74	735	0.24	0.42	4.59	0.46	2.70
Anagenesis (RC)	54	537	456.0	98.6	98.8	0.24	2.17	54	552	0.34	0.47	7.77	0.26	3.11
Anagenesis (AS)	37	252	392.5	73.8	88.9	0.21	2.72	37	253	0.35	0.47	5.64	0.23	3.06
Cladogenesis (RC)	35	516	472.6	66.6	91.6	0.16	2.16	35	583	0.27	0.32	3.78	0.17	2.29
Cladogenesis (AS)	36	411	337.8	69.3	68.8	0.21	0.73	37	482	0.17	0.38	3.90	0.61	2.46

For this article, to allow ease of visual comparisons of results among the species, emphasis has been placed on selected graphic presentations. SplitsTree Neighbor-Net was employed with the AFLP data, and the results are given in a series of graphs (Fig. 3). Neighbour-joining based on genetic distance was used for analysis of the SSRs, and simplified networks were used to show relationships among the populations (Fig. 4). For summary comparisons of genetic diversity among species, AGDOL was used with the AFLP data (Fig. 5). Not all calculated values for all original populations are presented or discussed in this review. The reader is referred to the original publications for additional methods and data.

Results

The results from the AFLP and SSR data analyses are given in Tables 1–4 and shown graphically in Figs 3–5. In general, the results from the two sources of genetic data are similar, with some exceptions, reinforcing confidence in the patterns seen. These data will be presented in context of the two modes of speciation, anagenesis and clado-genesis, but with attention also to the different ages of the islands. Robinson Crusoe Island is ~4 million years old and Alejandro Selkirk 1–2 million (Stuessy *et al.* 1984).

Anagenesis

The results from analysis of species that have evolved anagenetically include those from *Myrceugenia fernandeziana, M. schulzei, Robinsonia masafuerae, D. confertifolia* and *R. venustus*. The first species occurs only on the older island, the second and third species only on the younger island and the last two on both islands. A number of points seem evident. First, all anagenetically derived species show considerable levels of genetic diversity (Table 1, and Figs 3 and 5), and none of them shows geographic patterns over the island landscape (López-Sepúlveda *et al.* 2013*b*, 2014, P. López-Sepúlveda, K. Takayama, D. J. Crawford, J. Greimler, P. Peñailillo, M. Baeza, E. Ruiz, G. Kohl, K. Tremetsberger, A. Gatica, L. Letelier, P. Novoa, J. Novak, T. F. Stuessy, submitted for publication). This is what might be expected from the predictions regarding anagenesis based on previous studies. Even more interesting, perhaps, is that the amount of genetic diversity differs in species on the two islands of different ages. In *D. confertifolia*, and *R. venustus*, which occur on both islands, one sees in both cases more genetic diversity (SDI) in populations on the older island than on the younger island except for estimates of SSRs in *R. venustus* (Table 1). The explanation of these data may relate to the time available for a genetic change to take place. Because Alejandro Selkirk Island is no more than 1–2 million years old, this must be the maximum

Table 2. Summary of statistical tests based on Table 1. TNB, total number of bands (fragments); PPB, percentage of polymorphic bands; SDI, Shannon Diversity Index; AGDOL, average gene diversity over loci; RI, rarity index; H_O, observed proportion of heterozygotes; H_E, expected proportion of heterozygotes; N_A, number of alleles per locus; A_{R5}, allelic richness standardized by five individuals. Bold font indicates significant values ($P < 0.05$).

	High genetic diversity in anagenetically derived species	High genetic diversity in Robinson Crusoe Island species
AFLPs		
TNB	0.351	**0.024**
PPB	0.086	0.235
SDI	0.101	**0.045**
AGDOL	**0.050**	0.227
RI	**0.004**	0.085
SSRs		
H_O	**0.006**	0.132
H_E	0.061	0.236
N_A	**0.040**	0.308
A_{R5}	**0.038**	0.388

Table 3. Summary of two-way ANOVA based on Table 1. TNB, total number of bands (fragments); PPB, percentage of polymorphic bands; SDI, Shannon Diversity Index; AGDOL, average gene diversity over loci; RI, rarity index; H_O, observed proportion of heterozygotes; H_E, expected proportion of heterozygotes; N_A, number of alleles per locus; A_{R5}, allelic richness standardized by five individuals. For all F-values, the degree of freedom was 1. Bold font indicates significant values ($P < 0.05$).

Factor		F-value	P-value
AFLPs			
TNB	Island	4.78	**0.046**
	Speciation mode	0.22	0.645
	Island vs. speciation mode	0.51	0.489
PPB	Island	0.67	0.427
	Speciation mode	2.60	0.129
	Island vs. speciation mode	2.05	0.174
SDI	Island	3.61	0.078
	Speciation mode	2.36	0.147
	Island vs. speciation mode	0.47	0.504
AGDOL	Island	0.85	0.372
	Speciation mode	4.09	0.063
	Island vs. speciation mode	4.67	**0.048**
RI	Island	4.63	**0.049**
	Speciation mode	13.71	**0.002**
	Island vs. speciation mode	10.53	**0.006**
SSRs			
H_O	Island	2.03	0.176
	Speciation mode	11.65	**0.004**
	Island vs. speciation mode	1.64	0.221
H_E	Island	0.47	0.502
	Speciation mode	3.44	0.085
	Island vs. speciation mode	0.19	0.671
N_A	Island	0.47	0.502
	Speciation mode	3.44	0.085
	Island vs. speciation mode	0.19	0.671
A_{R5}	Island	0.10	0.752
	Speciation mode	4.54	0.051
	Island vs. speciation mode	0.11	0.744

time available for population divergence to take place. With anagenetically evolved species, all factors being equal, genetic variation increases through time, and this can be seen in the species investigated.

One case of anagenesis in the archipelago also merits comment. *Robinsonia masafuerae* is a species that appears to have speciated from *R. evenia*, with which it has been closely associated in all studies so far (Crawford *et al.* 1993a; Sang *et al.* 1995; Takayama *et al.* 2015). Previous investigations on ITS 1 and 2 in *Robinsonia* (Sang *et al.* 1995) have shown sequence divergence between *R. evenia* and *R. masafuerae* as only 0.0063 (two base substitutions). Although one cannot place an absolute time on this divergence, it is the lowest level among any pair of species in the genus, which correlates well with the youthful geological age of Alejandro Selkirk Island. Genetic variation in *R. masafuerae* is much lower from AFLP data than in *R. evenia* from Robinson Crusoe (Table 1 and Fig. 5), but in SSRs, the pattern reverses with the anagenetically derived species, *R. masafuerae*, showing more variation than any single one of the cladogenetically originated species on Robinson Crusoe (Table 1).

It is also possible to make comparisons between populations of continental progenitors with endemic island derivatives. In the case of *Myrceugenia schulzei*, the closest continental congener is *M. colchaguensis* (Landrum 1981a, b; Ruiz *et al.* 2004). Although the sampling of populations on the continent is limited to two populations,

the amount of genetic diversity is particularly low as shown by AFLP data, although somewhat higher with SSRs (López-Sepúlveda *et al.* 2013b). Although *M. schulzei* is known only on the younger island, it did not diverge from *M. fernandeziana* on the older island because the two are unrelated (Murillo-Aldana *et al.* 2012), so much so that the latter has now been transferred to another genus (*Nothomyrcia*; Murillo-Aldana and Ruiz 2011). With *D. confertifolia*, comparisons with *D. winteri* and *D. andina* show less genetic variation in the two latter species as

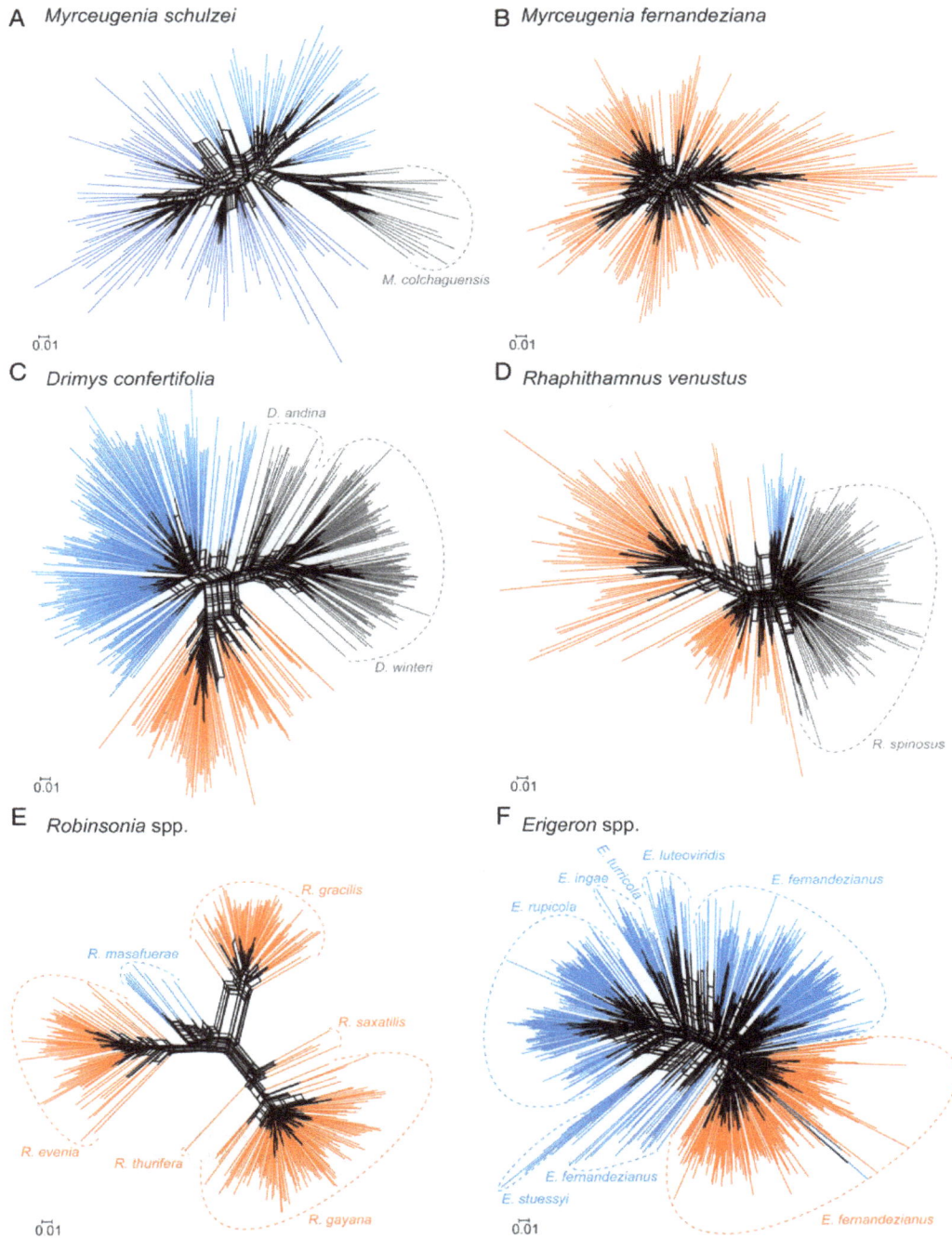

Figure 3. SplitsTree NeighborNet showing genetic relationships based on AFLPs among individuals in endemic species of *Myrceugenia* (A and B), *Drimys* (C), *Rhaphithamnus* (D), *Robinsonia* (E) and *Erigeron* (F) in the Juan Fernández Archipelago. Closely related continental relatives are also shown in A, C and D. Orange = species and populations on Robinson Crusoe Island; blue = on Alejandro Selkirk Island and black = on the or islands continent.

seen from AFLPs and SSRs (López-Sepúlveda *et al.* 2014). In *R. venustus*, which is a congener of *R. spinosus* (the only other known species in the genus; Moldenke 1937; Crawford *et al.* 1993*b*), the amount of genetic diversity is again greater in the population on Robinson Crusoe Island

than documented on the continent, although considerably lower in the population on Alejandro Selkirk (P. López-Sepúlveda, K. Takayama, D. J. Crawford, J. Greimler, P. Peñailillo, M. Baeza, E. Ruiz, G. Kohl, K. Tremetsberger, A. Gatica, L. Letelier, P. Novoa, J. Novak, T. F. Stuessy,

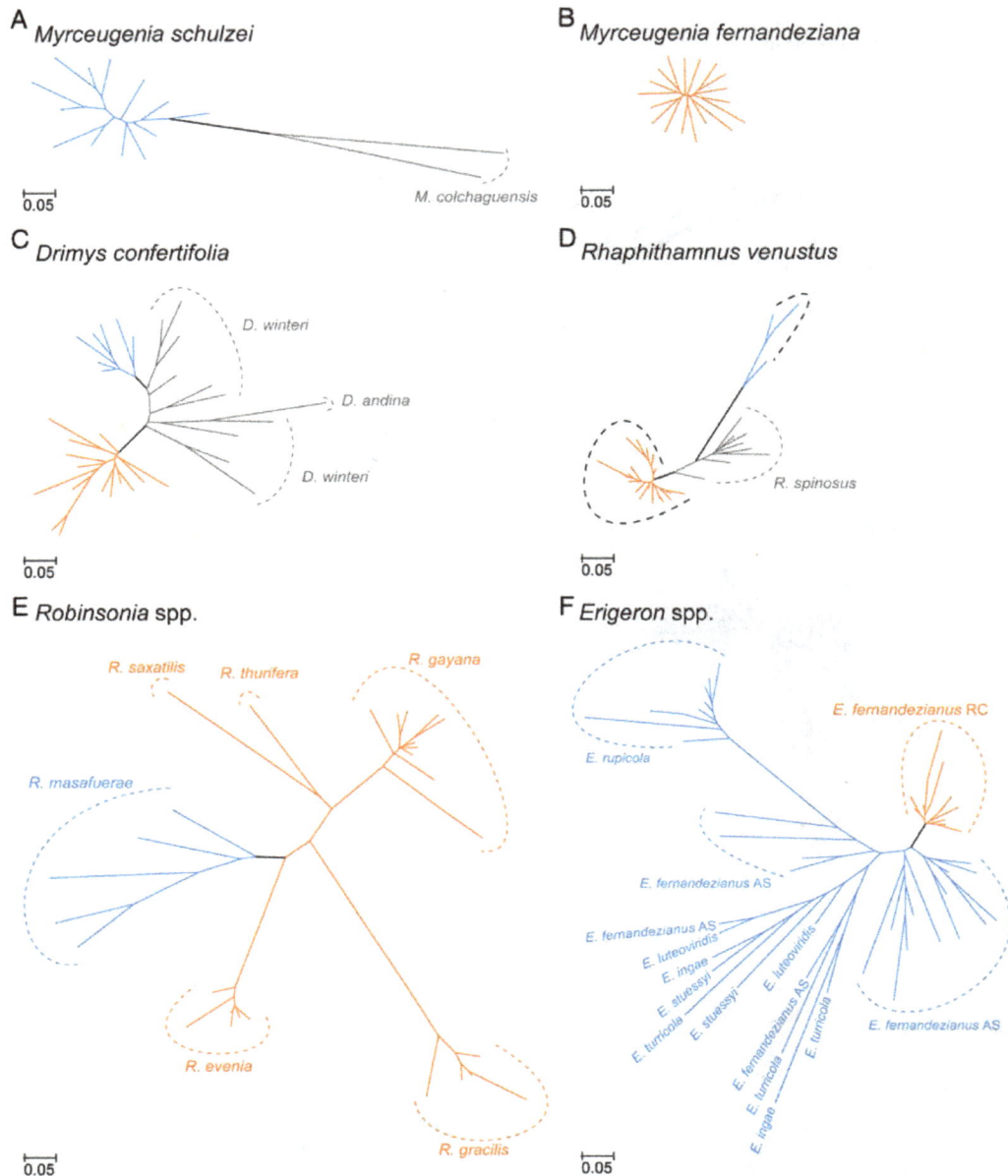

Figure 4. Neighbour-joining tree showing genetic relationships based on SSRs among populations in endemic species of *Myrceugenia* (A and B), *Drimys* (C), *Rhaphithamnus* (D), *Robinsonia* (E) and *Erigeron* (F) in the Juan Fernández Archipelago. Closely related continental relatives are also shown in A, C and D. Orange = species and populations on Robinson Crusoe Island; blue = on Alejandro Selkirk Island and black = on the continent.

submitted for publication). These results support the concept that over time, considerable genetic variation can accumulate in anagenetically derived populations, so much so that the degree of variation can approximate and even surpass that in the progenitor source populations.

Cladogenesis

Two of the largest genera of the archipelago are *Robinsonia* with eight endemic species and *Erigeron* with six. Both are in Asteraceae, although unrelated and placed in different tribes (Senecioneae vs. Astereae, respectively).

Robinsonia has adaptively radiated on Robinson Crusoe Island during the past 4 million years (maximum value) and *Erigeron* has done so on Alejandro Selkirk Island in the past 1–2 million years.

Robinsonia is the second largest genus in the archipelago. The largest is *Dendroseris*, also of Asteraceae but from still another tribe (Cichorieae). This latter genus is of interest as it has derived cladogenetically on the older island with three independent dispersals to the younger island and three anagenetic speciations there (Sanders *et al.* 1987; Pacheco *et al.* 1991; Sang *et al.* 1994). Most of

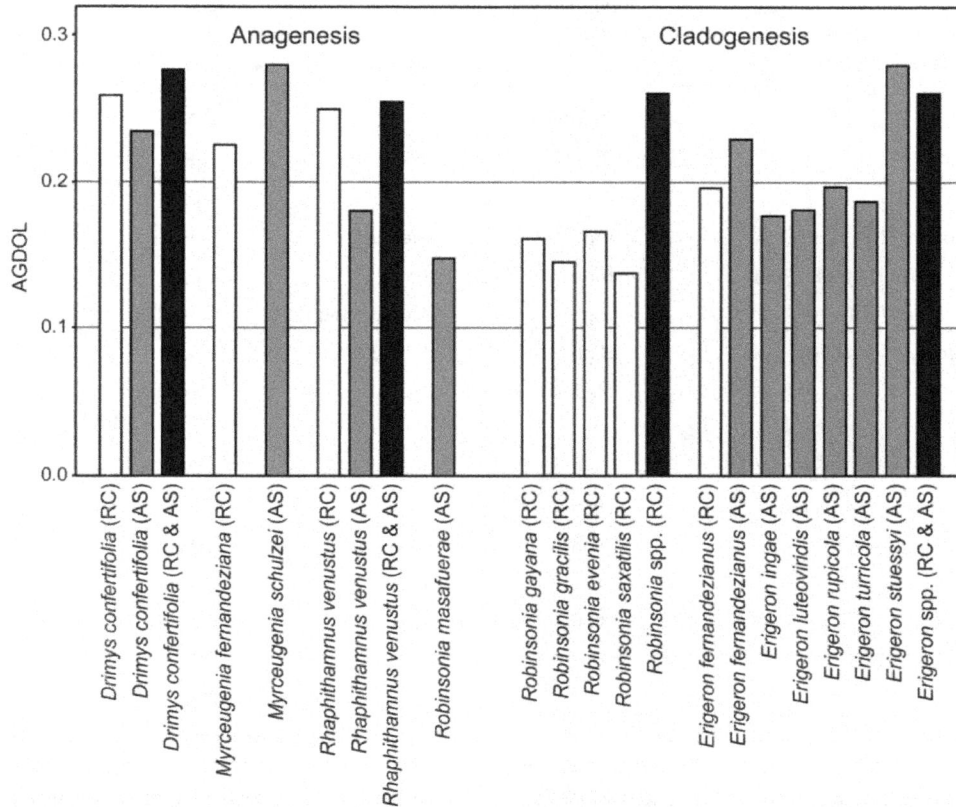

Figure 5. Summary of genetic diversities, AGDOL, within the endemic species of *Drimys*, *Myrceugenia* and *Rhaphithamnus* having originated by anagenesis, and *Robinsonia* and *Erigeron* having been derived through cladogenesis. *Robinsonia masafuerae* from the younger island is also an anagenetic derivative from the cladogenetic complex of *Robinsonia* on the older island. AS, Alejandro Selkirk Island; RC, Robinson Crusoe Island. White bar indicates an endemic species in RC, grey bar an endemic species in AS and black bar multiple species or islands combined.

these species are quite rare now, however, which precluded our being able to obtain sufficient population data for genetic evaluation. *Robinsonia* has eight species, but two are presumed extinct (*R. berteroi* and *R. megacephala*; Danton *et al.* 2006). Our studies have focussed on five species having originated cladogenetically on the older island. Comments have already been made regarding the one anagenetically derived species (*R. masafuerae*) on Alejandro Selkirk Island. The results from AFLP data are shown in Fig. 3 and from SSRs in Fig. 4. Most notable from the SplitsTree graph in Fig. 3 is that the different species of *Robinsonia* are very distinct genetically. Divergence has obviously taken place during adaptive radiation and also during a maximum time available of 4 million years. The species *R. gayana*, *R. thurifera* and *R. saxatilis* form an evolutionary complex, which taxonomically has been regarded as sect. *Robinsonia* (Skottsberg 1922, as sect. *Symphyolepis*; Takayama *et al.* 2015). *Robinsonia gracilis* ties with *R. evenia* and its close anagenetic relative *R. masafuerae* in sect. *Eleutherolepis* (Skottsberg 1922). With SSR data (Fig. 4), the species are also very distinct and genetically more cohesive, with the anagenetic species *R. masafuerae* showing the greatest genetic diversity.

Another important point seen clearly in Figs 3 and 4 is that the range of genetic diversity within each of these cladogenetic species is limited in comparison with the anagenetically derived species discussed above.

Although *Erigeron* is not an endemic genus in the archipelago, six endemic species occur there having evolved via cladogenesis and adaptive radiation. The origin of this complex is unusual in that the colonist(s) presumably arrived directly to the younger island (Valdebenito *et al.* 1992). Amplified fragment length polymorphism and SSR data (Figs 3 and 4) reveal considerable genetic diversity within these endemic species, and each species is reasonably distinct. An exception is the *Erigeron ingae* complex consisting of *E. ingae*, *E. luteoviridis* and *E. turricola*. These species are sometimes difficult to distinguish morphologically. Solbrig (1962) and Marticorena *et al.* (1998), for example, placed *E. turricola* into synonymy with *E. ingae*, but Danton *et al.* (2006) kept them distinct. The molecular data parallel this morphological inconsistency. This may be a population complex in early stages of speciation, now undergoing divergence from within a pool of morphological and genetic variation. All of these species grow in the 'alpine zone' on the younger island (Skottsberg 1922),

Table 4. Generalized comparison of the levels of genetic diversity obtained with AFLPs and SSRs from species that have originated via anagenesis and cladogenesis on the two islands of the Juan Fernández Archipelago. See Table 1 for the actual data. RC, Robinson Crusoe Island; AS, Alejandro Selkirk Island.

	Anagenesis		Cladogenesis	
	RC	AS	RC	AS
AFLPs				
Total number of bands (TNB)	High	Medium	High	Low
Percentage of polymorphic bands (PPB)	High	Low	Low	Low
Shannon Diversity Index (SDI)	High	Medium high	Medium high	Low
Average gene diversity over loci (AGDOL)	High	Medium high	Low	Medium high
Rarity index (RI)	Medium	High	Medium	Very low
Microsatellites (SSRs)				
Observed proportion of heterozygotes (H_O)	High	High	Medium	Low
Expected proportion of heterozygotes (H_E)	High	High	Medium	Medium
Number of alleles per locus (N_A)	High	Medium	Low	Low
Inbreeding coefficient (F_{IS})	Low	Low	Low	High
Allelic richness (A_{R5})	High	High	Low	Low

and we have not noticed any clear habitat differences among them. The species *E. rupicola* is confined to coastal rocks along the sea and also penetrates into the quebradas (ravines); its close relative, *E. stuessyi*, is also found on rocky ledges but residing inside the cool and deep ravines. *Erigeron fernandezianus* occurs in a broad altitudinal range (100–1200 m), and it inhabits mainly rocky areas in middle elevation plains, quebradas and ridges. This species also occurs on the older island, but it is found there in many plant communities and especially in disturbed sites. It appears, therefore, to be an example of back migration from the younger to the older island (Valdebenito *et al.* 1992; López-Sepúlveda *et al.* 2015).

Although most species of *Erigeron* on the younger island are distinct genetically, the degree of distinctness is much less than observed among species of *Robinsonia* on the older island (Figs 3 and 4). It may be that these species of *Erigeron* have had less time to diverge in comparison with those of *Robinsonia*. With the passage of time, therefore, the genetic profiles of species undergoing adaptive radiation may remain narrow due to strong directional selection in each different habitat. In both *Erigeron* and *Robinsonia*, however, the range of genetic variation seen is less than that in the anagenetically derived species.

Discussion

Comparison of anagenesis and cladogenesis

Predictions from theory (Stuessy 2007) would suggest that higher levels of genetic diversity should be found within the anagenetically derived species. This is because the founding population increases in size over time, accumulating genetic diversity mainly through mutation and recombination. One would expect no (or very little) geographic partitioning over the landscape. Likewise, due to a lack of strong selection, one would not expect to find high levels of private alleles or bands, nor a high RI. With cladogenetic speciation, on the other hand, one would expect less overall genetic diversity within each species, but with more private alleles due to strong directional selection. As for impact from the age of the islands, one would predict less total genetic diversity within anagenetically derived species on the younger island because diversity increases through time. As for the cladogenetic species, one would predict less genetic divergence (distinctness) on the younger island in comparison with species on the older island, because directional selection continues over time and refines the genetic profile of each species as it adapts to the particular ecological zone.

Results from genetic analyses of 5 anagenetic species and 10 cladogenetic species allow comparisons between the two modes of speciation and the two islands of differing ages (Tables 1–3). A number of general points can be observed (Table 4 and Fig. 5). First, in anagenetic species, the level of genetic diversity tends to be higher per species than in the cladogenetic species, especially on Robinson Crusoe Island. This can be seen in percentage of polymorphic bands, SDI, AGDOL, observed proportion of heterozygotes, expected proportion of heterozygotes, number of alleles per locus and allelic richness. Second, in the anagenetic species, the individuals on each island behave genetically as one large population, showing no genetic pattern over the landscape (López-Sepúlveda et al. 2013b, 2014; Takayama et al. 2015; P. López-Sepúlveda, K. Takayama, D. J. Crawford, J. Greimler, P. Peñailillo, M. Baeza, E. Ruiz, G. Kohl, K. Tremetsberger, A. Gatica, L. Letelier, P. Novoa, J. Novak, T. F. Stuessy, submitted for publication). This is true on both islands of differing ages. This suggests that this pattern can develop easily within 1–2 million years and that it can persist for up to 4 million. This is consistent with the results reported for Ullung Island, Korea, which is known to be 1.8 million years old (Pfosser et al. 2005; Takayama et al. 2012, 2013a). Third, the ability of an immigrant population to radiate adaptively has much to do with the properties of the colonists (and progenitors) and less with differences of habitat. Some colonists remain as a single larger population and are not responsive to adaptive change in different ecological zones, whereas others disperse well to micro-zones and quickly become modified morphologically and genetically. Fourth, perhaps most importantly, the total amount of genetic diversity within an anagenetically derived species in comparison with an entire adaptively radiating lineage is approximately the same (Fig. 5).

Genetics of speciation in endemic plants of oceanic islands

A number of previous studies have assessed levels of genetic variation within and among populations of endemic species of the Juan Fernández Archipelago with other markers such as isozymes, random amplified polymorphic DNA (RAPDs) and inter simple sequence repeats (ISSRs). Isozymes have been analysed in Dendroseris (Crawford et al. 1987), Chenopodium sanctae-clarae (Crawford et al. 1988), Wahlenbergia (Crawford et al. 1990), Robinsonia (Crawford et al. 1992), Lactoris (Crawford et al. 1994) and Myrceugenia (Jensen et al. 2002). RAPDs have been investigated in Dendroseris (Esselman et al. 2000) and Lactoris (Brauner et al. 1992), and ISSRs also in Lactoris (Crawford et al. 2001b).

Crawford et al. (2001a) summarized the results from isozyme studies on 29 endemic species of the Juan Fernández Archipelago, and this represents the best set of observations to compare with the AFLP and SSR data summarized here. The most conspicuous result is that the mean genetic diversities at the species level are low ($H_{es} = 0.065$). Higher levels of diversity were seen in larger populations or in many small populations and also in outcrossing species in contrast to selfers. Of relevance for comparisons to the present study, isozymes have been analysed from four species of Robinsonia and in M. fernandeziana, E. fernandezianus and R. venustus. It is difficult to compare the results of the isozymes because they provide less detailed genetic information than from AFLPs and SSRs. Isozyme studies on the endemic Lactoris fernandezianus, for example Crawford et al. (1994), revealed virtually no variation, but ISSRs showed considerable variation within and among populations (Crawford et al. 2001b). Studies on isozymes (Crawford et al. 1987) and RAPDs (Esselman et al. 2000) from Dendroseris showed greater resolution of relationships from the latter. The isozyme data for the four cladogenetically derived species of Robinsonia show higher levels of genetic variation than in the anagenetic R. venustus (Crawford et al. 1993b) and Myrceugenia (Jensen et al. 2002), which would be in contrast to the trends documented here. It is important, therefore, that for questions involving population genetics in endemic plants of oceanic islands, rapidly evolving markers need to be used.

The employment of AFLPs and SSRs in the present study from 15 species of the Juan Fernández Archipelago, therefore, does provide detailed genetic data at the population level for purposes of comparing consequences of different modes of speciation. A general review has recently been published on the general topic of interpretation of genetic variation within endemic species of oceanic islands (Stuessy et al. 2014), and the present data corroborate ideas summarized there. Clearly, the alternative modes of speciation, anagenesis and cladogenesis result in different genetic consequences. Interpretation of the evolutionary significance of levels of genetic diversity, therefore, must be done in context of type of speciation. As can be seen in the results of adaptive radiation in Erigeron and Robinsonia, on the young and older islands, respectively, the geological age of the island also matters, as this provides the time dimension in which the evolutionary processes unfold.

Another very significant impact on levels of genetic variation in populations of endemic plants of oceanic islands is that from human activity. Because oceanic islands often have agreeable climates and attractive beaches, people have come to live, play and build homes and apartments, all of which have caused pressures on the native vegetation. In the Juan Fernández Archipelago, for example, people have been living continuously on Robinson Crusoe Island for >300 years (Woodward 1969; Wester 1991). It is not impossible that the species of Robinsonia on the older island have suffered some genetic loss due to human activity. Although these species occur either on high ridges or in deep forests, far removed from most persons who live at sea level in the village (San Juan Bautista), incursions into the native forest must have taken place and some plants destroyed. It is known that two species of Robinsonia, both on Robinson Crusoe Island, are now extinct (R. berteroi and R. megacephala; Danton and Perrier 2005; Danton et al.

2006). Assessing the level of human impact on the vegetation of an oceanic island, therefore, is challenging. At least in the Juan Fernández Archipelago, there were no aboriginal peoples, and human activity could only have begun with discovery by Europeans (Juan Fernández; Medina 1974) at the end of the 16th century. Since that time, however, considerable negative impact from human activity has been documented in the archipelago (Wester 1991; Matthei et al. 1993; Stuessy et al. 1997; Swenson et al. 1997; Cuevas and Leersum 2001; Greimler et al. 2002; Dirnböck et al. 2003; Cuevas et al. 2004; Ricci 2006; Vargas et al. 2011), especially from introduced animals, such as rats, rabbits and goats (e.g. Camus et al. 2008). These combined activities have surely had some impact on the levels of genetic variation within and among populations.

Sources of Funding

This work was supported by an FWF (Austrian Science Fund) grant (P21723-B16) to T.F.S. and a Japan Society for the Promotion of Science (JSPS) Postdoctoral Fellowship for Research Abroad (grant 526) to Ko.T.

Contributions by the Authors

Ko.T. conceived the idea behind the article; all authors participated in the field work except G.K. and Ka.T.; J.N., P.L.-S., G.K. and Ko.T. completed the laboratory work; J.N. coordinated the NGS data acquisition; T.F.S. and Ko.T wrote the initial draft and all authors contributed to subsequent drafts and offered comments for improvement.

Acknowledgements

We very much appreciate the generous logistic and facility support of Sr Iván Leiva, Chief of the Robinson Crusoe Islands national park, administered by the Corporación Nacional Forestal (CONAF); the help and cooperation in fieldwork from the CONAF guides, especially Jorge Angulo, Danilo Arredondo, Danilo Arredondo, Jr, Oscar Chamorro, Michael González, Bernardo López, Eduardo Paredes, Ramon Schiller and Manuel Tobar; and the Armada de Chile for logistic support in transporting supplies from the continent to the islands. The results presented in this paper form part of an Open Partnership Joint Project of the JSPS Bilateral Joint Research program.

Literature Cited

Baldwin BG. 2003. A phylogenetic perspective on the origin and evolution of Madiinae. In: Carlquist S, Baldwin BG, Carr GD, eds. *Tarweeds & silverswords: evolution of the Madiinae (Asteraceae).* St. Louis: Missouri Botanical Garden Press, 193–228.

Baldwin BG, Crawford DJ, Francisco-Ortega J, Kim S-C, Sang T, Stuessy TF. 1998. Molecular phylogenetic insights on the origin and evolution of oceanic island plants. In: Soltis DE, Soltis PS, Doyle JJ, eds. *Molecular systematics of plants II: DNA sequencing.* Boston: Kluwer, 410–441.

Böhle U-R, Hilger HH, Martin WF. 1996. Island colonization and evolution of the insular woody habit in *Echium* L. (Boraginaceae). *Proceedings of the National Academy of Sciences of the USA* **93**: 11740–11745.

Bramwell D, Caujapé-Castells J. 2011. *The biology of island floras.* Cambridge: Cambridge University Press.

Brauner S, Crawford DJ, Stuessy TF. 1992. Ribosomal DNA and RAPD variation in the rare plant family Lactoridaceae. *American Journal of Botany* **79**:1436–1439.

Bryant D, Moulton V. 2004. Neighbor-Net: an agglomerative method for the construction of phylogenetic networks. *Molecular Biology and Evolution* **21**:255–265.

Camus P, Castro S, Jaksic F. 2008. El conejo europeo en Chile: Historia de una invasión histórica. *Historia* **41**:305–339.

Carlquist S. 1974. *Island biology.* New York: Columbia University Press.

Carlquist S, Baldwin BG, Carr GD, eds. 2003. *Tarweeds & silverswords: evolution of the Madiinae (Asteraceae).* St. Louis: Missouri Botanical Garden.

Crawford DJ. 2010. Progenitor-derivative species pairs and plant speciation. *Taxon* **59**:1413–1423.

Crawford DJ, Stuessy TF, Silva M. 1987. Allozyme divergence and the evolution of *Dendroseris* (Compositae: Lactuceae) on the Juan Fernandez Islands. *Systematic Botany* **12**:435–443.

Crawford DJ, Stuessy TF, Silva M. 1988. Allozyme variation in *Chenopodium sanctae-clarae*, an endemic species of the Juan Fernandez Islands, Chile. *Biochemical Systematics and Ecology* **16**: 279–284.

Crawford DJ, Stuessy TF, Lammers TG, Silva M, Pacheco P. 1990. Allozyme variation and evolutionary relationships among three species of *Wahlenbergia* (Campanulaceae) in the Juan Fernandez Islands. *Botanical Gazette* **151**:119–124.

Crawford DJ, Stuessy TF, Haines DW, Cosner MB, Silva M, Lopez P. 1992. Allozyme diversity within and divergence among four species of *Robinsonia* (Asteraceae: Senecioneae), a genus endemic to the Juan Fernandez Islands, Chile. *American Journal of Botany* **79**:962–966.

Crawford DJ, Stuessy TF, Cosner MB, Haines DW, Silva M. 1993a. Ribosomal and chloroplast DNA restriction site mutations and the radiation of *Robinsonia* (Asteraceae: Senecioneae) on the Juan Fernandez Islands. *Plant Systematics and Evolution* **184**: 233–239.

Crawford DJ, Stuessy TF, Rodriguez R, Rondinelli M. 1993b. Genetic diversity in *Rhaphithamnus venustus* (Verbenaceae), a species endemic to the Juan Fernandez Islands. *Bulletin of the Torrey Botanical Club* **120**:23–28.

Crawford DJ, Stuessy TF, Cosner MB, Haines DW, Wiens D, Peñaillo P. 1994. *Lactoris fernandeziana* (Lactondaceae) on the Juan Fernandez Islands: allozyme uniformity and field observations. *Conservation Biology* **8**:277–280.

Crawford DJ, Sang T, Stuessy TF, Kim S-C, Silva M. 1998. *Dendroseris* (Asteraceae: Lactuceae) and *Robinsonia* (Asteraceae: Senecioneae) on the Juan Fernandez Islands: similarities and differences in biology and phylogeny. In: Stuessy TF, Ono M, eds. *Evolution and speciation of island plants*. Cambridge: Cambridge University Press, 97–119.

Crawford DJ, Ruiz E, Stuessy TF, Tepe E, Aqueveque P, González F, Jensen RJ, Anderson GJ, Bernardello G, Baeza CM, Swenson U, Silva M. 2001a. Allozyme diversity in endemic flowering plant species of the Juan Fernandez Archipelago, Chile: ecological and historical factors with implications for conservation. *American Journal of Botany* **88**:2195–2203.

Crawford DJ, Tago-Nakazawa M, Stuessy TF, Anderson GJ, Bernardello G, Ruiz E, Jensen RJ, Baeza C, Wolfe AD, Silva M. 2001b. Intersimple sequence repeat (ISSR) variation in *Lactoris fernandeziana* (Lactoridaceae), a rare endemic of the Juan Fernández Archipelago, Chile. *Plant Species Biology* **16**:185–192.

Cuevas JG, van Leersum G. 2001. Project "Conservation, restoration, and development of the Juan Fernández Islands, Chile". *Revista Chilena de Historia Natural* **74**:899–910.

Cuevas JG, Marticorena A, Cavieres LA. 2004. New additions to the introduced flora of the Juan Fernández Islands: origin, distribution, life history traits, and potential of invasion. *Revista Chilena de Historia Natural* **77**:523–538.

Danton P, Perrier C. 2005. Note sur la disparition d'une espèce emblématique, *Robinsonia berteroi* (DC.) Sanders, Stuessy & Martic. (Asteraceae) dans l'ile Robinson Crusoe, archipel Juan Fernández (Chili). *Journal de Botanique de la Société Botanique de France* **31**:1–6.

Danton P, Perrier C, Martinez Reyes G. 2006. Nouveau catalogue de la flore vasculaire de l'archipel Juan Fernández (Chili). *Nuevo catálogo de la flora vascular del Archipiélago Juan Fernández (Chile)*. *Acta Botanica Gallica* **153**:399–587.

Darwin C. 1842. *The structure and distribution of coral reefs*. London: Smith, Elder and Co.

Dirnböck T, Greimler J, López P, Stuessy TF. 2003. Predicting future threats to the native vegetation of Robinson Crusoe Island, Juan Fernandez Archipelago, Chile. *Conservation Biology* **17**:1650–1659.

Ehrich D. 2006. AFLPdat: a collection of R functions for convenient handling of AFLP data. *Molecular Ecology Notes* **6**:603–604.

Eliasson U. 1974. Studies in Galapágos Plants XIV. The genus *Scalesia* Arn. *Opera Botanica* **36**:1–117.

Esselman EJ, Crawford DJ, Brauner S, Stuessy TF, Anderson GJ, Silva M. 2000. RAPD marker diversity within and divergence among species of *Dendroseris* (Asteraceae: Lactuceae). *American Journal of Botany* **87**:591–596.

Excoffier L, Laval G, Schneider S. 2005. Arlequin (version 3.0): an integrated software package for population genetics data analysis. *Evolutionary Bioinformatics Online* **1**:47–50.

Frankham R. 1997. Do island populations have less genetic variation than mainland populations? *Heredity* **78**:311–327.

Givnish TJ, Millam KC, Mast AR, Paterson TB, Theim TJ, Hipp AL, Henss JM, Smith JF, Wood KR, Sytsma KJ. 2009. Origin, adaptive radiation and diversification of the Hawaiian lobeliads (Asterales: Campanulaceae). *Proceedings of the Royal Society B Biological Sciences* **276**:407–416.

Gleiser G, Verdú M, Segarra-Moragues JG, González-Martínez SC, Pannell JR. 2008. Disassortative mating, sexual specialization, and the evolution of gender dimorphism in heterodichogamous *Acer opalus*. *Evolution* **62**:1676–1688.

Greimler J, Stuessy TF, Swenson U, Baeza CM, Matthei O. 2002. Plant invasions on an oceanic archipelago. *Biological Invasions* **4**:73–85.

Huson DH, Bryant D. 2006. Application of phylogenetic networks in evolutionary studies. *Molecular Biology and Evolution* **23**:254–267.

Jensen RJ, Schwoyer M, Crawford DJ, Stuessy TF, Anderson GJ, Baeza CM, Silva M, Ruiz E. 2002. Patterns of morphological and genetic variation among populations of *Myrceugenia fernandeziana* (Myrtaceae) on Masatierra Island: implications for conservation. *Systematic Botany* **27**:534–547.

Jorgensen TH, Olesen JM. 2001. Adaptive radiation of island plants: evidence from *Aeonium* (Crassulaceae) of the Canary Islands. *Perspectives in Plant Ecology, Evolution and Systematics* **4**:29–42.

Kikuchi S, Shibata M, Tanaka H, Yoshimaru H, Niiyama K. 2009. Analysis of the disassortative mating pattern in a heterodichogamous plant, *Acer mono* Maxim. using microsatellite markers. *Plant Ecology* **204**:43–54.

Kim YK. 1985. Petrology of Ulreung volcanic island, Korea—Part 1. Geology. *Journal of the Japanese Association of Mineralogists, Petrologists and Economic Geologists* **80**:128–135.

Landrum L. 1981a. A monograph of the genus *Myrceugenia* (Myrtaceae). Bronx, NY: New York Botanical Garden Press.

Landrum LR. 1981b. The phylogeny and geography of *Myrceugenia* (Myrtaceae). *Brittonia* **33**:105–129.

Langella O. 1999. *Populations, 1.2.30*. http://www.bioinformatics.org/~tryphon/populations/ (15 October 2014).

Liu H-Y. 1989. *Systematics of Aeonium (Crassulaceae)*. Special Publications of the National Museum of Natural Science (Taichung), No. 3, 1–102.

López-Sepúlveda P, Tremetsberger K, Ortiz MA, Baeza CM, Peñailillo P, Stuessy TF. 2013a. Radiation of the *Hypochaeris apargioides* complex (Asteraceae: Cichorieae) of southern South America. *Taxon* **62**:550–564.

López-Sepúlveda P, Takayama K, Greimler J, Peñailillo P, Crawford DJ, Baeza M, Ruiz E, Kohl G, Tremetsberger K, Gatica A, Letelier L, Novoa P, Novak J, Stuessy TF. 2013b. Genetic variation (AFLPs and nuclear microsatellites) in two anagenetically derived endemic species of *Myrceugenia* (Myrtaceae) on the Juan Fernández Islands, Chile. *American Journal of Botany* **100**:722–734.

López-Sepúlveda P, Takayama K, Greimler J, Crawford DJ, Peñailillo P, Baeza M, Ruiz E, Kohl G, Tremetsberger K, Gatica A, Letelier L, Novoa P, Novak J, Stuessy TF. 2014. Progressive migration and anagenesis in *Drimys confertifolia* of the Juan Fernández Archipelago, Chile. *Journal of Plant Research* **128**:73–90.

López-Sepúlveda P, Takayama K, Crawford DJ, Greimler J, Peñailillo P, Baeza M, Ruiz E, Kohl G, Tremetsberger K, Gatica A, Letelier L, Novoa P, Novak J, Stuessy TF. 2015. Speciation and biogeography of *Erigeron* (Asteraceae) endemic to the Juan Fernández Archipelago, Chile, based on AFLPs and SSRs. *Systematic Botany*, **40**: doi: 10.1600/036364415X689311.

Lowe A, Harris S, Ashton P. 2004. *Ecological genetics: design, analysis, and application*. Oxford: Blackwell.

Marticorena C, Stuessy TF, Baeza M. 1998. Catálogo de la flora vascular del Archipiélago de Juan Fernández, Chile. *Gayana Botánica* **55**:187–211.

Matthei O, Marticorena C, Stuessy TF. 1993. La flora adventicia del archipiélago de Juan Fernández. *Gayana Botánica* **50**:69–102.

Medina JT. 1974. *El piloto Juan Fernández, descubridor de las islas que llevan su nombre, y Juan Jufre, armador de la expedición que hizo en busca de otras en el Mar del Sur*. Santiago: Editora Nacional Gabriela Mistral.

Moldenke HN. 1937. A monograph of the genus *Rhaphithamnus*. *Repertorium Novarum Specierum Regni Vegetabilis* **42**:62–82.

Murillo-Aldana J, Ruiz E. 2011. Revalidación de *Nothomyrcia* (Myrtaceae), un género endémico del Archipiélago de Juan Fernández. *Gayana Botánica* **68**:129–134.

Murillo-Aldana J, Ruiz-P E, Landrum LR, Stuessy TF, Barfuss MHJ. 2012. Phylogenetic relationships in *Myrceugenia* (Myrtaceae) based on plastid and nuclear DNA sequences. *Molecular Phylogenetics and Evolution* **62**:764–776.

Nei M, Tajima F, Tateno Y. 1983. Accuracy of estimated phylogenetic trees from molecular data II. Gene frequency data. *Journal of Molecular Evolution* **19**:153–170.

Pacheco P, Crawford DJ, Stuessy TF, Silva OM. 1991. Flavonoid evolution in *Dendroseris* (Compositae, Lactuceae) from the Juan Fernandez Islands, Chile. *American Journal of Botany* **78**:534–543.

Peakall R, Smouse PE. 2006. GENALEX 6: genetic analysis in Excel. Population genetic software for teaching and research. *Molecular Ecology Notes* **6**:288–295.

Perugganan MD, Remington DL, Robichaux RH. 2003. Molecular evolution of regulatory genes in the silversword alliance. In: Carlquist S, Baldwin BG, Carr GD, eds. *Tarweeds & silverswords: evolution of the Madiinae (Asteraceae)*. St. Louis: Missouri Botanical Garden Press, 171–182.

Pfosser M, Jakubowsky G, Schlüter PM, Fer T, Kato H, Stuessy TF, Sun B-Y. 2005. Evolution of *Dystaenia takesimana* (Apiaceae), endemic to Ullung Island, Korea. *Plant Systematics and Evolution* **256**:159–170.

Raymond M, Rousset F. 1995. GENEPOP (version 1.2): population genetics software for exact test and ecumenicism. *Journal of Heredity* **86**:248–249.

R Core Team. 2013. *A language and environment for statistical computing*. Vienna: R Foundation for Statistical Computing.

Rensch B. 1959. *Evolution above the species level*. New York: Columbia University Press.

Ricci M. 2006. Conservation *status* and *ex situ* cultivation efforts of endemic flora of the Juan Fernández Archipelago. *Biodiversity and Conservation* **15**:3111–3130.

Ruiz E, Crawford DJ, Stuessy TF, González F, Samuel R, Becerra J, Silva M. 2004. Phylogenetic relationships and genetic divergence among endemic species of *Berberis, Gunnera, Myrceugenia* and *Sophora* of the Juan Fernández Islands (Chile) and their continental progenitors based on isozymes and nrITS sequences. *Taxon* **53**:321–332.

Rundell RJ, Price TD. 2009. Adaptive radiation, nonadaptive radiation, ecological speciation and nonecological speciation. *Trends in Ecology and Evolution* **24**:394–399.

Sanders RW, Stuessy TF, Marticorena C, Silva M. 1987. Phytogeography and evolution of *Dendroseris* and *Robinsonia*, tree-Compositae of the Juan Fernandez Islands. *Opera Botanica* **92**:195–215.

Sang T, Crawford DJ, Kim S-C, Stuessy TF. 1994. Radiation of the endemic genus *Dendroseris* (Asteraceae) on the Juan Fernandez Islands: evidence from sequences of the ITS regions of nuclear ribosomal DNA. *American Journal of Botany* **81**:1494–1501.

Sang T, Crawford DJ, Stuessy TF, Silva M. 1995. ITS sequences and the phylogeny of the genus *Robinsonia* (Asteraceae). *Systematic Botany* **20**:55–64.

Schluter D. 2001. Ecology and the origin of species. *Trends in Ecology and Evolution* **16**:372–380.

Schlüter PM, Harris SA. 2006. Analysis of multilocus fingerprinting data sets containing missing data. *Molecular Ecology Notes* **6**:569–572.

Skottsberg C. 1922. The phanerogams of the Juan Fernandez Islands. In: Skottsberg C, ed. *The natural history of Juan Fernandez and Easter Island*, Vol. 2. Uppsala: Almqvist & Wiksells, 95–240.

Solbrig OT. 1962. The South American species of *Erigeron*. *Contributions from the Gray Herbarium of Harvard University* **191**:3–79.

Stuessy TF. 1995. Juan Fernández Islands. In: Davis SD, Heywood VH, Hamilton AC, eds. *Centres of plant diversity: a guide and strategy of their conservation*. Cambridge: IUCN Publications Unit, 565–568.

Stuessy TF. 2007. Evolution of specific and genetic diversity during ontogeny of island floras: the importance of understanding process for interpreting island biogeographic patterns. In: Ebach MC, Tangney RS, eds. *Biogeography in a changing world*. Boca Raton: CRC Press, 117–133.

Stuessy TF, Foland KA, Sutter JF, Sanders RW, Silva M. 1984. Botanical and geological significance of potassium-argon dates from the Juan Fernandez Islands. *Science* **225**:49–51.

Stuessy TF, Crawford DJ, Marticorena C. 1990. Patterns of phylogeny in the endemic vascular flora of the Juan Fernandez Islands, Chile. *Systematic Botany* **15**:338–346.

Stuessy TF, Swenson U, Marticorena C, Mathei O, Crawford DJ. 1997. Loss of plant diversity and extinction on Robinson Crusoe Island, Chile. In: Peng C-I, ed. *Rare, threatened, and endangered floras of Asia and the Pacific Rim*. Taipei: Academia Sinica (Monograph Series No. 16), 147–257.

Stuessy TF, Crawford DJ, Marticorena C, Rodriguez R. 1998. Island biogeography of angiosperms of the Juan Fernandez archipelago. In: Stuessy TF, Ono M, eds. *Evolution and speciation of island plants*. Cambridge: Cambridge University Press, 121–138.

Stuessy TF, Jakubowsky G, Gómez RS, Pfosser M, Schluter PM, Fer T, Sun B-Y, Kato H. 2006. Anagenetic evolution in island plants. *Journal of Biogeography* **33**:1259–1265.

Stuessy TF, Takayama K, López-Sepúlveda P, Crawford DJ. 2014. Interpretation of patterns of genetic variation in endemic plant species of oceanic islands. *Botanical Journal of the Linnean Society* **174**:276–288.

Swenson U, Stuessy TF, Baeza M, Crawford DJ. 1997. New and historical plant introductions, and potential pests in the Juan Fernández Islands, Chile. *Pacific Science* **51**:233–253.

Takayama K, López P, König C, Kohl G, Novak J, Stuessy TF. 2011. A simple and cost-effective approach for microsatellite isolation in non-model plant species using small-scale 454 pyrosequencing. *Taxon* **60**:1442–1449.

Takayama K, Sun B-Y, Stuessy TF. 2012. Genetic consequences of anagenetic speciation in *Acer okamotoanum* (Sapindaceae) on Ullung Island, Korea. *Annals of Botany* **109**:321–330.

Takayama K, Sun B-Y, Stuessy TF. 2013a. Anagenetic speciation in Ullung Island, Korea: genetic diversity and structure in the island endemic species, *Acer takesimense* (Sapindaceae). *Journal of Plant Research* **126**:323–333.

Takayama K, López-Sepúlveda P, Kohl G, Novak J, Stuessy TF. 2013b Development of microsatellite markers in *Robinsonia* (Asteraceae) an endemic genus of the Juan Fernández Archipelago, Chile. *Conservation Genetics Resources* **5**:63–67.

Takayama K, López-Sepúlveda P, Greimler J, Crawford DJ, Peñailillo P, Baeza M, Ruiz E, Kohl G, Tremetsberger K, Gatica A, Letelier L,

Novoa P, Novak J, Stuessy TF. 2015. Relationships and genetić consequences of contrasting modes of speciation among endemic species of *Robinsonia* (Asteraceae, Senecioneae) of the Juan Fernández Archipelago, Chile, based on AFLPs and SSRs. *New Phytologist* **205**:415–428.

Tremetsberger K, Stuessy TF, Guo Y-P, Baeza CM, Weiss H, Samuel RM. 2003. Amplified fragment length polymorphism (AFLP) variation within and among populations of *Hypochaeris acaulis* (Asteraceae) of Andean southern South America. *Taxon* **52**:237–245.

Valdebenito H, Stuessy TF, Crawford DJ, Silva M. 1992. Evolution of *Erigeron* (Compositae) in the Juan Fernandez Islands, Chile. *Systematic Botany* **17**:470–480.

van Oosterhout C, Hutchinson WF, Wills DPM, Shipley P. 2004. MICRO-CHECKER: software for identifying and correcting genotyping errors in microsatellite data. *Molecular Ecology Notes* **4**:535–538.

Vargas R, Reif A, Faúndez MJ. 2011. The forests of Robinson Crusoe Island, Chile: an endemism hotspot in danger. *Bosque* **32**:155–164.

Vos P, Hogers R, Bleeker M, Reijans M, van de Lee T, Hornes M, Friters A, Pot J, Paleman J, Kuiper M, Zabeau M. 1995. AFLP: a new technique for DNA fingerprinting. *Nucleic Acids Research* **23**:4407–4414.

Wallace AR. 1881. *Island life*. London: Macmillan and Co.

Wester L. 1991. Invasions and extinctions on Masatierra (Juan Fernández Islands): a review of early historical evidence. *Journal of Historical Geography* **17**:18–34.

Whittaker RJ, Fernández-Palacios JM. 2007. *Island biogeography*, 2nd edn. Oxford: Oxford University Press.

Whittaker RJ, Triantis KA, Ladle RJ. 2008. A general dynamic theory of oceanic island biogeography. *Journal of Biogeography* **35**: 977–994.

Woodward RL. 1969. Robinson Crusoe's Island: a history of the Juan Fernandez Islands. Chapel Hill: University of North Carolina Press.

Yim Y-J, Lee E-B, Kim S-H. 1981. *Vegetation of Ulreung and Dogdo Islands*. A Report on the Scientific Survey of the Ulreung and Dogdo Islands. Seoul: The Korean Association for Conservation of Nature.

13

Long-term ecology resolves the timing, region of origin and process of establishment for a disputed alien tree

Janet M. Wilmshurst[1,2]*, Matt S. McGlone[1] and Chris S.M. Turney[3]

[1] Landcare Research, PO Box 69040, Lincoln 7640, New Zealand
[2] School of Environment, The University of Auckland, Private Bag 92019, Auckland 1142, New Zealand
[3] School of Biological, Earth and Environmental Sciences, University of New South Wales, NSW 2052, Australia

Guest Editor: Jose Maria Fernandez-Palacios

Abstract. Alien plants are a pervasive environmental problem, particularly on islands where they can rapidly transform unique indigenous ecosystems. However, often it is difficult to confidently determine whether a species is native or alien, especially if establishment occurred before historical records. This can present a management challenge: for example, should such taxa be eradicated or left alone until their region of origin and status are clarified? Here we show how combining palaeoecological and historical records can help resolve such dilemmas, using the tree daisy *Olearia lyallii* on the remote New Zealand subantarctic Auckland Islands as a case study. The status of this tree as native or introduced has remained uncertain for the 175 years since it was first discovered on the Auckland Islands, and its appropriate management is debated. Elsewhere, *O. lyallii* has a highly restricted distribution on small sea bird-rich islands within a 2° latitudinal band south of mainland New Zealand. Analysis of palaeoecological and historical records from the Auckland Islands suggest that *O. lyallii* established there c. 1807 when these islands were first exploited by European sealers. Establishment was facilitated by anthropogenic burning and clearing and its subsequent spread has been slow, limited in distribution and probably human-assisted. *Olearia lyallii* has succeeded mostly in highly disturbed sites which are also nutrient enriched from nesting sea birds, seals and sea spray. This marine subsidy has fuelled the rapid growth of *O. lyallii* and allowed this tree to be competitive against the maritime communities it has replaced. Although endemic to the New Zealand region, our evidence suggests that *O. lyallii* is alien to the Auckland Islands. Although such 'native' aliens can pose unique management challenges on islands, in this instance we suggest that ongoing monitoring with no control is an appropriate management action, as *O. lyallii* appears to pose minimal risk to ecological integrity.

Keywords: Alien; Asteraceae; dispersal; facilitation; historical ecology; invasion; *Olearia lyallii*; palaeoecology; pollen; subantarctic islands.

* Corresponding author's e-mail address: wilmshurstj@landcareresearch.co.nz

Introduction

Alien plant species can pose a major threat to indigenous species, habitats and ecosystem function on islands, particularly if they become invasive (Vitousek 1988; Whittaker and Fernández-Palacios 2007). Although eradication or control is the common conservation response to invasive taxa (Simberloff et al. 2013), this requires a confident assessment that the targeted species is in fact alien. However, exactly what circumstances make a plant 'alien', let alone 'invasive' is unclear. Webb (1985) defines 'native' as a plant that has either evolved in a given place, or arrived at that place 'entirely independently of human activity'. This definition is somewhat problematical, because a plant that can disperse long distances and arrive on an island without human assistance will be classified as a 'native', whereas another plant introduced by humans to the same place will be classified as 'alien'. Both may be equivalent in terms of their ecological impact on that island. This can make it difficult to know how to manage the human-assisted movement and naturalization of plant species away from their natural biogeographic range, either between islands, or within larger islands, in an archipelago. For example, in New Zealand, there has been a tendency to regard any plant that is native within the archipelago to be native throughout, which defines the flora on the basis of political boundaries rather than biogeography. More recently, plants native to the New Zealand archipelago but growing out of their natural range are increasingly recognized as undesirable aliens that require control in conservation planning (e.g. Sawyer et al. 2003; DOC 2008).

Problems arise with regard to the practical application of the twin criteria (unaided movement and natural range) for native status in the face of climate change, and the biogeographic reality that many plants could have had a much larger 'natural' range before long-past events, such as the Last Glacial Maximum. Definitions that link alien status to anthropogenic dispersal can create conflicts for active management and global change mitigation strategies. The 'Projected Dispersal Envelope' concept of Webber and Scott (2012) argues for a definition that is based instead on the potential limit of a species, the distribution margin being determined by (i) the natural mechanisms that could move the dispersal unit the furthest distance within its native range and (ii) the time period regarded as relevant (e.g. post-glacial). Thus, if movement by natural mechanisms within the given time period is deemed impossible, then the organism is regarded as an alien within its new location. Another issue to consider is the ecological impact of the alien on its new range (Davis et al. 2011). In natural ecosystems, the problem with alien plant invasions is usually replacement, exclusion or suppression of native plants

and detrimental changes to ecosystem function. However, where the alien is closely related to the species or even genotypes in the host area (as is often the case in interachipelago invasions), hybridization or genetic pollution is seen as the major threat (Godley 1972; Petit 2004).

In many cases, the region of origin for an alien species is clear, but often there is scope for confusion (Willis and Birks 2006). Palaeoecological records are increasingly recognized as a way to help determine a species' region of origin and native/alien status with more confidence, by reconstructing the history of taxa over longer timescales than is possible through direct observations alone (Gillson et al. 2008). This approach works particularly well when fossil evidence for a species is morphologically unique, and species-specific baselines can be reconstructed with confidence (van Leeuwen et al. 2005). For example, pollen and macrofossil records have helped to resolve uncertainty around the native/alien status of numerous taxa on islands (e.g. van Leeuwen et al. 2005, 2008; Connor et al. 2012; Schofield et al. 2013).

Well-dated, long-term and high-temporal resolution reconstructions of former vegetation composition can also show how, and under what ecological and environmental conditions, a species manages to invade and establish, and can determine the subsequent speed and spatial extent of spread (Gillson et al. 2008). The entire process of establishment and expansion can be documented through to the present, and then integrated with botanical or historical observations to develop a rich temporal and spatial perspective on an invasion, providing valuable insights for management practice and policy. We use this approach here to address the controversial status of a tree daisy Olearia lyallii (Asteraceae) on the Auckland Islands, a subantarctic island group in the New Zealand archipelago (Fig. 1). This tree is endemic to the New Zealand flora, but its origin and appropriate management on the Auckland Islands remains uncertain (Campbell and Rudge 1976; Lee et al. 1991; DOC 1998). By integrating palaeoecological records with historical evidence (written and photographic) and previous ecological investigations, we establish the history of O. lyallii arrival, establishment and subsequent spread on these islands. We also address the unresolved status of O. lyallii on the Auckland Islands according to the Projected Dispersal Envelope concept of Webber and Scott (2012) and determine whether its history and ecological role suggests that it poses a threat to the ecological integrity of the Auckland Island ecosystems.

Background of O. lyallii on the Auckland Islands

The remote, uninhabited Auckland Islands (50.5°S) are afforded the highest level of protection status by the New Zealand Department of Conservation (DOC 1998).

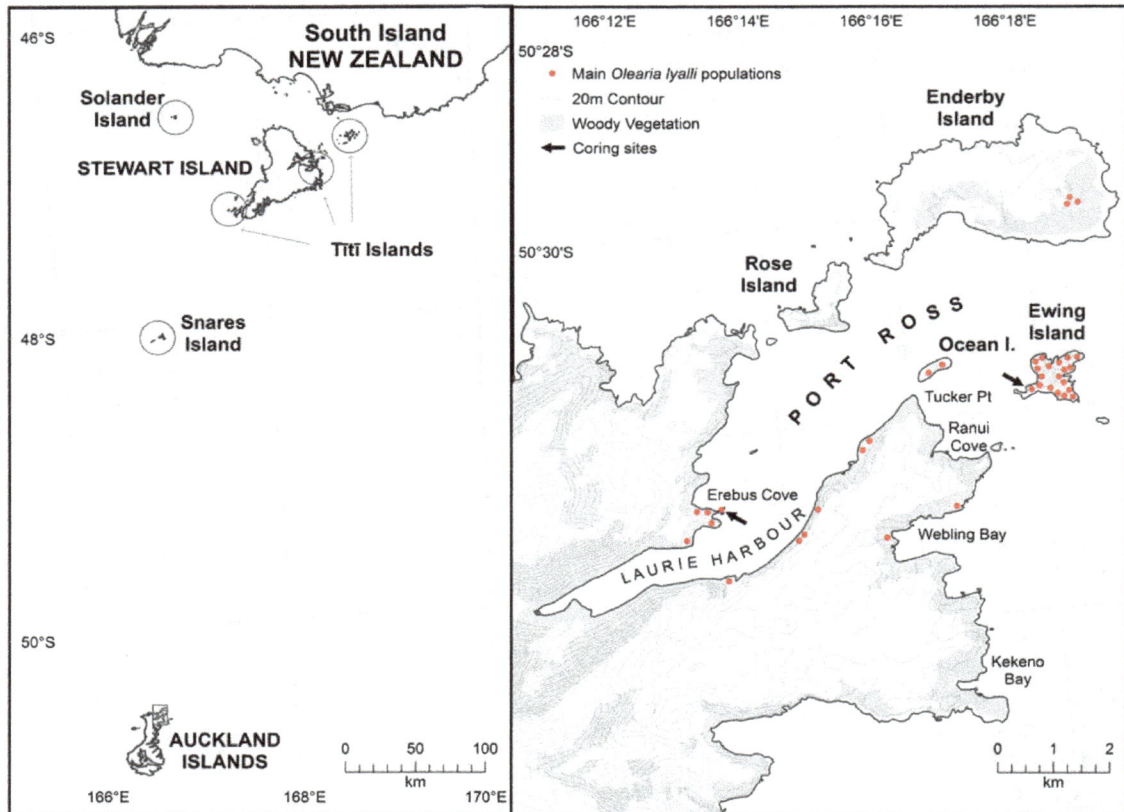

Figure 1. Map (left) showing the location of subantarctic Auckland Islands in relation to the South Island of New Zealand, and islands (circled) where *O. lyallii* currently occurs in New Zealand, and (right) the main *O. lyallii* populations on the north-eastern Auckland Islands (boxed area enlarged from map on left).

Olearia lyallii has a highly restricted distribution on small islands and adjacent coastal habitat in the northernmost Port Ross region of the Auckland Islands (Fig. 1). This includes all but the central parts of Ewing Island, where *O. lyallii* is thought to have initially established (Godley 1965), and at the short-lived Enderby Settlement site in Erebus Cove (Fig. 1) on the main Auckland Island, where canopy heights can reach 10 m. Elsewhere, there are isolated stands of a few trees or saplings scattered down the eastern side of Port Ross, Webling Bay and on Enderby and Ocean Island (Lee *et al.* 1991). A single tree has been recorded from Adams Island, the southernmost island of the Auckland Island archipelago, but has since been removed (Walls 2009). Lee *et al.* (1991), in their survey along the Laurie Harbour coastline, recorded ~50 *O. lyallii* trees at nine sites. These scattered stands can be distinguished by their distinctive pale silvery foliage in Google Earth satellite imagery [see Supporting Information—Figs S1 and S2].

Elsewhere in the New Zealand region *O. lyallii* is confined to islands in a narrow 2° latitudinal belt, including the Snares Islands ~270 km north of the Auckland Islands; coastal patches on Stewart Island; and on the Tītī and Solander Islands which are scattered around the coast of Stewart Island, ~440 km to the north of the Auckland Islands (Fig. 1). Generally, *O. lyallii* dominated forests form dense, tangled and darkly shaded stands that suppress many lower-statured and light-demanding species leaving the sub-canopy almost devoid of any other plant species (Fig. 2). The tree has thick coriaceous leaves, is fast growing, nutrient demanding, flowers in profusion, is insect-pollinated and the seeds are wind-dispersed (Lee *et al.* 1991).

The status of *O. lyallii* on the Auckland Islands has been controversial since its discovery on Ewing Island off the main Auckland Islands in 1840 (Hooker 1844). It was first thought to be a remnant of a once more widely distributed forest (Hooker 1844; Cockayne 1909), but a more recent view is that it is an alien translocated in historic times from islands to the north (Godley 1965; Campbell and Rudge 1976). Campbell and Rudge (1976) saw *O. lyallii* as a threat to the native Auckland Island vegetation and argued for its control. In contrast, Lee *et al.* (1991) suggested that *O. lyallii* was causing minimal ecological impact and that regardless of how it arrived on the Auckland Islands, the tree was well within its natural

Figure 2. Photo (taken in 2013) of *O. lyallii* forest on the south west coast of Ewing Island, showing typically dense, tangled and shaded understory, with a ground cover of bare peat and ferns *Asplenium obtusatum* and *Blechnum durum*.

dispersal limits and would have established on the islands eventually. On these grounds Lee *et al.* (1991) considered *O. lyallii* an acceptable biogeographic addition to the Auckland Island flora. The Department of Conservation's (DOC) management strategy (DOC 1998) for *O. lyallii* is cautious because of its unresolved native status. However, it is recognized that without any control, the current distribution of *O. lyallii* is likely to continue spreading slowly into, and reducing the area of, the maritime tussock–scrub–herbfield community that lies between the shore and the *Metrosideros* forest, particularly on the leeward side of the island (Lee *et al.* 1991). Department of Conservation's management strategy (DOC 1998) is for minimal control, but accompanied by monitoring of its distribution.

The Auckland Islands are well-suited to palaeoecological research. The islands are completely covered with thick (up to 12 m) organic peat deposits which have developed under a regime of high rainfall (2000 mm year^{-1}) and a cool, cloudy and humid climate (McGlone 2002). Pollen and spores are well-preserved in the peats and allow detailed high-resolution vegetation reconstructions (McGlone *et al.* 2000; McGlone 2002), and current vegetation communities and species well-characterized by the modern pollen spectra (McGlone and Moar 1997).

The Auckland Islands are also well-documented from an historical point of view, with published observations of *O. lyallii* from 1840 onwards providing a chronological description of its habitat, stature and distribution (Hooker 1844; Cockayne 1909; Godley 1965; Campbell and Rudge 1976; Lee *et al.* 1991). The Auckland Islands were initially discovered by Polynesian voyagers in the 13th century, which is evidenced by earth ovens, shell and bone middens, stone flakes and scrapers and charcoal preserved

in the sand dunes on Enderby Island (Fig. 1) (Anderson 2005). However, these early visitors did not settle on the islands permanently or leave any trace of their presence in the palaeo-vegetation or charcoal records (McGlone *et al.* 2000; Anderson 2005). Rediscovery of the islands by Europeans in 1806 marked the beginning of a short but intense period of disturbance and exploitation including: sealing and whaling from 1807; burning and small clearances, especially at Erebus Cove (Fig. 1) for the Enderby Settlement of 300 people between 1850 and 1852 and subsequent sheep grazing between 1874 and 1877 (Dingwall 2009); six major shipwrecks between 1833 and 1907 (Egerton *et al.* 2009); the introduction of alien mammals (pigs, cattle, rabbits, cats, mice, sheep, dogs and goats—of which only cats, pigs and mice now remain on the main Auckland Island); and the introduction of 37 plant species most of which are short-statured herbs and grasses of little threat to ecosystems or ecological processes (DOC 1998). The last ecological investigation of *O. lyallii* distribution and rate of spread was carried out in 1982, and predictions made of its likely trajectory (Lee *et al.* 1991).

The flora of the Auckland Islands is well-described (Cockayne 1909; Johnson and Campbell 1975). In the Port Ross region (Fig. 1), *Metrosideros umbellata* (Myrtaceae) forms a low forest with canopy heights ranging from 6 to 14 m where it is not exposed to strong winds or poor drainage **[see Supporting Information—Fig. S2]**. The small trees *Raukaua simplex*, *Myrsine divaricata*, *Dracophyllum longifolium* and *Coprosma foetidissima* are subdominant throughout the *Metrosideros* forest, reaching the canopy in tree-fall gaps, clearances or slips. They also occur on exposed coastal areas or in the upper forest-grassland ecotone. On coasts exposed to strong wind and salt spray, a maritime community of shrubland-grassland forms, including woody species: *Veronica elliptica*, *D. longifolium* and *Coprosma* spp.; graminoids: *Poa litorosa*, *P. foliosa* and *Carex appressa*; and ferns: *Polystichum vestitum* and *Asplenium obtusatum*. Where introduced pigs cannot get access, large-leaved forbs such as *Stilbocarpa polaris* and *Anistome latifolia* are prominent. *Metrosideros* forest is stunted or absent from the windward areas of Enderby Island and confined to the centre of Rose Island, Ocean Island and Ewing Island (Fig. 1).

Methods

Coring

We collected three peat cores from the Port Ross area, one from Ewing Island and two from Erebus Cove [**see Supporting Information—Figs S1 and S2**]. We used a hand-operated D-section corer to collect the Ewing Island core in 2013, which was taken from the southern end of the island to a maximum depth of sediment at 1.65 m (50°31′50.17″S; 166°17′50.43″E). The core site (Fig. 1) was under a mature monotypic coastal forest of *O. lyallii*, ~2 m above sea level and ~10 m from one of the few protected boat landings on the island.

At Erebus Cove, two short peat cores were collected in 2008 by digging pits and pushing a half-drainpipe into the wall of the peat sections. One of these Erebus Cove cores, labelled 'coastal Erebus Cove' (51 cm deep), was collected ~5 m from the shoreline under a forest co-dominated by mature *O. lyallii* and *M. umbellata* trees ~2 m above sea level (50°32′45.76″S; 166°12′55.22″E). This coring site is adjacent to the Enderby Settlement flagstaff (later replaced with a signal mast), which can be seen in many early historic photos and paintings of Erebus Cove (e.g. Figs 3 and 4).

The second Erebus Cove core, labelled 'inland Erebus Cove' (53 cm deep), was taken from under *M. umbellata* forest canopy with a *Dracophyllum*, *R. simplex* and *Coprosma* understory. The site was ~500 m inland from the coastal *O. lyallii* site (50°32′48.94″S; 166°12′46.09″E). The approximate locations of both Erebus Cove cores are marked on a historic painting from 1850 of the Enderby Settlement (Fig. 3).

All cores comprised coarse, fibrous, poorly humified, highly organic, red-brown peat. The cores were wrapped in the field, and sub-sampled in a clean laboratory environment. We sampled for pollen and charcoal to a depth of 100 cm in the Ewing Island core, and to the base of the Erebus Cove cores.

Microscopic pollen and charcoal analyses

We used standard treatments of highly organic peats (KOH, acetolysis, and filtering through a 100 μm mesh

Figure 3. Painting by Charles Enderby, 1850–52, showing the clearing at Erebus Cove, the extent of the Enderby Settlement building, and the approximate locations of our coastal (c) and inland (i) core sites at Erebus Cove. The settlement flagstaff (later replaced with a signal mast) can be seen to the right of the red flowering *Metrosideros* tree and also marks the location of the coastal core site. The distinctive outcrop of Mt Eden can be seen on the hills to the south in the background. McNulty, Dorothy (Mrs), fl 1961. [Enderby, Charles] 1797–1876. Attributed works. :[Port Ross, Auckland Islands, Between 1850 and 1852?]. Ref: A-093-008. Alexander Turnbull Library, Wellington, New Zealand. http://natlib.govt.nz/records/23243247.

sieve) to prepare microscopic pollen slides (Moore *et al.* 1991). We counted pollen and spores on each slide until we had recorded at least 250 grains from terrestrial plants (the pollen sum) from which percentages were calculated. We have used the recommended nomenclature for New Zealand pollen taxonomic groups (Moar *et al.* 2011). Statistical differences in composition between pollen zones were estimated with a non-parametric permutational multivariate analysis of variance (PERMANOVA) (Anderson 2001) using adonis in the vegan library with default settings, Bray's distance measure, and 9999 permutations. We log-transformed the pollen data while preserving zero values following (McCune *et al.* 2002). Results were compared with those using analyses of multivariate abundance using the function many.lm in the R library mvabund (Wang *et al.* 2012).

For the Ewing Island core we reconstructed local fire history following standard charcoal-analysis procedures (Whitlock and Larsen 2001), counting all charcoal particles present in a 1 mL sample that were retained on nested sieves of 125 and 250-μm mesh size. For the Erebus Cove cores, we counted microscopic charcoal particles on the pollen slides (Clark 1982), which we expressed as a percentage of the total number of pollen grains counted. Although we have used two different techniques to record charcoal presence in the Ewing

Figure 4. Historic photograph of Erebus Cove taken by G. Wolfram in 1874, showing the former site (abandoned and dismantled) of the Enderby Settlement; the signal mast (the same one shown in Fig. 3) that replaced the Enderby Settlement flagstaff; and the tussock grasses which were grazed by Monckton's sheep between 1874 and 1877. An abundant patch of herbaceous *Acaena* can be seen in the foreground, and the Mt Eden outcrop on the hills in the background (between the supporting wires on the left). Photo one of many taken by G. Wolfram in 1874, courtesy of State Library of Victoria, Melbourne (H86.2/9).

Island and Erebus Cove cores, our work from subantarctic Campbell Island has shown that microscopic and macroscopic charcoal records are highly correlated (McGlone *et al.* 2007).

Pollen identification

Pollen analysis of surface samples taken under different vegetation types on the Auckland Island show that broad communities can be distinguished by characteristic pollen taxa (McGlone and Moar 1997). *Olearia lyallii* is closely related to three Asteraceae herb species in the *Pleurophyllum* genera found on the Auckland Islands. While *O. lyallii* pollen can easily be distinguished from *P. speciosum*, it cannot always be reliably distinguished from *P. hookeri* type (including *P. hookeri* and *P. criniferum*) (Moar *et al.* 2011) **[see Supporting**

Information—Fig. S3]. However, on the Auckland Islands *P. hookeri* is rarely found at sea level, is only common above 450 m in mountain tundra communities where is rarely exceeds 5 % of the pollen sum; and *P. criniferum*, although found in maritime communities, rarely makes up >0.5 % of the pollen sum. We use these modern abundances as a guide to provide a level of confidence on the *O. lyallii* pollen curves, placing a 5 % reference baseline on our pollen diagrams. None of the *Pleurophyllum* species occurs under a forest canopy, and the palatable *P. criniferum* is scarce in the presence of pigs, which are common at Erebus Cove. In contrast to *Pleurophyllum* pollen representation, surface samples taken from under an *O. lyallii*-dominant forest on the Snares Island and Ewing Island show that *O. lyallii* pollen makes up 50 and 80 % of the pollen sum, respectively **[see Supporting Information—Table S1]**. The greater number of flowers and therefore pollen produced by *O. lyallii* compared with herbaceous *Pleurophylum* spp. per unit area sampled results in significantly higher pollen percentages under an *O. lyallii* canopy.

Radiocarbon dating

Peat samples were taken from the cores (1 cm vertical thickness) and submitted for Accelerator Mass Spectrometry (AMS) radiocarbon dating at the Waikato Radiocarbon and Beta Analytic Dating Laboratories (Table 1), with eight dates from the Ewing Island core, and two from each of the Erebus Cove cores. Radiocarbon ages were calibrated using OxCal (Ramsey 2008) using the SHCal13 calibration dataset (Hogg *et al.* 2013). Modern radiocarbon ages (i.e. post 1950 AD) were calibrated using Calibomb (http://calib.qub.ac.uk/ CALIBomb (accessed 2015); using SHCAL 13 and SHZ1-2 bomb extension zone options). We calculated an age-depth model for the Ewing Island core using the P_sequence option in OxCal **[see Supporting Information—Table S2 and Fig. S4]**. Using Bayes theorem, the algorithms employed possible solutions with a probability that is the product of the prior and likelihood probabilities. The posterior probability densities quantify the most likely age distributions. The OxCal outlier model (A_{model}= 98.9; $A_{overall}$= 99.5) identified one date (BETA-395476) as an outlier that was removed from the model. All calibrated ages are reported here as calendar (cal) years AD (Table 1). We estimated the time for the first appearance of *O. lyallii* pollen in the coastal Erebus Cove core using linear interpolations between the two calibrated dates from this core **[see Supporting Information—Table S3]**.

For the historical ecology, we examined published accounts of various botanical excursions to the islands (including: Hooker 1844; Chapman 1891; Cockayne 1903, 1905, 1909; Godley 1965; Campbell and Rudge 1976;

Table 1. Radiocarbon dates from Ewing Island and Erebus Cove peat cores, Auckland Islands. Calibrations based on Southern Hemisphere Calibration Curve (SHCAL13) from Hogg *et al.* (2013). **Identified as an outlier in age-depth model **[see Supporting Information—Fig. S4]** and *modern dates on Calibomb (http://calib.qub.ac.uk/CALIBomb).

Core site (and laboratory code)	^{14}C Lab code	Depth (cm)	Conventional C^{14} age	Dated material	AD calibrated years 1 sigma calibration (with relative area) and most likeliest age with probability highlighted
Ewing Island south, Coastal *Olearia* (site X13/84)	BETA-395475	10	116.5 ± 0.3	Peat	1959 (0.06)
					1960 (0.02)
					1963 (0.002)
					1988 (0.08)
					1989–91 (0.7)
					1991 (0.09)
					1992 (0.04)
	BETA-400420	25	130 ± 30	Plant remains	1705–21 (0.13)
					1810–37 (0.24)
					1845–66 (0.16)
					1879–1931 (0.46)
					1939–42 (0.01)
	**BETA-395476	33	670 ± 30	Peat	Not calibrated (age inversion)
	BETA-400421	45	290 ± 30	Plant remains	1518–38 (0.16)
					1626–68 (0.81)
					1788–92 (0.03)
	BETA-395477	75	750 ± 30	Peat	**1274–1302 (0.82)**
					1365–75 (0.18)
	BETA-400422	80	720 ± 30	Plant remains	**1286–1312 (0.56)**
					1359–80 (0.45)
	BETA-400423	148	3790 ± 30	Peat	**BC 2205–2129 (0.73)**
					2087–48 (0.27)
	Wk-38432	165	8768 ± 27	Peat	**BC 7789–7648 (1)**
Auckland Is, Inland Erebus Cove *Metrosideros* (site X08/22)	*Wk-31424	5	100.3 ± 0.4	Peat	1955 (0.29)
					1955–56 (0.70)
					1956 (0.009)
	Wk-31425	50	275 ± 27	Peat	**1635–70 (0.84)**
					1749–52 (0.03)
					1784–94 (0.13)
Auckland Is, Coastal Erebus Cove, *Olearia* (site X08/23)	*Wk-31426	5	116.50 ± 0.4	Peat	1958–59 (0.1)
					1995–95 (0.2)
					1996–98 (0.7)
	Wk-31427	50	31 ± 27	Peat	1890–1910 (0.36)
					1815–30 (0.33)

Smith 2002) and examined photos and paintings of the Port Ross area of Auckland Island from electronic archives, including the Museum of New Zealand Te Papa Tongarewa, Alexander Turnbull Library (New Zealand) and State Library of Victoria (Australia).

Results

For ease of interpretation, we have divided the Ewing Island and coastal Erebus Cove profiles into two zones: the uninvaded zone and *Olearia* zone, the latter defined by the first presence of *O. lyallii* pollen. The inland Erebus cove pollen profile is divided into the pre- and post-Enderby Settlement zones, the latter defined by the decline and subsequent regeneration of *Metrosideros* forest.

Ewing Island core

The radiocarbon dates (Table 1) and age-depth model for this core [see Supporting Information—Table S2 and Fig. S4] indicate peat accumulation began on Ewing Island c. 10 000 cal year before the present. The base of our pollen record (Fig. 5) starts at c. 1600 cal year before the present (c. 400 cal year AD), at which time the site was covered with a coastal maritime community dominated by the shrub *V. elliptica*, with grasses, macrophyllous forbs *S. polaris* and *A. latifolia*, sedges, and abundant ground ferns. The low levels of *Metrosideros* pollen throughout the core suggests that this forest was limited to the more protected interior of the island behind the coastal belt of maritime vegetation. Low counts of charcoal (<5 fragments per 1 mL of peat) are first recorded in the peat profile at the top of the uninvaded zone, just prior to c. 1800 cal year AD but probably reflect reworking as a consequence of site disturbance during the European era (also supported by age inversion at 33 cm—Table 1).

At the base of the *Olearia* zone at 32 cm, further charcoal particles, and the first trace of *O. lyallii* occur at an estimated age of c. 1800 cal year AD [see Supporting Information—Table S2]. At the same time, there is an increase of the herbs *Callitriche antarctica* and *Urtica australis*, and a decline of grasses, *S. polaris* and *A. latifolia*. Pollen of *O. lyallii* remains below 5 % of the pollen sum until 1870 cal year AD after which time it increases towards the top of the core, while ground ferns, herbs and grasses decline. As the *O. lyallii* canopy matured and closed creating a dense shade, it suppressed many lower-statured and light-demanding plants, leaving the sub-canopy and canopy floor almost bare. The pollen composition in the uninvaded and *Olearia* zone was significantly different ($F_{1,28} = 27.4$, $P < 0.001$). Similar, statistically strong differences in composition were detected using an analysis of multivariate abundance using the function many.lm in the R library mvabund (Wald statistic $= 26.15$, $P = 0.001$).

Erebus Cove cores

The uninvaded and pre-Enderby Settlement zones of both the coastal and inland Erebus Cove cores, respectively, record a *Metrosideros*-dominated forest (constituting ~40–50 % of the pollen sum). The coastal site (Fig. 6) has a greater representation of ground ferns, *S. polaris*, *Acaena* and grass, reflecting its more open canopy. In contrast, the inland site (Fig. 7) has a substantial representation of the small trees *R. simplex*, *D. longifolium* and *M. divaricata*.

Both the coastal and inland Erebus Cove pollen profiles show a rapid and marked reduction of *Metrosideros* at 45 and 36 cm, respectively, accompanied by an abundance of charcoal fragments at the coastal site. This forest decline reflects the burning and cutting of trees

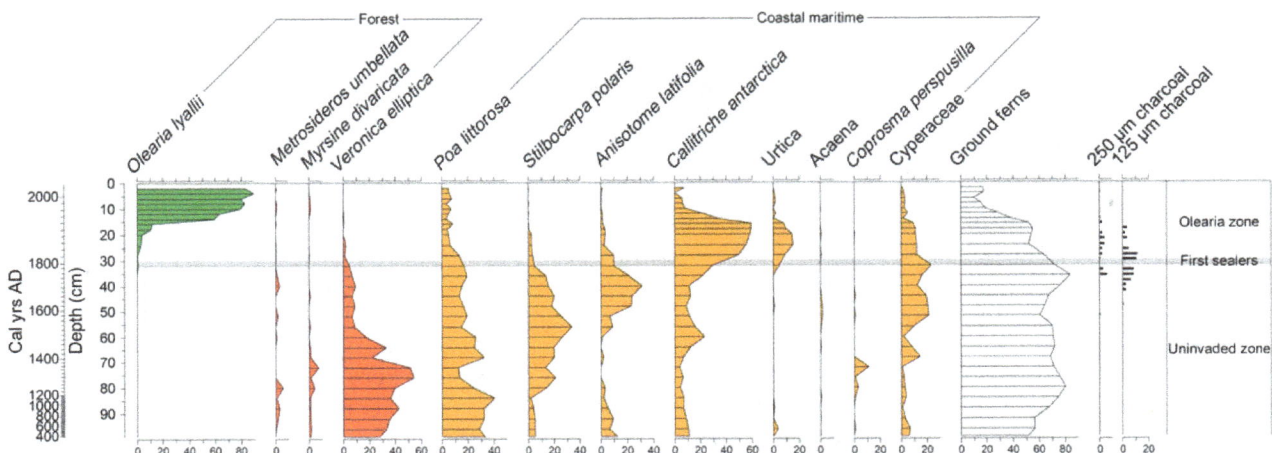

Figure 5. Summary percentage pollen record from Ewing Island, with pollen taxa plotted against depth, with the calibrated age scale in years AD shown on the secondary axis. Grey zone shows time of earliest sealing activity in the region (1807–10).

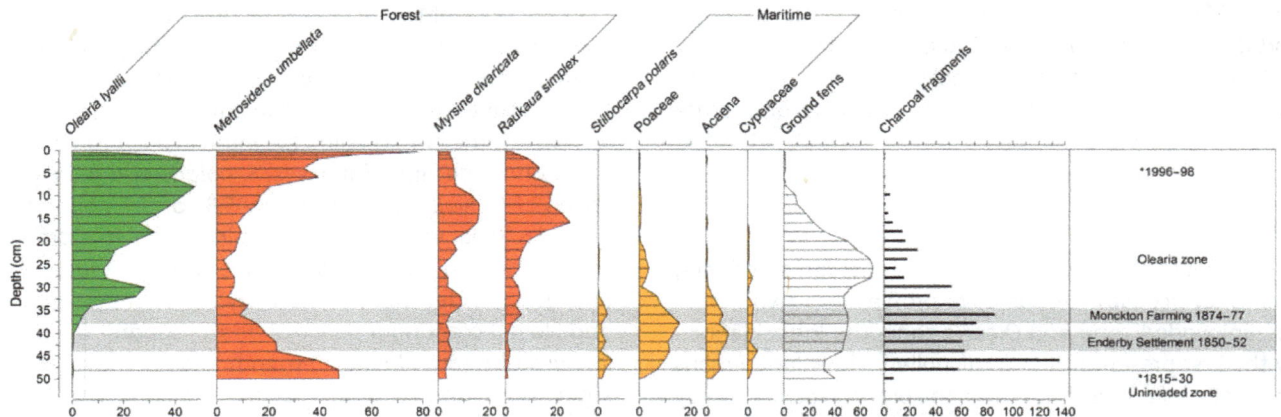

Figure 6. Summary percentage pollen record from the coastal Erebus Cove site (taken under an *O. lyallii* canopy), main Auckland Island. Grey bars show the time of the Enderby Settlement (1850–52) and Monckton Farming (1874–77) periods according to age-depth model. *Position and age of calibrated radiocarbon (cal year AD) dates from Table 1.

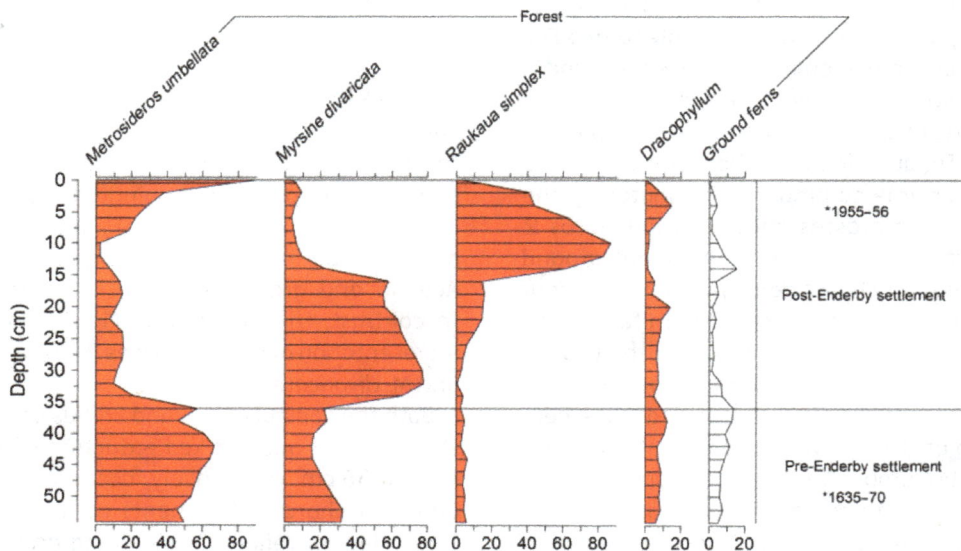

Figure 7. Summary percentage pollen record from the inland Erebus Cove site (taken under a *M. umbellata* canopy 500 m inland from the coastal Erebus Cove site) main Auckland Island. *Position and age of calibrated radiocarbon dates (cal year AD) from Table 1.

at the coastal site, and felling of trees at the inland site to make way for the Enderby Settlement in 1850, as shown in paintings and photos from this period (e.g. Figs 3 and 4).

The calibrated date from the base of the coastal Erebus Cove core provides two age ranges (1910–1890 and 1830–15 cal year AD) with equal probability distributions (Table 1). As botanists did not record *O. lyallii* trees at Erebus Cove during a visit in 1890 (Chapman 1891), we take the older of the two solutions to create an age-depth model (using linear interpolation between this date and the 1955–56 cal year AD date at 5 cm [see Supporting Information—Table S3]). This model provides an estimate for the first trace of *O. lyallii* pollen (0.5 % at 48 cm)

of 1823–37 cal year AD. *Olearia lyallii* pollen is not recorded again until 42 cm (c. 1847–60 cal year AD) after the *Metrosideros* has been burnt and cleared. After this, *O. lyallii* is consistently recorded, increasing to a peak of 47 % (Fig. 6) and later declining as *Metrosideros* begins to recover at the site.

At the inland Erebus Cove site (Fig. 7) charcoal and *O. lyallii* pollen are not recorded following the decline of *Metrosideros*. Instead, after forest clearance, a woody succession takes place from *M. divaricata* through *R. simplex* to *Metrosideros*, the latter recovering mainly through sprouting from cut stumps which are still visible in the forest today.

Pollen composition was significantly different in the pre- and post-settlement zones of the inland Erebus

Cove core ($F_{1,27} = 7.33$, $P < 0.0001$). Similar, statistically strong differences in composition were detected using an analysis of multivariate abundance using the function many.lm in the R library mvabund (Wald statistic = 116.9, $P < 0.0001$). However, as there was only one sample from the uninvaded period from the coastal Erebus Cove core, statistical comparisons of compositional variance in the uninvaded and *Olearia* zones are very weak and non-significant (PERMANOVA: $F_{1,25} = 1.47$, $P = 0.18$ and multivariate generalized linear models Wald statistic = 7.38, $P = 0.22$).

Earliest historical documentation of *O. lyallii* on the Auckland Islands

Written and photographic records from the 19th century to the mid-20th century provide a surprisingly detailed account of *O. lyallii* on the Auckland Islands (e.g. Hooker 1844; Chapman 1891; Cockayne 1903, 1905, 1909; Godley 1965; Campbell and Rudge 1976). The first record of the tree was the type specimen collected from Ewing Island by David Lyall in 1840, a botanist on the Ross Expedition. Hooker (1844:38) described the specimen as follows: '. . .a short stout trunk rises a few inches above the ground, and then sends off horizontally patent branches, which radiate as from a common centre for 10 to 12 feet on all sides, a little above the surface of the earth.' Hooker (1844) further describes *O. lyallii* as rare on the Auckland Islands, and McCormick (1884), who also landed on the island at the same time as Lyall, did not mention seeing this species. Godley (1965) remarked that these descriptions typically matched stunted plants in exposed locations, but a canopy of at least 6 m in diameter would suggest the specimen described by Hooker (1844) had been growing for some years. However, as the *O. lyallii* specimen was not flowering (despite being collected in early summer) and low-growing, Campbell and Rudge (1976) suggested on the basis of observations of *O. lyallii* growing on the Snares that this tree may have been <20 years old when first seen by Lyall in 1840.

Discussion

Timing, dispersal and origin of *O. lyallii* on Auckland Islands

The estimated ages for the first appearance of *O. lyallii* pollen in our dated pollen profiles indicate that this tree daisy established on Ewing Island ~1800, and then later at Erebus Cove c. 1823–37, exceeding >5 % of the pollen sum at both sites <60 years later. Despite the limited precision expected of radiocarbon dates from fibrous peat, this timing is consistent with historical observations (Godley 1965; Smith 2002; Prickett 2009) that strongly suggest initial establishment on Ewing Island c. 1807–10.

Sealers were active at this time within the natural range of *O. lyallii* on the Snares Island and the 36 smaller Tītī islands around Stewart Island, as well as on the northern Auckland Islands (Smith 2002). Sealing activities caused localized disturbance to the mainly coastal flora and fauna (Smith 2002), and undoubtedly increased the possibilities for seed translocation between islands. Our age estimates are also consistent with age–diameter relationships made in 1982 of one of the largest erect *O. lyallii* trees on Ewing Island located close to an old whaling boat shed. This sampled tree had a trunk diameter of 110 cm indicating establishment ~1820 (Lee et al. 1991).

It has been proposed that *O. lyallii* had the capacity to eventually disperse to the Auckland Islands through natural agencies and fill a previously vacant niche (Lee et al. 1991). *Olearia lyallii* seeds are adapted for transport by wind and can also potentially attach to feathers of ground nesting sea birds; the 4-mm long achenes have 6-mm long fluffy pappus hairs (Allan 1982) and are produced in abundance. The Auckland Islands are also only 270 km from the source islands of *O. lyallii* which also harbour large populations of nesting sea birds. Thus, this tree daisy has had numerous opportunities during the Holocene to disperse naturally to the Auckland Islands. However, our pollen records, and other pollen records from the islands (McGlone et al. 2000; McGlone 2002), show that *Metrosideros* forest, and coastal maritime communities have dominated sheltered and exposed coastal habitats, respectively, on the Auckland Islands for at least 12 000 years. *Olearia lyallii* has only managed to establish in a few scattered places in the northern Auckland Islands in the last 200 years, coincident with the earliest European exploitation of the region. We conclude from this evidence that there is a high probability that the Auckland Islands lie outside of the natural distribution range of *O. lyallii*, and require human-assisted seed dispersal and/or new niches in order to establish.

Process of initial *O. lyallii* establishment on Ewing Island

The pollen and charcoal record from Ewing Island (Fig. 5) shows that the initial establishment of *O. lyallii* on this island was into an anthropogenically disturbed habitat, not into a pristine coastal maritime community. Fire-induced changes to the coastal vegetation preceded the establishment of *O. lyallii*, suggesting facilitation by anthropogenic disturbance. Charcoal is almost absent from all subantarctic island Holocene peat records until the arrival of Europeans (McGlone et al. 2000; McGlone 2002; Wilmshurst et al. 2004; Bestic et al. 2005).

Sealers often used overland routes to access sealing spots as the seas were rough and dangerous, and New

Zealand fur seals (*Arctocephalus forsteri*) and Hooker's sea lions (*Phocarctos hookeri*) dispersed along the coast (Smith 2002). Walking through the dense and tangled vegetation of the Auckland Islands is notoriously slow and arduous, and fire was used liberally by 19th century travellers to clear the way for easier travel. For instance, officers from the 1840 Terror and Erebus expedition to the islands set fire to forest and scrub in the hills immediately above Erebus Cove. Robert McCormick from the same expedition, in his excursion from Terror Cove to Matheson Bay and around the peninsula to Deas Head, noted that there was extensive burnt grassland on the cliffs (McCormick 1884). The earliest charcoal presence in the Ewing Island core precedes the earliest shipwrecks in the Auckland Islands (from 1888 to 1907) (Egerton *et al.* 2009) and suggests that the Ewing Island coring area had been burnt by sealing gangs to ease their passage through dense coastal vegetation to reach seal haul-out sites. The pollen and charcoal record from Ewing Island also shows that during the first few decades of *O. lyallii* establishment, the herbs *U. australis* and *C. antarctica* became abundant, indicating a succession similar to that recorded on the Snares Island following dieback of *Veronica* and subsequent abandonment of penguin colonies (Hay *et al.* 2004). *Callitriche antarctica* commonly colonizes abandoned penguin colonies on the Snares Islands (Hay *et al.* 2004), and *U. australis* responds favourably to disturbance and high-light (Allan 1982). These nutrient-demanding herbs likely established on the disturbed and marine-enriched patches of bare peat on Ewing Island that were previously maintained by seals in haul-out areas along the coast. As vast numbers of seal carcasses were usually left to rot *in situ* where the animals were slaughtered on these coastal habitats (Smith 2002) their decomposing bodies may have provided a pulse of nutrient enrichment during the time of *O. lyallii* establishment. Outside of current *O. lyallii* forest patches on the Auckland Islands, *O. lyallii* seedlings are most commonly recorded on recently abandoned sea lion haul-outs with bare patches of peat (Lee *et al.* 1991).

The presence of ground nesting and burrowing sea birds on Ewing Island may also explain why *O. lyallii* has been so successful on Ewing Island compared with elsewhere on the Auckland Islands, as they also provide substantial and continuing sources of disturbance and nutrients to the Ewing Island peats. On the main Auckland Islands and other smaller islands in the Port Ross area, introduced pigs, cattle, cats (Challies 1975; Lee *et al.* 1991) and potentially mice (c.f. Cuthbert and Hilton 2004), have almost completely eliminated nesting sea bird populations (through trampling and predation), and therefore associated marine nutrient transfer. In contrast, Ewing Island has remained free of introduced mammals. The continued input of marine-derived nutrients from sea birds during the early human disturbance phase, and later recovery of seal populations, has likely promoted *O. lyallii* establishment on Ewing Island, and fuelled its rapid growth rates.

On Ewing Island, as the *O. lyallii* canopy became taller and more open as it matured, and bulky leaf litter built up thick peat deposits, these conditions would have become increasingly attractive for nesting sea birds where they could burrow and land/take-off easily (see Whitehead *et al.* 2014). The input of marine nutrients by seals and sea birds, and the dark shade and rapid growth of *O. lyallii* forests on enriched soils may allow this tree to exclude the former coastal maritime communities indefinitely. *Olearia lyallii* benefits from marine-enriched soils in its natural range on the Snares, Solander and Tītī islands (Fig. 1). These islands are wind-swept, drenched with salt spray during storms and largely covered with organic peat deposits which are extensively burrowed and disturbed by nesting sea birds and seal activity. The strong fertilizing effect of marine animals on the soils is reflected in the *O. lyallii* leaves from the Snares Islands which have yielded some of the highest leaf ^{15}N enrichment levels ever recorded for plants (Martinelli *et al.* 1999; Hawke and Newman 2007). However, away from the smaller islands, on nearby Stewart Island nesting sea bird densities are much lower and *O. lyallii* has a very limited distribution (Wilson 1987). *Olearia lyallii* distribution is almost certainly dependent on marine subsidies introduced by sea birds and seals.

Rates of *O. lyallii* spread on Ewing Island

Historical observations and the pollen record from Ewing Island suggests that it took *O. lyallii* ~80 years to shade out the tall tussock-and herbaceous maritime communities (Fig. 5). In 1840, 20 years after its establishment, *O. lyallii* was described as quite rare and stunted among the maritime tussock and scrub on Ewing Island (Hooker 1844). By 1890, substantial trees of *O. lyallii* were present along the sea shore, but large tussocks remained common (Chapman 1891). In 1907, a low *O. lyallii* forest appears to have extended over most of the island aside from the central *Metrosideros* core, but patches of tall tussocks apparently were still present (Godley 1965). By the 1960s the tree daisy had formed a coastal fringe around the island (Godley 1965). The slow spread of *O. lyallii* into the *Metrosideros* forest over the last 50 years suggests a superior competitiveness of *O. lyallii* on nutrient-enriched soils in exposed locations. Although *Metrosideros* is a long-lived tree, and can resprout or layer after damage, it is shade-intolerant and slow growing (Wardle 1971) and vulnerable to over-topping by *O. lyallii* where wind and salt exposure causes shorter-statured canopies.

Timing of *O. lyallii* establishment at Erebus Cove

Historical observations also suggest *O. lyallii* has made a slow and limited spread away from Ewing Island to other sites on the main Auckland Island and smaller islands in the Port Ross area (Campbell and Rudge 1976; Lee *et al.* 1991). The age estimate for the first trace of *O. lyallii* pollen in the dated coastal Erebus Cove pollen profile is 1823–37 cal year AD (Fig. 6), some 15–40 years after it is recorded at our site on Ewing Island. These first traces of *O. lyallii* may reflect pollen contributed from scattered individuals in the coastal communities at Erebus Cove that did not initially succeed, or possibly *P. criniferum*, which can occur in coastal communities at trace levels (McGlone and Moar 1997). The pollen type is not recorded again until 42 cm in the profile (c. 1847–60) after which it steadily increases to a peak of 47 %. However, the precision of these age estimates is limited, as they are only based on linear interpolation between two dates, the lowest of which has a spread of calibrated age ranges spanning 75 years, therefore we use the historical evidence to refine the timing of *O. lyallii* establishment at Erebus Cove.

The conjecture has been that colonists at the Enderby Settlement transplanted *O. lyallii* from Ewing Island to this site, either accidentally or as an ornamental, at some time between 1850 and 1852 (Godley 1965; Campbell and Rudge 1976). The extent of clearance for the Enderby Settlement (1850–52) can be seen in the painting of Erebus Cove in c. 1850 by Charles Enderby (Fig. 3). In 1865, Captain Musgrave visited the Enderby Settlement site at Erebus Cove and made reference to 'two trees' (Musgrave 1866) which have been interpreted by Campbell and Rudge (1976) as being *O. lyallii* specimens. However, shortly after the Enderby Settlement dismantled, there was a brief farming episode (September 1874–May 1877) by Monckton, a lease-holder at Erebus Cove (Dingwall 2009), when 56 sheep were grazed on the cleared site, scrub and grass was burnt, and grass and oats were planted (Dingwall 2009). A photograph in 1874 of the signal mast that replaced the original settlement flagstaff on Davis Point (the coastal Erebus core site) shows a landscape dominated by tall grass, with a mat of the herb *Acaena* in the foreground, but with no sign of any trees (Fig. 4). This corresponds well with the *Acaena* pollen recorded from this site, after the forest clearance but before the *O. lyallii* invasion (Fig. 6). Chapman and Kirk visited Erebus Cove in 1890, and they did not record *O. lyallii* (Chapman 1891). Only in 1907, some 55 years after the Enderby Settlement, and 30 years after the Monckton lease, there is evidence for *O. lyallii* at Davis Point near the flagstaff (Godley 1965). Cockayne (1909) mentions 'a few trees in the neighbourhood of the Port Ross depot'. If *O. lyallii* had been introduced by the Enderby Settlement colonists to Erebus Cove, these trees had made little growth. From these observations it seems more likely that *O. lyallii* established during or after the Monckton farming interval (1874–77).

Process of establishment at Erebus Cove

The pollen record from the coastal Erebus Cove site (Fig. 6) is similar in one key respect to the Ewing Island record, in that *O. lyallii* did not establish into a pristine *Metrosideros* forest but into a burned and disturbed coastal maritime community dominated by grasses and the herbs *Acaena* and *S. polaris*, and with low levels of the successional small trees *M. divaricata* and *R. simplex*. However, it differs significantly from Ewing Island, in that as the *O. lyallii* canopy matured towards the present, it became co-dominant with the recovering *Metrosideros*. As there are almost no nesting sea birds at Erebus Cove and limited seal presence, *O. lyallii* lacks its preferred enriched peat. This sheltered site favours *Metrosideros* which is now co-dominant with *O. lyallii* and, being longer-lived, will eventually over-top and replace it.

There is no evidence from the pollen record for *O. lyallii* presence at the inland Erebus Cove site following abandonment of the Enderby Settlement (Fig. 7), despite being only <500 m away from the well-established population at the coastal site. Instead, the pollen record shows a succession from *Myrsine* to *Raukaua* and back to the pre-settlement *Metrosideros* forest. Seed dispersal by wind, people or animals over such a short distance cannot have limited *O. lyallii* establishment at the inland site or indeed elsewhere in the Port Ross area. However, nutrients derived from marine aerosols drop off rapidly with distance from the shore (Meurk *et al.* 1994), and without marine nutrient subsidies and disturbance, *O. lyallii* loses its competitive advantage.

Conclusions

Despite having ample opportunity to disperse to the Auckland Islands from the small island groups to the north (Snares, Solander, Tītī and Stewart Islands), a combination of palaeoecological and historical observations suggests that *O. lyallii* is only a recent addition to the flora of the Auckland Islands. It was most likely introduced by sealers between 1807 and 1810. Under the Projected Dispersal Envelope concept of Webber and Scott (2012) regarding natural dispersal and time, it is unambiguously an alien plant on the Auckland Islands. Despite its alien status, our observations indicate that *O. lyallii* is not highly invasive, and poses little threat to the ecological integrity of the island, in agreement with Lee *et al.* (1991). Detailed palaeoecological records have shown that the establishment of *O. lyallii* on the Auckland Islands was facilitated by human disturbance; that its spread has been slow; and its distribution limited to exposed coastal habitats where peats have been enriched by sea birds, seals and salt spray. Given the limited distribution of anthropogenically disturbed and enriched habitats on the islands, *O. lyallii* is unlikely to pose a significant threat to the existing maritime habitat on the uninhabited islands, and no threat to the *Metrosideros* forest.

Climate change and the inevitable human-assisted movement of propagules across landscapes will ensure that the issue of 'native' aliens will arise repeatedly. However, palaeoecological and historical research such as presented here and elsewhere (e.g. van Leeuwen *et al.* 2008) can help conservation agencies make considered decisions regarding the management and status of such plants (Gillson *et al.* 2008). We support the dynamic and pragmatic 'monitor and see' approach for *O. lyallii* (Davis *et al.* 2011) that balances what appears to be a limited loss of ecological integrity with the high cost and low probability of successful control.

Sources of Funding

Our study was supported by Core Funding for Crown Research Institutes, from the New Zealand Ministry of Business, Innovation and Employment's Science and Innovation Group.

Contributions by the Authors

J.M.W. and M.S.M. conceived the study, collected cores, analysed the cores and wrote the paper. C.S.M.T. contributed to the writing of the paper and calculated the age-depth models.

Acknowledgements

We thank the Department of Conservation (M. Carruthers and L. Chilvers) for collecting soil cores from Erebus Cove, and for their support of the Australasian Antarctic Expedition 2013–14 when we cored Ewing Island (DOC research permit National Authorisation # 37687-FAU). We also thank B. Beaven, J. McDiarmid, V. Flett and V. Meduna for field assistance in 2013; R. Price for help with Fig. 1; K. Boot for laboratory sampling and macroscopic charcoal counts; A. Watkins and N. Bolstridge for preparing pollen slides; and S. Richardson, J. Wood and three anonymous reviewers for helpful comments on an earlier draft.

Supporting Information

The following additional information is available in the online version of this article –

Figures S1 and S2. Location and vegetation cover of Ewing Island and Erebus Cove coring sites.

Figure S3. Microphotographs of *Olearia lyallii* and *Pleurophyllum* spp. pollen grains.

Figure S4. Output graph for Ewing Island age-depth model.

Table S1. Modern pollen percentages from surface samples under *Olearia lyalli*-dominated canopies.

Table S2. Age-depth model for Ewing Island core.

Table S3. Linear interpolation between calibrated radiocarbon ages in coastal Erebus Cove core.

Literature Cited

Allan HH. 1982. *Flora of New Zealand Volume 1*. Wellington: P.D. Hasselberg, Government Printer.

Anderson A. 2005. Subpolar settlement in South Polynesia. *Antiquity* **79**:791–800.

Anderson MJ. 2001. A new method for non-parametric multivariate analysis of variance. *Austral Ecology* **26**:32–46.

Bestic KL, Duncan RP, McGlone MS, Wilmshurst JM, Meurk CD. 2005. Population age structure and recent *Dracophyllum* spread on subantarctic Campbell Island. *New Zealand Journal of Ecology* **29**:291–297.

Campbell DI, Rudge MR. 1976. A case for controlling the distribution of the tree daisy (*Olearia lyallii*) Hook.F. in its type locality, Auckland Islands. *Proceedings of the New Zealand Ecological Society* **23**:109–115.

Challies CN. 1975. Feral pigs (*Sus scrofa*) on Auckland Island: status, and effects on vegetation and nesting sea birds. *New Zealand Journal of Zoology* **2**:479–490.

Chapman FR. 1891. The outlying islands of New Zealand. *Transactions of the New Zealand Institute* **23**:491–522.

Clark RL. 1982. Point count estimation of charcoal in pollen preparations and thin sections. *Pollen et Spores* **24**:523–535.

Cockayne L. 1903. *Botanical excursion during midwinter to the southern islands of New Zealand*. New Zealand Institute.

Cockayne L. 1905. Notes on a brief botanical visit to the Poor Knights Islands. *Transactions and Proceedings of the New Zealand Institute* **38**:351–360.

Cockayne L. 1909. The ecological botany of the subantarctic islands of New Zealand. In: Chilton C, ed. *The Subantarctic Islands of New Zealand*. Christchurch: Philosophical Institute of Canterbury.

Connor SE, van Leeuwen JFN, Rittenour TM, van der Knaap WO, Ammann B, Björck S. 2012. The ecological impact of oceanic island colonization–a palaeoecological perspective from the Azores. *Journal of Biogeography* **39**:1007–1023.

Cuthbert R, Hilton G. 2004. Introduced house mice *Mus musculus*: a significant predator of threatened and endemic birds on Gough Island, South Atlantic Ocean? *Biological Conservation* **117**: 483–489.

Davis MA, Chew MK, Hobbs RJ, Lugo AE, Ewel JJ, Vermeij GJ, Brown JH,

Rosenzweig ML, Gardener MR, Carroll SP, Thompson K, Pickett STA, Stromberg JC, Tredici PD, Suding KN, Ehrenfeld JG, Grime JP, Mascaro J, Briggs JC. 2011. Don't judge species on their origins. *Nature* **474**:153–154.

Dingwall PR. 2009. Pastoral farming at the Auckland Islands. In: Dingwall PR, Jones KL, Egerton R, eds. *In care of the Southern Ocean: an archaeological and historical survey of the Auckland Islands.* Auckland: New Zealand Archaeological Association.

DOC. 1998. Management Strategy Subantarctic Islands 1998–2008, Southland Conservancy Conservation Management Planning Series Number 10.

DOC. 2008. *Abel Tasman National Park Management Plan.* Management Plan Series. Nelson: New Zealand Department of Conservation.

Egerton R, Burgess S, Petchey P, Dingwell PR. 2009. The Auckland Islands shipwreck era. In: Dingwall PR, Jones KL, Egerton R, eds. *In care of the Southern Ocean: an archaeological and historical survey of the Auckland Islands.* Auckland: New Zealand Archaeological Association.

Gillson L, Ekblom A, Willis KJ, Froyd C. 2008. Holocene palaeoinvasions: the link between pattern, process and scale in invasion ecology? *Landscape Ecology* **23**:757–769.

Godley EJ. 1965. Notes on the vegetation of the Auckland Islands. *Proceedings of the New Zealand Ecological Society* **12**:69–72.

Godley EJ. 1972. Does planting achieve its purpose? *Forest and Bird Journal* **185**:25–26.

Hawke DJ, Newman J. 2007. Carbon-13 and nitrogen-15 enrichment in coastal forest foliage from nutrient-poor and seabird-enriched sites in southern New Zealand. *New Zealand Journal of Botany* **45**:309–315.

Hay CH, Warham J, Fineran BA. 2004. The vegetation of The Snares, islands south of New Zealand, mapped and discussed. *New Zealand Journal of Botany* **42**:861–872.

Hogg AG, Hua Q, Blackwell PG, Niu M, Buck CE, Guilderson TP, Heaton TJ, Palmer JG, Reimer PJ, Reimer RW, Turney CSM, Zimmerman SRH. 2013. SHCal13 Southern Hemisphere calibration, 0–50,000 Years cal BP. *Radiocarbon* **55**:1889–1903.

Hooker JD. 1844. *The Botany of the Antarctic Voyage of H.M. Discovery Ships 'Erebus' and 'Terror' in the years 1839–1843, under the Command of Captain Sir James Clark Ross. Vol. 1 Flora Antarctica. Part 1. Botany of Lord Auckland's Group and Campbell's Island.* London: Reeve.

Johnson PN, Campbell DJ. 1975. Vascular plants of the Auckland Islands. *New Zealand Journal of Botany* **13**:665–720.

Lee WG, Wilson JB, Meurk CD, Kennedy PC. 1991. Invasion of the subantarctic Auckland Islands, New Zealand, by the asterad tree *Olearia lyallii* and its interaction with a resident myrtaceous tree *Metrosideros umbellata*. *Journal of Biogeography* **18**:493–508.

Martinelli LA, Piccolo MC, Townsend AR, Vitousek PM, Cuevas E, McDowell W, Robertson GP, Santos OC, Treseder K. 1999. Nitrogen stable isotopic composition of leaves and soil: tropical versus temperate forests. In: Townsend A, ed. *New perspectives on nitrogen cycling in the temperate and tropical Americas.* The Netherlands: Springer.

McCormick R. 1884. *Voyages of discovery in the Arctic and Antarctic seas, and round the world, Volume 1.* London: Sampson Low, Marston, Seale, and Rivington.

McCune B, Grace JB, Urban DL. 2002. *Analysis of ecological communities.* Gleneden Beach, OR: MjM Software Design.

McGlone M, Wilmshurst J, Meurk C. 2007. Climate, fire, farming and the recent vegetation history of subantarctic Campbell Island. *Earth and Environmental Science Transactions of the Royal Society of Edinburgh* **98**:71–84.

McGlone MS. 2002. The late Quaternary peat, vegetation and climate history of the Southern Oceanic Islands of New Zealand. *Quaternary Science Reviews* **21**:683–707.

McGlone MS, Moar NT. 1997. Pollen-vegetation relationships on the subantarctic Auckland Islands, New Zealand. *Review of Palaeobotany and Palynology* **96**:317–338.

McGlone MS, Wilmshurst JM, Wiser SK. 2000. Lateglacial and Holocene vegetation and climatic change on Auckland Island, subantarctic New Zealand. *The Holocene* **10**:719–728.

Meurk CD, Foggo MN, Thomson BM, Bathurst ETJ, Crompton MB. 1994. Ion-rich precipitation and vegetation pattern on subantarctic Campbell Island. *Arctic and Alpine Research* **26**:281–289.

Moar NT, Wilmshurst JM, McGlone MS. 2011. Standardizing names applied to pollen and spores in New Zealand Quaternary palynology. *New Zealand Journal of Botany* **49**:201–229.

Moore PD, Webb JA, Collinson ME. 1991. *Pollen analysis.* Oxford, UK: Blackwell Scientific.

Musgrave T. 1866. *Castaway on the Auckland Islands.* London: Lockwood and Co.

Petit RJ. 2004. Biological invasions at the gene level. *Diversity and Distributions* **10**:159–165.

Prickett N. 2009. Sealing in the Auckland Islands. In: Dingwall PR, Jones KL, Egerton R, eds. *In care of the Southern Ocean: an archaeological and historical survey of the Auckland Islands.* Auckland: New Zealand Archaeological Association.

Ramsey CB. 2008. Radiocarbon dating: revolutions in understanding. *Archaeometry* **50**:249–275.

Sawyer J, McFadgen B, Hughes P. 2003. Karaka (Corynocarpus laevigatus J.R. et G. Forst.) in Wellington Conservancy (excluding Chatham Islands). DOC Internal Science Series. Wellington: New Zealand Department of Conservation.

Schofield JE, Edwards KJ, Erlendsson E, Ledger PM. 2013. Palynology supports 'Old Norse' introductions to the flora of Greenland. *Journal of Biogeography* **40**:1119–1130.

Simberloff D, Martin J-L, Genovesi P, Maris V, Wardle DA, Aronson J, Courchamp F, Galil B, García-Berthou E, Pascal M, Pyšek P, Sousa R, Tabacchi E, Vilà M. 2013. Impacts of biological invasions: what's what and the way forward. *Trends in Ecology and Evolution* **28**:58–66.

Smith I. 2002. *The New Zealand Sealing Industry.* Wellington: Department of Conservation.

van Leeuwen JF, Schäfer H, Van der Knaap W, Rittenour T, Björck S, Ammann B. 2005. Native or introduced? Fossil pollen and spores may say. An example from the Azores Islands. *Neobiota* **6**:27–34.

van Leeuwen JFN, Froyd CA, van der Knaap WO, Coffey EE, Tye A, Willis KJ. 2008. Fossil pollen as a guide to conservation in the Galapagos. *Science* **322**:1206.

Vitousek PM. 1988. Diversity and biological invasions of oceanic islands. *Biodiversity* **20**:181–189.

Walls G. 2009. Picking up the plant trail: botanical evidence of people in the Auckland Islands. In: Dingwall PR, Jones KL, Egerton R, eds. *In care of the Southern Ocean.* Auckland: New Zealand Archaeological Association.

Wang Y, Naumann U, Wright ST, Warton DI. 2012. mvabund–an R package for model-based analysis of multivariate abundance data. *Methods in Ecology and Evolution* **3**:471–474.

Wardle P. 1971. Biological Flora of New Zealand 6. *Metrosideros umbellata* Cav. [Syn. M. lucida (Forst.f.) A. Rich.] (Myrtaceae) Southern Rata. *New Zealand Journal of Botany* **9**:645–671.

Webb D. 1985. What are the criteria for presuming native status? *Watsonia* **15**:231–236.

Webber BL, Scott JK. 2012. Rapid global change: implications for defining natives and aliens. *Global Ecology and Biogeography* **21**:305–311.

Whitehead AL, Lyver POB, Jones CJ, Bellingham PJ, MacLeod CJ, Coleman M, Karl BJ, Drew K, Pairman D, Gormley AM, Duncan RP. 2014. Establishing accurate baseline estimates of breeding populations of a burrowing seabird, the grey-faced petrel (*Pterodroma macroptera gouldi*) in New Zealand. *Biological Conservation* **169**:109–116.

Whitlock C, Larsen C. 2001. Charcoal as a Fire Proxy. In: Smol JP, Birks HJB, Last WM, eds. *Tracking environmental change using lake sediments: volume 3 terrestrial, algal, and siliceous indicators*. Dordrecht: Kluwer Academic Publishers.

Whittaker RJ, Fernández-Palacios JM. 2007. *Island biogeography: ecology, evolution and conservation*. Oxford, UK: Oxford University Press.

Willis KJ, Birks HJB. 2006. What is natural? The need for a long-term perspective in biodiversity conservation. *Science* **314**:1261–1265.

Wilmshurst JM, Bestic KL, Meurk CD, McGlone MS. 2004. Recent spread of Dracophyllum scrub on subantarctic Campbell Island, New Zealand: climatic or anthropogenic origins? *Journal of Biogeography* **31**:401–413.

Wilson HD. 1987. *Vegetation of Stewart Island, New Zealand*. Department of Scientific and Industrial Research.

14

Origins and diversity of a cosmopolitan fern genus on an island archipelago

Paul G. Wolf[1,2]*, Carol A. Rowe[1], Joshua P. Der[3], Martin P. Schilling[1], Clayton J. Visger[4] and John A. Thomson[5]

[1] Department of Biology, Utah State University, Logan, UT 84322, USA
[2] Ecology Center, Utah State University, Logan, UT 84322, USA
[3] Department of Biological Science, California State University, Fullerton, CA 92834, USA
[4] Department of Biology, University of Florida, Gainesville, FL 32611, USA
[5] National Herbarium of NSW, Royal Botanic Gardens and Domain Trust, Mrs Macquaries Road, Sydney, NSW 2000, Australia

Associate Editor: Chelsea D. Specht

Abstract. Isolated oceanic islands are characterized by patterns of biological diversity different from that on nearby continental mainlands. Isolation can provide the opportunity for evolutionary divergence, but also set the stage for hybridization between related taxa arriving from different sources. Ferns disperse by haploid spores, which are produced in large numbers and can travel long distances in air currents, enabling these plants to become established on most oceanic islands. Here, we examine the origins and patterns of diversity of the cosmopolitan fern genus *Pteridium* (Dennstaedtiaceae; bracken) on the Galapagos Islands. We use nucleotide sequences from two plastid genes, and two nuclear gene markers, to examine phylogeography of *Pteridium* on the Galapagos Islands. We incorporate data from a previous study to provide a worldwide context. We also sampled new specimens from South and Central America. We used flow cytometry to estimate genome size of some accessions. We found that both plastid and nuclear haplotypes fall into two distinct clades, consistent with a two-diploid-species taxonomy of *P. aquilinum* and *P. esculentum*. As predicted, the allotetraploid *P. caudatum* possesses nuclear haplotypes from both diploid species. Samples from the Galapagos include *P. esculentum* subsp. *arachnoideum*, *P. caudatum* and possible hybrids between them. Multiple *Pteridium* taxa were also observed growing together at some sites. We find evidence for multiple origins of *Pteridium* on the Galapagos Islands and multiple origins of tetraploid *P. caudatum* throughout its range in Central and South America. We also posit that *P. caudatum* may include recent diploid hybrids, backcrosses to *P. esculentum*, as well as allotetraploid plants. The Galapagos Islands are positioned close to the equator where they can receive dispersing propagules from both hemispheres. This may partly explain the high levels of diversity found for this cosmopolitan fern on these islands.

Keywords: Biogeography; bracken; ferns; Galapagos; hybridization; islands; nuclear genes; phylogeny; *Pteridium*.

* Corresponding author's e-mail address: paul.wolf@usu.edu

Introduction

Oceanic islands provide an ideal biological setting for evolutionary change and thus for the study of evolutionary processes. On islands that are distantly isolated from continents, there can be an increased opportunity for organisms to diverge genetically from those in the original source populations. Furthermore, remote islands can act as a sink for individuals of the same species (or closely related species) arriving from more than one original source, thereby setting the stage for hybridization, increased genetic diversity, or both. Angiosperms and gymnosperms colonize islands most often via seeds, which contain diploid embryos. This is contrasted in ferns (monilophytes) and lycophytes, which are dispersed by haploid spores. In many cases, spore-bearing plants require two spores, and subsequently two gametophytes, for successful establishment (Soltis and Soltis 1987; Soltis et al. 1988). Moreover, spore-dispersed plants appear to have high levels of gene flow relative to seed plants (Soltis and Soltis 1987), presumably because spores are smaller and more easily transported in the air than most seeds. An increased dispersal potential can result in colonization of new areas from multiple spore sources, and may involve an increased opportunity for hybridization relative to seed plants. Here, we explore this possibility as we assess the origins and diversity of the cosmopolitan fern genus *Pteridium* on the Galapagos Islands.

The Galapagos Islands are a group of ~14 main islands and ~100 rocks or islets, ~1000 km west of mainland Ecuador (Snell et al. 1996). The islands vary in age ranging from ~0.7 to ~4.2 million years since emergence above sea level (White et al. 1993). However, additional evidence points to a much older archipelago existing at the same spot prior to the current emergence (Christie et al. 1992). Rassmann (1997) cited this idea to explain why estimates of some lineage ages are more than 5 million years old. The relative isolation of the islands from the nearest mainland coasts (Ecuador and Costa Rica) has resulted in a high level of endemism. There are estimated to be 236 endemic plant species on the islands (Tye and Francisco-Ortega 2011). Origins of the Galapagos flora are now thought to be quite diverse and include northern and southern Andes and other parts of South America, as well as Central America and the Caribbean (Tye and Francisco-Ortega 2011). The wide range of elevations and rainfall patterns results in a diversity of ecological habitats (Itow 2003).

Pteridium is a worldwide genus that has been treated from as few as one species to >20. There are several taxonomic challenges in the genus, one of which is the high level of variability and phenotypic plasticity for morphological characters, including those that are used for taxonomic treatments. Furthermore, regional and local treatments often do not incorporate the context of variation that is seen at the global scale. A few authors have examined *Pteridium* in a worldwide context. For example, Ching (1940) considered the genus to comprise 6 species, whereas a year later, Tryon (1941) treated *Pteridium* as a single species with 2 subspecies and 12 varieties. Page (1976) reviewed information on geographic variation and concluded that there is probably more than one species, but he made no formal taxonomic changes. More recently, in a series of articles (Thomson 2000, 2012; Thomson et al. 2008), Thomson has recognized two main diploid species: *P. aquilinum* (corresponding to Tryon's subspecies *aquilinum*) from Europe, North America, Asia and Africa and *P. esculentum* (corresponding approximately to Tryon's subspecies *caudatum*). *Pteridium esculentum* is treated by some authors as two species: *P. esculentum* in Australia and New Zealand and *P. arachnoideum* in South America (see, for example, Schwartsburd et al. 2014), whereas others treat *esculentum* and *arachnoideum* as subspecies of *P. esculentum* (Thomson 2012; Zhou et al. 2014), a system we follow here. Regardless of rank assignment, evidence for two main clades of *Pteridium* includes analyses of plastid DNA variation (Der et al. 2009; Zhou et al. 2014). Further, several hybrids and allotetraploids have been examined (Thomson and Alonso-Amelot 2002; Zhou et al. 2014). Der et al. (2009) noted that development of nuclear genomic markers would be critical for establishing the origins of hybrid taxa and for other systematic studies of *Pteridium*.

South America is home to two main *Pteridium* taxa: diploid *P. esculentum* subsp. *arachnoideum* and allotetraploid *P. caudatum*, the latter a hybrid between *P. esculentum* from South America and *P. aquilinum* from North America (Thomson and Alonso-Amelot 2002). Tetraploidy was inferred on the basis of Feulgen cytometry (Tan and Thomson 1990) and spore size and guard cell length (Thomson and Alonso-Amelot 2002). The hybrid origin of *P. caudatum* is further supported by the additive pattern of DNA markers from *P. aquilinum* and *P. esculentum* (Thomson 2000; Thomson and Alonso-Amelot 2002). Additional characters that can be used to distinguish *P. caudatum* from *P. esculentum* subsp. *arachnoideum* in South America include the presence of gnarled trichomes between veins abaxially (Thomson and Martin 1996) and free laminar lobes on *P. esculentum*. An additional taxon was recently described from north eastern Brazil (Schwartsburd et al. 2014).

Most chromosome counts of *Pteridium* show $2n = 104$ (Page 1976; Sheffield et al. 1989; Thomson 2000; Tindale and Roy 2002; Bainard et al. 2011), with other complements, such as triploidy (Sheffield et al. 1993), assumed to be rare. One count of $2n = 52$ from Spain (Löve and Kjellqvist 1972) has not been corroborated despite resampling from the same area (Sheffield *et al.* 1989). Jarrett *et al.* (1968) reported the first observation of cytological variation in the genus, with a count of $2n = 208$ (tetraploid) for one sporophyte of *Pteridium* from the Galapagos Islands. This report has been the motivation for previous as well as the current focus on *Pteridium* from these islands. Klekowski (1973) used gametophytes grown from spores collected from the islands to examine the ability to self-fertilize and cross with *Pteridium* from other sources. The results demonstrated that bracken from Hawaii (*P. aquilinum*) and samples from the Galapagos Islands were interfertile with *Pteridium* from Central and South America. However, Hawaiian and Galapagos *Pteridium* were intersterile with each other.

Recent examination of *Pteridium* collections from the Galapagos islands suggests that more than one taxon is present on Galapagos. We set out to examine Galapagos *Pteridium* with the following objectives:

(1) To examine the origins of *P. caudatum* on both islands and mainland.

(2) To determine how many *Pteridium* taxa are on the Galapagos Islands.
(3) To examine whether different *Pteridium* taxa occupy different islands, or elevations on the Galapagos Islands.
(4) To examine the possible mainland origins of Galapagos *Pteridium*.

Methods

We sampled 17 *Pteridium* from three of the Galapagos Islands (Fig. 1; Table 1): Santa Cruz, Isabela and San Cristobal. We also scouted on Floriana, but were unable to locate any *Pteridium* on that island. At each site, we collected expanding frond segments onto silica gel, and collected intact fronds for herbarium specimens, deposited at the herbarium of the Charles Darwin Research Station (CDS). In addition, we included a selection of DNA samples from the plastid gene study of Der *et al.* (2009) to obtain nuclear gene sequences. We selected representatives of each of the major plastid clades of Der *et al.* (2009). We also included additional *Pteridium* samples

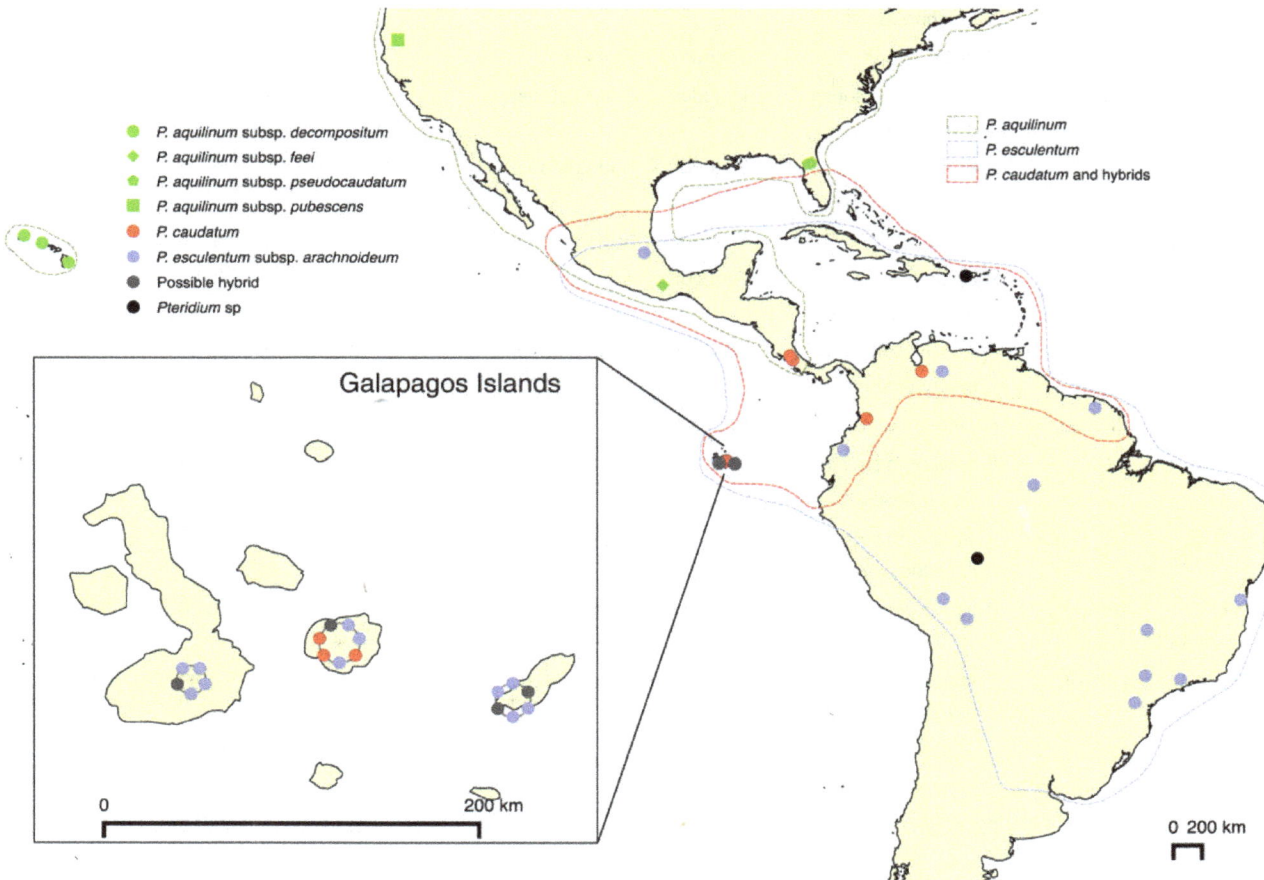

Figure 1. Map of America showing locations of *Pteridium* sampled for this study. Inset shows details of Galapagos Islands. Approximate taxon boundaries are based on Tryon (1941), Tryon and Tryon (1982) and Mickel and Smith (2004).

Table 1. Voucher and locality information for samples used in this study. Code (as used in tree figures) indicates collector and number, with full name in parentheses when abbreviated in code. 'Possible hybrids' are likely to be between *P. esculentum* subsp. *arachnoideum* and *P. caudatum*.

Code	Herbarium	Taxon	Country	Island/province/state	Latitude (°)	Longitude (°)	Elevation (m)
Wolf 1001	CDS	*P. esculentum* subsp. *arachnoideum*	Ecuador	Santa Cruz	−0.63	−90.38	592
Wolf 1002	CDS	*P. caudatum*	Ecuador	Santa Cruz	−0.66	−90.40	420
Wolf 1003	CDS	*P. esculentum* subsp. *arachnoideum*	Ecuador	Santa Cruz	−0.64	−90.33	874
Wolf 1004	CDS	*P. esculentum* subsp. *Arachnoideum*	Ecuador	Santa Cruz	−0.65	−90.33	732
Wolf 1005a	CDS	Possible hybrid	Ecuador	Santa Cruz	−0.66	−90.33	580
Wolf 1005c	CDS	*P. caudatum*	Ecuador	Santa Cruz	−0.66	−90.33	580
Wolf 1006	CDS	*P. caudatum*	Ecuador	Santa Cruz	−0.67	−90.32	476
Wolf 1007	CDS	*P. esculentum* subsp. *arachnoideum*	Ecuador	Isabela	−0.81	−91.09	1009
Wolf 1008	CDS	*P. esculentum* subsp. *arachnoideum*	Ecuador	Isabela	−0.83	−91.09	1006
Wolf 1009	CDS	*P. esculentum* subsp. *arachnoideum*	Ecuador	Isabela	−0.84	−91.09	822
Wolf 1010	CDS	*P. esculentum* subsp. *arachnoideum*	Ecuador	Isabela	−0.84	−91.07	627
Wolf 1011	CDS	*Possible hybrid*	Ecuador	Isabela	−0.85	−91.04	405
Wolf 1012	CDS	*P. esculentum* subsp. *arachnoideum*	Ecuador	San Cristobal	−0.91	−89.55	381
Wolf 1013	CDS	*P. esculentum* subsp. *arachnoideum*	Ecuador	San Cristobal	−0.90	−89.48	683
Wolf 1014	CDS	Possible hybrid	Ecuador	San Cristobal	−0.90	−89.48	676
Wolf 1015	CDS	*P. esculentum* subsp. *arachnoideum*	Ecuador	San Cristobal	−0.90	−89.52	739
Wolf 1016	CDS	*P. esculentum* subsp. *arachnoideum*	Ecuador	San Cristobal	−0.90	−89.53	544
Wolf 1017	CDS	Possible hybrid	Ecuador	San Cristobal	−0.90	−89.53	544
Wolf 1018	UTC	*P. aquilinum* subsp. *decompositum*	USA	Hawaii	19.43	−155.28	1247
Wolf 1019	UTC	*P. aquilinum* subsp. *pseudocaudatum*	USA	Florida	29.63	−81.92	42
AL 147 (A. Larsson)	DUKE	*P. aquilinum* subsp. *feei*	Mexico	Oaxaca	17.17	−96.60	2660
IJ 786 (Jiménez)	LPB, UC	*P. esculentum* subsp. *arachnoideum*	Bolivia	Franz Tamayo	−14.62	−68.95	2350
IJ 1245 (Jiménez)	LPB, UC	*P. esculentum* subsp. *arachnoideum*	Bolivia	Ayopaya	−16.65	−66.62	2750
IJ 2048 (Jiménez)	LPB, UC	*Pteridium* sp.	Bolivia	Federico Román	−10.48	−65.57	140

Continued

Table 1. *Continued*

Code	Herbarium	Taxon	Country	Island/province/state	Latitude (°)	Longitude (°)	Elevation (m)
Wolf 1020	UTC	*P. caudatum*	Costa Rica	San Jose	9.56	−83.80	2270
Wood 15788	HAW	*P. aquilinum* subsp. *decompositum*	USA	Hawaii	22.15	−159.65	1280
Wolf 1023	HAW	*P. aquilinum* subsp. *decompositum*	USA	Hawaii	21.40	−157.89	419
Worthington 35231	DUKE	*Pteridium* sp.	Puerto Rico	Ponce	18.13	−66.68	792
JJdG 14388 (de Granville)	NSW 729390	*P. esculentum* subsp. *arachnoideum*	French Guiana	Saint-Laurent-du-Maroni	4.70	−53.97	480
Matos 231	NY 01198119	*P. esculentum* subsp. *arachnoideum*	Brazil	Bahia	−14.71	−39.60	700
Ortiz 497	NY 00089157	*P. esculentum* subsp. *arachnoideum*	Ecuador	Esmeraldas	0.40	−78.80	1925
Delprete 10293	NY 01019119	*P. esculentum* subsp. *arachnoideum*	Brazil	Goias	−17.80	−48.82	1150
Prado 2351	SP	*P. esculentum* subsp. *arachnoideum*	Brazil	Paranà	−25.14	−50.03	1000
Prado 2337	SP	*P. esculentum* subsp. *arachnoideum*	Brazil	São Paulo	−22.77	−45.53	1888
Wolf 795	UC 1622577	*H. incisa*					
Wolf 387	UTC	*P. scaberula*					
Wolf 376	UTC	*B. pubescens*	La Réunion				

from the mainland of Central and South America, and outgroups *Histiopteris*, *Blotiella* and *Paesia* (Table 1).

Morphology

Preliminary morphological analysis of samples was undertaken independently of molecular studies. Photographs of fresh unpressed pinnae supplemented by macrophotographs of the abaxial surface of individual ultimate segments were used for the examination of gross features of laminal dissection, presence/absence of free laminal lobes on pinna and pinnule axes, and abaxial laminal indumentum.

We used small subsamples of each accession comprising one or two dried pinnules for more detailed microscopic study following Thomson and Martin (1996). We examined the indument of abaxial pinnulet and segment midveins, determined presence versus absence of gnarled intervein trichomes, measured false-indusial width, estimated the number of cells per millimetre along the outer margin of the false indusium and measured stomatal guard cell length. Taxonomic designation was based on previous descriptions of the characters we

used (Thomson 2000, 2012; Thomson and Alonso-Amelot 2002; Thomson et al. 2008).

DNA sequencing

DNA was extracted from fresh, desiccated or herbarium tissue using the DNeasy Plant Mini kit (QIAGEN, Valencia, CA, USA), following the manufacturer's protocol.

The plastid markers *trnS–rpS4* (spacer + gene) and *rpL16* intron were amplified in 25 μL polymerase chain reactions (PCRs) using the fern-specific primers published in Small et al. (2005). However, the complete plastid genome sequence of *P. aquilinum* (GenBank accession NC_014348) was used to redesign the *rpl16* reverse and the *trnS* primers (names now with ptaq suffix; see Table 2).

Nuclear primers were based on those of Rothfels et al. (2013). We chose two nuclear genes *SQD1* (Region 1) and *ApPEFP_C* (Region 2), redesigning (now with suffix _ptaq) the forward *ApPEFP_C* and *SQD1* reverse, based on *Pteridium* sequences from a study of the transcriptome (Der et al. 2011). *SQD1* encodes sulfoquinovosyldiacylglyerol 1 involved in the biosynthesis of sulfolipids and

Table 2. Primer sequences for PCR and DNA sequencing. Suffix 'ptaq' denotes primers designed in this study.

Primer name	Primer sequence, 5'–3'
rpl16_r_ptaq	TCCTCTATGTTGCTTACGATAT
trns_gga_ptaq	CTACCGAGGGTTCAAATCCCTC
SQD_r2_ptaq	CCTTTGCCATAAACTGTAAGGGGGTG
EMSQD1E1F6	GCAAGGGTACHAAGGTHATGATCATAGG
ApPEFP_f25_ptaq	AATGCTCTAAGTCATTGTTACCGATC
ApPEFP_C4218_r7	TTGTAAATCTCTGTRTCRGATGYYGT
rps4_int_f1	CAGATTACTGAAAAACTAGC
rps4_int_r1	AGAAGAGCGAAAGGGTTC
rpl16_int_f1	GCGAAGCTGAAAACGATGCC
rpl16_int_r1	GTTCCATTTCTAAATAGCGG

ApPEFP_C encodes an appr-1-p processing enzyme family protein, ADP-ribose-1-monophosphatase (Appr-1-pase), a ubiquitous cellular processing enzyme. The PCR primer sequences (Table 2) used were as follows: SQD_r2_ptaq combined with EMSQD1E1F6, and ApPEFP_f25_ptaq combined with ApPEFP_C4218_r7. Polymerase chain reaction conditions followed Der *et al.* (2009), annealing at 56.5 °C for the plastid and nuclear genes. For sequencing, we used all PCR primers plus new internal primers (Table 2) for the two plastid genes: rps4_int_f1, rps4_int_r1, rpl16_int_f1 and rpl16_int_r1. In many samples, the two nuclear gene amplicons contained multiple haplotypes. To sequence each haplotype separately, we cloned the PCR products using the StrataClone PCR Cloning Kit (Agilent Technologies, Santa Clara, CA, USA). DNA sequences were assembled and edited using Sequencher 4.6 (Gene Codes Corporation, Ann Arbor, MI, USA).

Phylogenetic analysis

Sequences were aligned with MAFFT version 7.215 using the L-INS-i algorithm for accurate alignments. Newly generated sequences for the plastid genes *rps4* and *rpl16* were combined with data from Der *et al.* (2009; GenBank accession numbers FJ177158–FJ177206 for the *trnS*–*rps4* spacer + gene and FJ177239–FJ177287 for the rpL16 intron) and concatenated for phylogenetic analysis. Maximum likelihood (ML) phylogenetic inference was performed separately for each nuclear gene and the plastid data with RAxML version 8.1.17 using 100 rapid bootstrap replicates followed by a ML search under the GTRGAMMA model of evolution. Trees were rooted with the three outgroups.

Flow cytometry

Genome size was determined using flow cytometry. Approximately 0.75 cm^2 of fresh leaf tissue and 0.5 cm^2

of standard, *Vicia faba* (26.9 pg; Doležel *et al.* 1998), were co-chopped on a chilled surface using a fresh razor blade in 500 µL of ice-cold extraction buffer (0.1 M citric acid, 0.5 % v/v Triton X-100) (Hanson *et al.* 2005), with 1 % w/v PVP-40 (Yokoya *et al.* 2000). Tissue was chopped into a semi-fine slurry, and the suspension was swirled by hand until the liquid reached a light green tinge. The suspension was poured through a cell strainer (BD Falcon; Becton, Dickinson and Company, Franklin Lakes, NJ, USA). RNaseA (1 mg mL^{-1}) and 350 µL of propidium iodide staining solution (0.4 M NaPO$_4$, 10 mM sodium citrate, 25 mM sodium sulfate, 50 µg mL^{-1} propidium iodide) were added to 140 µL of filtrate, incubated at 25 °C for 30 min, followed by up to 2 h on ice. The stained solutions were analysed with an Accuri C6 using a 488 nm laser, and 10 000 events were captured per sample. The relative genome size was calculated using the ratio of the mean fluorescent peak of the sample to the internal standard multiplied by the genome size of the standard.

Results

In general, we found Galapagos *Pteridium* to be highly variable for both morphological and molecular characters. We find evidence of two *Pteridium* taxa plus possible hybrids, and multiple colonization events from different mainland sources.

Morphology

Pteridium caudatum and *P. esculentum* can be distinguished morphologically by a combination of characters (Thomson 2000; Thomson and Alonso-Amelot 2002; Table 3). We inferred that 11 of our samples were clearly *P. e.* subsp. *arachnoideum*, 2 were clearly *P. caudatum* and 4 were difficult to determine and inferred to be possible hybrids. The two *P. caudatum* samples were found at the two lowest sites on Santa Cruz. The possible hybrids were found at the lowest sites on Isabela and San Cristobal and mid-elevation sites on Santa Cruz and San Cristobal. We found different *Pteridium* taxa growing within a kilometre of each other on Santa Cruz and San Cristobal. Samples from one site included one frond that was *P. e. arachnoideum* (Wolf 1005a) and another frond (collected within 1 m of the other) was *P. caudatum* (Wolf 1005c) or a hybrid.

Most of our Galapagos samples of *Pteridium* fell into one of the two distinct categories for stomatal guard cell length: those with a mean below 40 µm and those above 40 µm. Wolf 1002 and Wolf 1006 fall within the range expected for tetraploid *P. caudatum* (Thomson and Alonso-Amelot 2002) and close to the guard cell length (46.5 µm, Thomson 2000) for the Galapagos plant showing 4n = 208 (K Sheet H2146/97/1, Jarrett

Table 3. Typical morphologies for *P. caudatum*, *P. esculentum* subsp. *arachnoideum*, and possible hybrids (or introgressants) between them. Information based on Tryon (1941) and Thomson and Alonso-Amelot (2002).

Determination	Wolf ID #	Free lobes on segment axes	False indusium: width (mm)	False indusium: cells/mm length along margin	Stomatal guard cell length (μm)	Abaxial surface between veins: gnarled trichomes	Abaxial surface: vein indumentum
P. caudatum	1002, 1006	Absent	0.3–0.5	~31	>40	Absent	Glabrous
P. esculentum subsp. *arachnoideum*	1001, 1003, 1004, 1007, 1008, 1009, 1010, 1012, 1013, 1015, 1016	Present	0.1–0.3	~48	<40	Present	Dense fine acicular white hairs, some twisted; fine white arachnoid hairs
Indeterminate: possibly introgressant	1005a	Absent	0.2	64	32.3	Present	Vein hairs less dense than for typical subsp. *arachnoideum*
	1011	Absent	0.4	56	39.1	Present	As for 1005a
	1014	Absent	0.15	48	39.3	Present	As for 1005a
	1017	Absent	0.15–0.2	40	34.3	Absent	As for 1005a

Table 4. Gene statistics and GenBank accession information.

Gene	Number of characters	Variable characters	Parsimony-informative characters	Differences between aquilinum and esculentum haplotypes	GenBank accession numbers
ApPEFP_C	785	51	28	17	KT345729–KT345821
SQD1	752	36	26	15	KT345856–KT345898
rps4	1036	33	25	11 + 1 indel	KT345822–KT345855
rpl16	792	27	21	9 + 2 indels	KT345899–KT345934
Total	3365	147	100	55	

et al. 1968), corroborating our morphology-based determination of these samples (Table 3). Ploidy level of the other Galapagos samples studied cannot be determined due to the extended, apparently continuous, series of guard cell lengths represented (Table 3). The wide range of lengths observed suggests that both diploid and triploid levels might be represented. On the basis of morphology, 11 of our samples are *P. esculentum* subsp. *arachnoideum* and 4 may be hybrids carrying genomic elements from outside *arachnoideum*, and may be triploid (Table 3).

DNA

Overall, nucleotide data from the four genes contained 3365 characters of which 100 were phylogenetically informative, and 55 distinguished the *P. aquilinum* clade from the *P. esculentum* clade (Table 4). Phylogenetic analyses of the two plastid genes were congruent, as found previously (Der *et al.* 2009). Thus, alignments of the two plastid genes were concatenated for a combined analysis (Fig. 2). In the three analyses (two plastid genes concatenated, *SQD1* and *ApPEFP_C*), the aquilinum and esculentum clades of *Pteridium* were sister to each other. Table 4 provides the ranges of GenBank accession numbers for each gene and **Supporting Information—File S1** lists the GenBank accession number for each sequence. All trees and associated nucleotide alignments are deposited in TreeBASE (http://purl.org/phylo/treebase/phylows/study/TB2:S18018). We include outgroups for phylogenetic analysis of the nuclear genes. However,

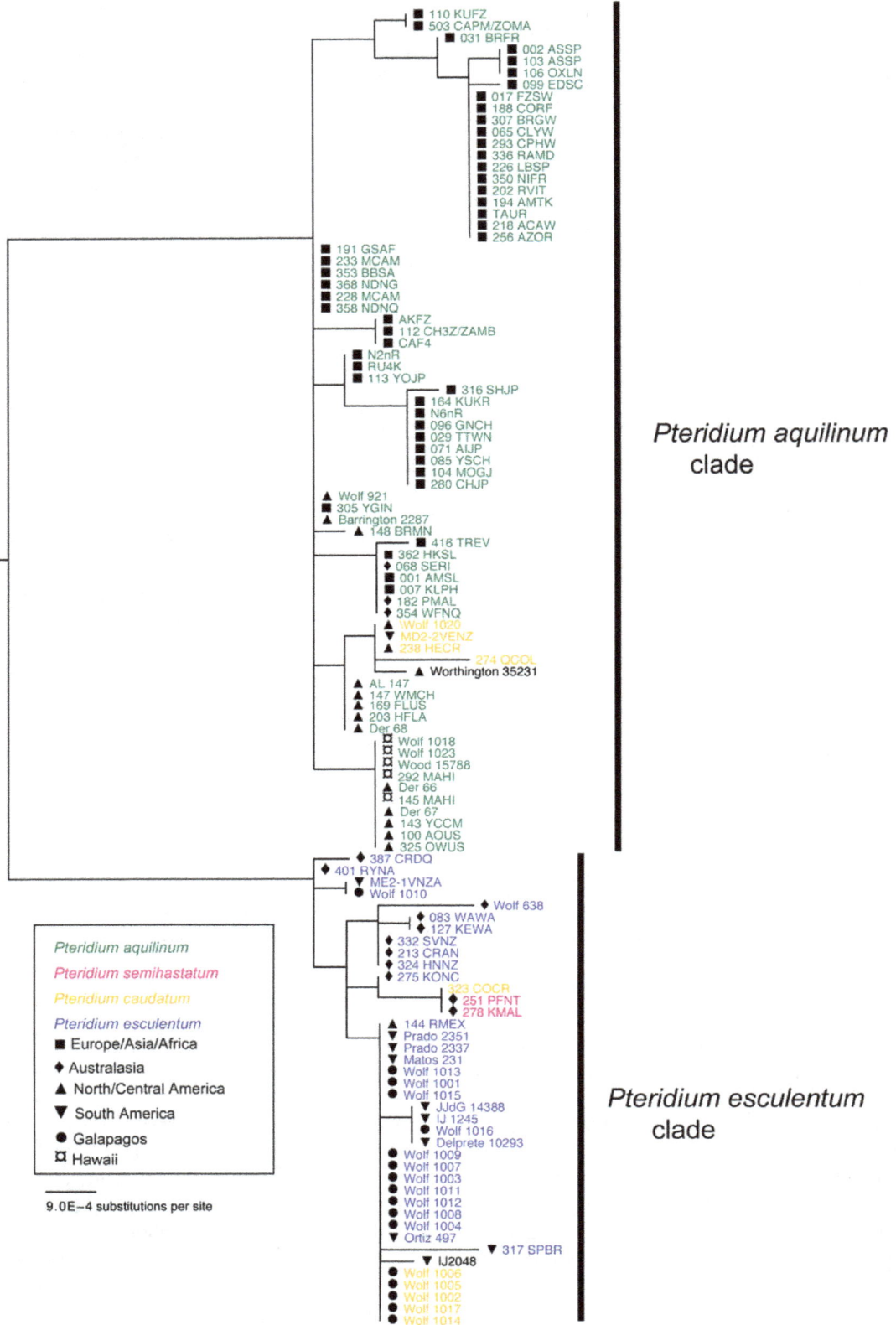

Figure 2. Phylogenetic tree based on the combined plastid gene data set. The tree was rooted with *Blotiella pubescens*, *Paesia scaberula* and *Histiopteris incisa*.

inclusion of outgroups in the plastid trees resulted in very short ingroup branch lengths, and is therefore not shown. The tree topology without outgroups is the same as that with outgroups.

We detected a total of 19 aquilinum and 31 esculentum plastid haplotypes. In samples from the Galapagos Islands, we detected 12 plastid haplotypes, one of which has been sampled previously in *P. e. arachnoideum* from Venezuela and Mexico (Der *et al.* 2009). The remaining Galapagos plastid haplotypes were nested within a clade that included haplotypes from Mexico and South America (Fig. 2). All samples of *P. aquilinum* had the expected plastid haplotype, and all new samples from South American *Pteridium*, including those from the Galapagos Islands, had esculentum haplotypes. A specimen from Costa Rica (Wolf 1020), which appears to be *P. caudatum*, had an aquilinum plastid haplotype, as did the single Mexican sample of *P. a. feei* (A. Larsson 147), a taxon not previously included in molecular studies. Previous studies (Thomson *et al.* 2008; Der *et al.* 2009) have noted two 5-bp polymorphic repeats in the *trnS–rps4* spacer. Together, these polymorphisms account for three haplotypes: haplotype C in outgroups and the *P. esculentum* clade, haplotype A in *P. aquilinum* and haplotype B in only European and African *P. aquilinum*.

The two nuclear genes showed a similar pattern of differentiation as the plastid genes: a set of distinct nucleotide differences (15 for *SQD1* and 17 for *ApPEFP_C*) distinguished aquilinum from esculentum haplotypes. For *SQD1*, samples of *P. caudatum* were heterozygous for the above nucleotide positions indicating that they were additive for *P. aquilinum* and *P. esculentum* haplotypes. However, although we were able to sequence a few distinct haplotypes from heterozygous plants, this was largely unsuccessful. Bacterial cells carrying the *SQD1* PCR product appeared to be clumping so that single colonies were usually not single clones and therefore remained heterozygous, despite re-streaking of colonies. We suspect that clumping was a function of partial expression of the gene. Conversely, we were able to clone several haplotypes of *ApPEFP_C* from heterozygous individuals and we found that *P. caudatum* plants indeed possessed both aquilinum and esculentum haplotypes. We detected a total of 12 aquilinum and 13 esculentum haplotypes for *SQD1* (Fig. 3), and 32 aquilinum and 30 esculentum haplotypes for *ApPEFP_C* (Fig. 3). In the samples from the Galapagos Islands, we detected 7 aquilinum and 14 esculentum *ApPEFP_C* haplotypes, and 1 esculentum and 9 aquilinum haplotypes for *SQD1*. Of the six samples that were heterozygous for *SQD1*, we were able to sequence five haplotypes from three individuals. Of the 17 samples that were heterozygous for *ApPEFP_C*, 8 had both aquilinum and esculentum haplotypes, 5 heterozygous samples had only aquilinum

haplotypes and 4 samples had only esculentum haplotypes. All Galapagos specimens have esculentum nuclear haplotypes, 11 with an esculentum haplotype only, 10 of which were *P. e. arachnoideum*. Five Galapagos samples have aquilinum and esculentum haplotypes, two of which were *P. caudatum*, two appeared to be hybrids and one was *P. e. arachnoideum*. One specimen (Wolf 1004) appeared to have aquilinum and esculentum haplotypes for *SQD1*, but only aquilinum haplotypes for *ApPEFP_C*. All mainland samples of *P. caudatum* have both aquilinum and esculentum haplotypes. Diploid individuals should have no more than two haplotypes at a locus, whereas tetraploids are expected to have no more than four haplotypes. However, 8 of the 13 heterozygous individuals had extra haplotypes: 3 *P. caudatum* samples with 5, 5 and 7 haplotypes, 1 *P. esculentum* with 6 haplotypes and 4 *P. aquilinum* samples with 3, 3, 3 and 5 haplotypes. In most cases, extra haplotypes differed from others from the same individual by one nucleotide, and at the most three nucleotides. All extra haplotypes appear to be the result of single nucleotide autapomorphies, and we cannot account for extra haplotypes by recombination, whether in plant cells, PCR tubes or during cloning.

Flow cytometry

We were able to estimate c-values for fresh *Pteridium* fronds on a consistent basis, but we were unable to do so for any dry samples, including those in silica gel and herbarium specimens. We estimate haploid genome size (mean $\pm$ coefficient of variation) of 15.88 ($\pm$ 0.67) pg for *P. a. decompositum* from Hawaii (Wolf 1018), 16.13 ($\pm$ 0.67) pg for *P. a. pseudocaudatum* from Florida (Wolf 1019) and 29.2 ($\pm$ 2.1) pg for *P. caudatum* from Costa Rica (Wolf 1020). These are consistent with previous estimates for diploid and tetraploid *Pteridium*, respectively (Tan and Thomson 1990; Bainard *et al.* 2011).

Discussion

In this article, we examined morphological and molecular variation in *Pteridium* from Galapagos Islands. To make inferences about taxonomic variation and possible origins on the islands, we first provided context with other mainland and worldwide samples, including sequences from a previous study (Der *et al.* 2009). We first discuss variation for morphological and molecular characters, followed by the implications for the origins of Galapagos *Pteridium*.

Molecular data

A growing body of research has examined variation for nuclear-encoded genes within species of ferns (for

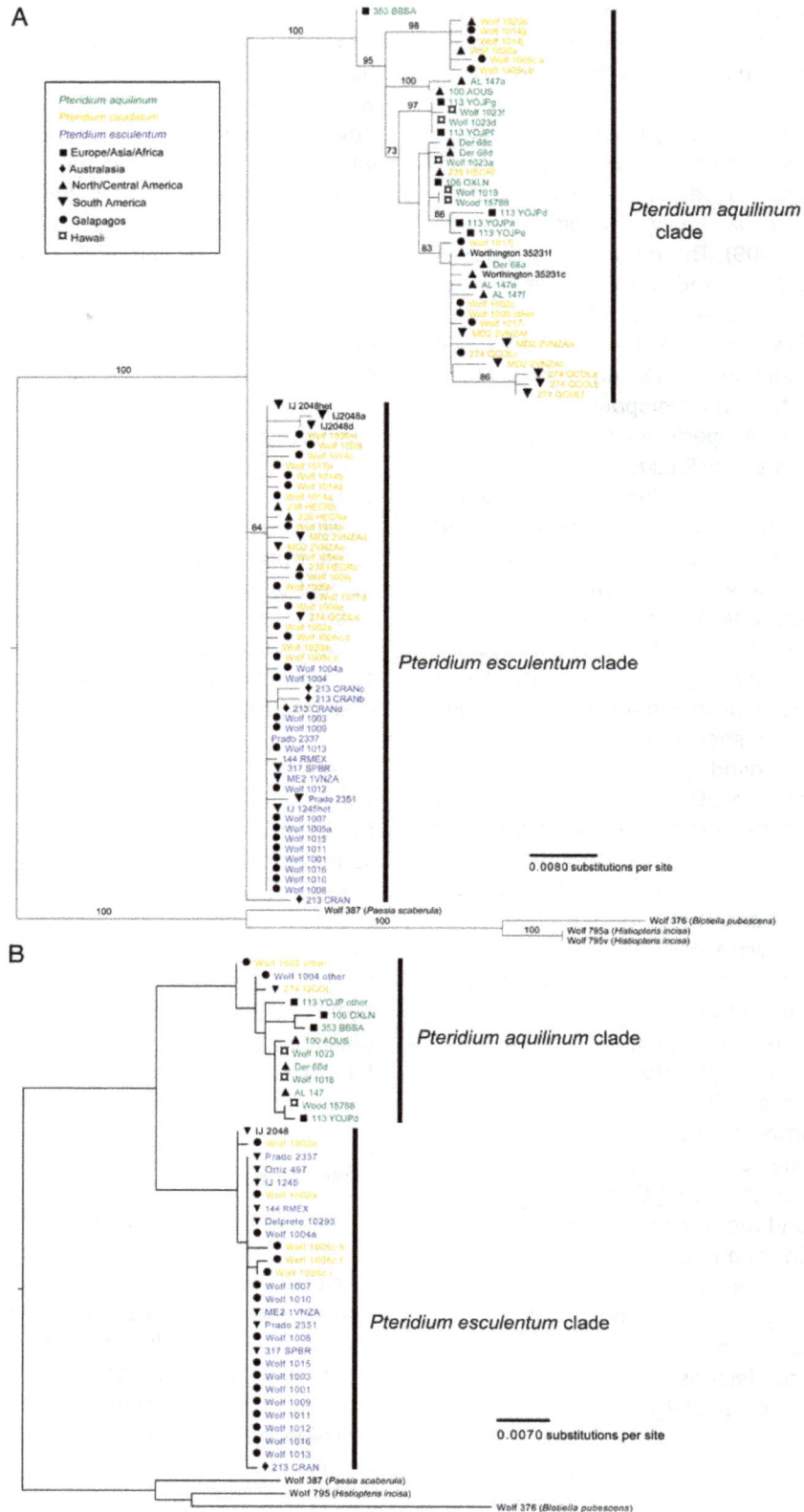

Figure 3. Phylogenetic tree based on the nuclear genes *ApPEFP_C* (A) and *SQD1* (B). Trees were rooted with *B. pubescens*, *P. scaberula* and *H. incisa*.

example see Grusz *et al.* 2009; Nitta *et al.* 2011; Sigel *et al.* 2014). Many studies use nucleotide information from the plastid genome, which appears to be effectively haploid and non-recombining in most plants. Nuclear genes, however, are subject to different processes as a result of chromosomal behaviour at a range of genomic scales. For a single-copy gene, a diploid individual should carry one or two haplotypes (alleles), and a tetraploid can have up to four. Eight of our samples carried more than the expected number based on our estimate of ploidy. Several explanations can account for these results. Polymerase chain reaction and sequencing error could manifest as extra haplotypes within an individual. It is also possible that *SQD1* and *ApPEFP_C* are not strictly single copy in *Pteridium*. Extra haplotypes can occur through several processes including segmental duplication of the chromosomal region carrying the gene, aneuploidy and polyploidy. All measures of genome size and chromosome number in *Pteridium* point to diploid and tetraploid being the most common arrangement. But samples sizes are small and we would benefit from population-level estimates of genome size, especially in areas with multiple species such as Galapagos Islands and areas of Central America. Regardless of the causes of extra haplotypes, they should be explored further. Meanwhile, because the extra haplotypes possessed only a few autapomorphic differences from others, they do not affect the phylogenetic inferences or estimates of origin numbers in this study.

Our only sample from the Caribbean was from a specimen represented by a young frond, and therefore, difficult to identify morphologically. Most descriptions indicate that *P. e. arachnoideum* is throughout the Caribbean, but this sample has aquilinum haplotypes. Future studies would benefit from increase sampling in the Caribbean.

Morphological variation

Pteridium is notorious for its phenotypic plasticity, including morphological variation among fronds within an individual clone and between pinnae on a frond (Tryon 1941; Sheffield *et al.* 1989; Ashcroft and Sheffield 1999; Thomson 2000). Wide morphological variability in *P. caudatum* led Tryon (1941) to recognize several 'phases'. Ortega (1990) found in Venezuela both typical *P. e. arachnoideum* and a second more compact form lacking free lobes between ultimate segments, while Schwartsburd *et al.* (2014) recognized three morphotypes of *P. e. arachnoideum* from Brazil. The significance and genetic basis of these character suites is yet to be established, but their variation has led to many reports of apparent intermediates between *P. caudatum*, *P. e. arachnoideum* and other taxa (Tryon 1941; Mickel and Beitel 1988; Ortega 1990; Mickel and Smith 2004; Schwartsburd *et al.* 2014).

The relationship between stomatal guard cell length and ploidy level was clearly documented for ferns by Barrington *et al.* (1986) and later established for *Pteridium* (Tan and Thomson 1990; Sheffield *et al.* 1993; Thomson 2000; Thomson and Alonso-Amelot 2002). Guard cell length is quite variable within and between *Pteridium* taxa at the subspecies level (Thomson 2000; Thomson and Alonso-Amelot 2002), and its relationship with ploidy, therefore, requires calibration and validation for each particular comparison, which we followed here.

Most of our samples from the Galapagos Islands (Table 3) had distinct morphological signatures of *P. e. arachnoideum* or *P. caudatum*. However, four samples did not fall clearly into either morphological category (Table 3). Therefore, we infer that the latter samples are possible hybrids between *P. caudatum* and *P. e. arachnoideum*, or the result of a ploidy level other than diploid or tetraploid.

The origins of *P. caudatum* in South America

Thomson (2000) and Thomson and Alonso-Amelot (2002) first outlined *P. caudatum* as one of the fertile allotetraploids between *P. aquilinum* and *P. esculentum*. Furthermore, these authors speculated that *P. caudatum* has had multiple origins in Central and South America. This hypothesis was supported by analysis of plastid DNA (Der *et al.* 2009), which showed that some *P. caudatum* samples had the *P. aquilinum* plastid DNA, whereas others had that of *P. esculentum*. Here, we provide additional evidence for the hybrid origin of *P. caudatum*; all samples had the additive pattern with both *P. aquilinum* and *P. esculentum* nuclear gene haplotypes. As for many allotetraploids, multiple origins can be inferred (Soltis and Soltis 1991; Ranker *et al.* 1994; Meimberg *et al.* 2009). We detected three *P. aquilinum* plastid haplotypes and seven *P. esculentum* plastid haplotypes among *P. caudatum* accessions. Even more haplotypes are seen in the nuclear DNA, but that is expected because two haplotypes can be transferred in a single origin involving a heterozygous plant. We detected 7 *P. aquilinum* and 12 *P. esculentum* nuclear DNA haplotypes across our *P. caudatum* accessions. Inferring the minimum number of origins is difficult because we do not know how much nucleotide change has occurred since the origin. But given the range of variation found in *P. caudatum*, we can infer at least 8 separate origins among the 11 accessions sampled here. From examination of the phylogenetic trees, it seems that the *P. aquilinum* parent could be *P. a. pseudocaudatum* (Florida and Caribbean), *P. a. latiusculum* (eastern North America) or *P. a. feei* (Mexico). All have similar plastid and nuclear haplotypes so that distinguishing the *P. aquilinum* parent further is challenging. The esculentum parent of *P. caudatum* includes only *P. e. arachnoideum*. However, this taxon is highly variable and probably includes several taxa (Schwartsburd *et al.* 2014). Future sampling should aim to include more samples of *P. e. arachnoideum* from western South America as well as Brazil.

We sampled *P. caudatum* more densely on the Galapagos Islands than the mainland, so it is difficult to determine whether *Pteridium* on the Galapagos Islands is more variable than for an equivalent area on the mainland. However, the variation that we detected indicates that *P. caudatum*, *P. e. arachnoideum* and possible backcrosses can be found in close proximity. Thus, it is possible that plants referred to *P. caudatum* include stable fertile allotetraploids, recently formed allotetraploids, homoploid hybrids between *P. aquilinum* and *P. e. arachnoideum*, and even possible hexaploid hybrids between *P. caudatum* and *P. e. arachnoideum*. Given this level of possible hybridization, we suggest that treating New World *Pteridium* as three species—the diploids *P. esculentum* and *P. aquilinum*, and hybrid *P. caudatum* (in all its manifestations)—represents best the biological situation in the genus (Thomson and Alonso-Amelot 2002; Zhou *et al.* 2014).

Pteridium on Galapagos Islands

We found evidence of *P. e. arachnoideum*, *P. caudatum* and their possible hybrids inhabiting three Galapagos Islands (Fig. 1), often with more than one taxon in close proximity. There was a tendency for *P. caudatum* to be found in lower elevation agricultural areas, but it is not clear if this is because of a habitat preference of *P. caudatum*, or if *P. caudatum* has been introduced with agricultural material. Long distance dispersal via spores is the most likely explanation for colonization of *Pteridium*. In fact, there is evidence of *Pteridium* hybrids in Scotland involving a parent from North America, suggesting transatlantic dispersal of *Pteridium* spores (Rumsey *et al.* 1991). *Pteridium* is highly variable within a relatively small area on the Galapagos Islands, a pattern that has not been observed, to our knowledge, to this extent on the mainland. However, this could be because collections have not been made at the scale used here, or collectors tend to favour specimens that key more easily to one taxon or another, rather than hybrids. If the high variability on the Galapagos Islands is not an artefact, then it could be explained by the location of the Islands. Because the closest mainland areas include both South and Central America, if spores are continually being introduced, then they could easily be coming from multiple sources. This is certainly consistent with the high number of haplotypes on the islands. Introduction from multiple sources has also been inferred for other plant species on the Galapagos Islands (Andrus *et al.* 2009) and for ferns on other island systems (Shepherd *et al.* 2009).

Earlier descriptions of the origin of Galapagos flora attributed the majority of the flora to have a Caribbean origin (Duncan and Hargraves 1984) as the result of ancient vicariance events. Conversely, Porter (1979) hypothesized that the Galapagos flora as a whole was mostly of South American origin. More recently, Tye and Francisco-Ortega (2011) compiled phylogenetic data to infer origins and showed that whereas the largest source (45 % of documented colonization events) was South American, other significant sources included Central America and the Caribbean (12 %), and North America (5 %). *Pteridium* adds an interesting twist to the data because all the above regions appear to be involved, although we do not yet have conclusive evidence for exact sources from North America; they could be from anywhere from Mexico to Florida.

It is unfortunate that we were unable to determine *c*-values for any of our Galapagos samples because this could have been used to test the prediction that putative hybrids between *P. caudatum* and *P. e. arachnoideum* are triploid. Future efforts will be made to sample appropriately for flow cytometry.

Conclusions

The most striking pattern across worldwide *Pteridium* is the morphological and molecular distinction between the *P. aquilinum* and *P. esculentum* clades. About half of the parsimony-informative molecular characters account for this difference between these two diploid species. How might speciation have occurred in the face of gene flow? One possibility is that initial divergence coincided with the separation of the southern landmasses, which was initiated about 180 million years ago (Scotese 2001) and continued until about 30 million years ago (McLoughlin 2001). However, additional factors would be required to maintain such a pattern of separation between species. One factor evident today is that mostly easterly and westerly prevailing winds operate at the equatorial regions (Oort and Yienger 1996). This would explain the similarities within *P. aquilinum* and within *P. esculentum* (Thomson 2012). However, northerly or southerly wind patterns crossing the equator, such as the inter-tropical convergence zone (Oort and Yienger 1996; Wright *et al.* 2001), are relatively rare. Yet this phenomenon could provide the means necessary for gene flow across the equator to enable hybridization between *P. aquilinum* and *P. esculentum*, thus forming the hybrids *P. caudatum* in South America, and *P. semihastatum* in Australia and tropical Asia (Thomson and Alonso-Amelot 2002).

Our new evidence for *P. caudatum* as a hybrid between diverged diploid species adds to previous examples from studies on ferns. Nitta *et al.* (2011) found evidence of hybrids between geographically distinct clades in the filmy fern genus *Crepidomanes*. Sessa *et al.* (2012) found evidence of rampant hybridization in *Dryopteris*. Furthermore, Rothfels *et al.* (2015) reported hybridization between fern taxa diverged for approximately 60 million years ago. This ability to form hybrids has been attributed to a slower 'speciation clock' in plants that lack pre-mating isolation mechanisms that involve biotically mediated fertilization (Rothfels *et al.* 2015). Such patterns are consistent with our findings for *Pteridium* in South America, and particularly on the Galapagos Islands.

In order to gain more resolution on origins of *Pteridium* and its hybrids on the Galapagos Islands, we would need additional genetic resolution. This could best be achieved by sampling at a finer geographic scale, and with many plants per site. In addition to using phylogenetic analysis of nucleotide sequences, it would be useful to include microsatellite loci or single nucleotide polymorphisms (Miller *et al.* 2007; Hohenlohe *et al.* 2010). Given our current data, we found evidence for multiple taxa, multiple origins and likely hybridization on an oceanic archipelago. The results from *Pteridium* add to a growing body of work on the origins of ferns on oceanic islands (Geiger *et al.* 2007) as well as the origins of the general flora of the Galapagos Islands (Tye and Francisco-Ortega 2011). More studies are needed to test whether these results for *Pteridium* extend to spore-bearing plants in general.

Accession Numbers

All nucleotide sequences used in this manuscript study have been deposited in (and released by) GenBank. GenBank accession numbers are provided in Table 4 and in more detail in **Supporting Information—File S1**. All phylogenetic trees and associated nucleotide alignments are deposited in TreeBASE. These data, including the actual nucleotide sequences, can be accessed by reviewers at http://purl.org/phylo/treebase/phylows/study/TB2:S18018?x-access-code=e0885899a1b8bff199f59defe19bb535&format=html.

Sources of Funding

This research project was supported by the Mary Gunson Memorial Bequest. C.J.V. was supported by the National Science Foundation Graduate research fellowship programme: DGE-1315138.

Contributions by the Authors

P.G.W. and J.T. conceived the project, planned the sampling and wrote the manuscript. J.T. conducted morphological and anatomical analyses. M.P.S. led the fieldwork part of the project. C.A.R. conducted all lab work and compiled the data. J.P.D. conducted all phylogenetic analyses and archiving of data. All authors contributed to and approved the manuscript.

Acknowledgements

We thank Tom Ranker, Mark Ellis, Alan Smith, Kent Perkins (FLAS), Robbin Moran (NYBG), Michael Kessler, Jefferson Prado and Layne Huiet (DUKE) for help obtaining and identifying specimens. We also thank the staff at CDF (Lenyn Betancourt, Frank Bungartz, Patricia Jaramillo, Sonia Cisneros and Solanda Rea) and the Park Nationale Galapagos for assistance in obtaining permits. We thank Margaret Mayger and Erick Yucailla for field assistance in the Galapagos. Two reviewers made helpful suggestions that improved the manuscript. The lab part of this research was performed while P.G.W. was on sabbatical in the labs of Kathleen Pryer (Duke University) and Pamela and Douglas Soltis (University of Florida). We thank all respective lab members for their hospitality.

Literature Cited

Andrus N, Tye A, Nesom G, Bogler D, Lewis C, Noyes R, Jaramillo P, Francisco-Ortega J. 2009. Phylogenetics of *Darwiniothamnus* (Asteraceae: Astereae)—molecular evidence for multiple origins in the endemic flora of the Galápagos Islands. *Journal of Biogeography* **36**:1055–1069.

Ashcroft CJ, Sheffield E. 1999. Rejection of *Pteridium aquilinum* subspecies *atlanticum* (C. N. Page). *Botanical Journal of the Linnean Society* **130**:157–170.

Bainard JD, Henry TA, Bainard LD, Newmaster SG. 2011. DNA content variation in monilophytes and lycophytes: large genomes that are not endopolyploid. *Chromosome Research* **19**:763–775.

Barrington DS, Paris CA, Ranker TA. 1986. Systematic inferences from spore and stomate size in the ferns. *American Fern Journal* **76**: 149–159.

Ching RC. 1940. On natural classification of the family "Polypodiaceae". *Sunyatsenia* **5**:201–268.

Christie DM, Duncan RA, Mcbirney AR, Richards MA, White WM, Harpp KS, Fox CG. 1992. Drowned islands downstream from the Galapagos hotspot imply extended speciation times. *Nature* **355**:246–248.

Der JP, Thomson JA, Stratford JK, Wolf PG. 2009. Global chloroplast phylogeny and biogeography of bracken (*Pteridium*; Dennstaedtiaceae). *American Journal of Botany* **96**:1041–1049.

Der JP, Barker MS, Wickett NJ, Depamphilis CW, Wolf PG. 2011. De novo characterization of the gametophyte transcriptome in bracken fern, *Pteridium aquilinum*. *BMC Genomics* **12**:99.

Doležel J, Greilhuber J, Lucretti S, Meister A, Lysák MA, Nardi L, Obermayer R. 1998. Plant genome size estimation by flow cytometry: inter-laboratory comparison. *Annals of Botany* **82**(Suppl A):17–26.

Duncan RA, Hargraves RB. 1984. Plate tectonic evolution of the Caribbean region in the mantle reference frame. *Memoirs of the Geological Society of America* **162**:81–94.

Geiger JMO, Ranker TA, Neale JMR, Klimas ST. 2007. Molecular biogeography and origins of the Hawaiian fern flora. *Brittonia* **59**:142–158.

Grusz AL, Windham MD, Pryer KM. 2009. Deciphering the origins of apomictic polyploids in the *Cheilanthes yavapensis* complex (Pteridaceae). *American Journal of Botany* **96**:1636–1645.

Hanson L, Boyd A, Johnson MAT, Bennett MD. 2005. First nuclear DNA C-values for 18 eudicot families. *Annals of Botany* **96**:1315–1320.

Hohenlohe PA, Bassham S, Etter PD, Stiffler N, Johnson EA, Cresko WA. 2010. Population genomics of parallel adaptation in threespine stickleback using sequenced RAD tags. *PLoS Genetics* **6**:e1000862.

Itow S. 2003. Zonation pattern, succession process and invasion by aliens in species-poor insular vegetation of the Galapagos Islands. *Global Environmental Research* **7**:39–58.

Jarrett FM, Manton I, Roy SK. 1968. Cytological and taxonomic notes on a small collection of living ferns from Galapagos. *Kew Bulletin* **22**:475–480.

Klekowski EJ Jr. 1973. Genetic endemism of Galapagos *Pteridium*. *Biological Journal of the Linnean Society* **66**:181–188.

Löve A, Kjellqvist E. 1972. Cytotaxonomy of Spanish plants. *Lagascalia* **2**:23–35.

Mcloughlin S. 2001. The breakup history of Gondwana and its impact on pre-Cenozoic floristic provincialism. *Australian Journal of Botany* **49**:271–300.

Meimberg H, Rice KJ, Milan NF, Njoku CC, Mckay JK. 2009. Multiple origins promote the ecological amplitude of allopolyploid *Aegilops* (Poaceae). *American Journal of Botany* **96**:1262–1273.

Mickel JT, Beitel JM. 1988. Pteridophyte flora of Oaxaca, Mexico. *Memoirs of the New York Botanical Garden* **46**:1–568.

Mickel JT, Smith AR. 2004. The pteridophytes of Mexico. *Memoirs of the New York Botanical Garden* **88**:1–1055.

Miller MR, Dunham JP, Amores A, Cresko WA, Johnson EA. 2007. Rapid and cost-effective polymorphism identification and genotyping using restriction site associated DNA (RAD) markers. *Genome Research* **17**:240–248.

Nitta JH, Ebihara A, Ito M. 2011. Reticulate evolution in the *Crepidomanes minutum* species complex (Hymenophyllaceae). *American Journal of Botany* **98**:1782–1800.

Oort AH, Yienger JJ. 1996. Observed interannual variability in the Hadley circulation and its connection to ENSO. *Journal of Climate* **9**:2751–2767.

Ortega FJ. 1990. El genero Pteridium en Venezuela: taxonomia y distribucion geografica. *Biollania* **7**:45–54.

Page CN. 1976. The taxonomy and phytogeography of bracken—a review. *Botanical Journal of the Linnean Society* **73**:1–34.

Porter DM. 1979. Endemism and evolution in Galapagos Islands vascular plants. In: Bramwell D, ed. *Plants and islands*. London: Academic Press.

Ranker TA, Floyd SK, Trapp PG. 1994. Multiple colonizations of *Asplenium adiantum-nigrum* onto the Hawaiian archipelago. *Evolution* **48**:1364–1370.

Rassmann K. 1997. Evolutionary age of the Galápagos iguanas predates the age of the present Galápagos Islands. *Molecular Phylogenetics and Evolution* **7**:158–172.

Rothfels CJ, Larsson A, Li FW, Sigel EM, Huiet L, Burge DO, Ruhsam M, Graham SW, Stevenson DW, Wong GKS, Korall P, Pryer KM. 2013. Transcriptome-mining for single-copy nuclear markers in ferns. *PLoS ONE* **8**:e76957.

Rothfels CJ, Johnson AK, Hovenkamp PH, Swofford DL, Roskam HC, Fraser-Jenkins CR, Windham MD, Pryer KM. 2015. Natural hybridization between genera that diverged from each other approximately 60 million years ago. *The American Naturalist* **185**:433–442.

Rumsey FJ, Sheffield E, Haufler CH. 1991. A re-assessment of *Pteridium aquilinum* (L.) Kuhn in Britain. *Watsonia* **18**:297–301.

Schwartsburd PB, De Moraes PLR, Lopes-Mattos KLB. 2014. Recognition of two morpho-types in eastern South American brackens (*Pteridium*—Dennstaedtiaceae—Polypodiopsida). *Phytotaxa* **170**:103–117.

Scotese CR. 2001. *Atlas of Earth history*. Arlington, TX: PALEOMAP Project.

Sessa EB, Zimmer EA, Givnish TJ. 2012. Reticulate evolution on a global scale: A nuclear phylogeny for New World *Dryopteris* (Dryopteridaceae). *Molecular Phylogenetics and Evolution* **64**:563–581.

Sheffield E, Wolf PG, Haufler CH, Ranker TA, Jermy AC. 1989. A re-evaluation of plants referred to as *Pteridium herediae* (Colmeiro) Löve and Kjellqvist. *Botanical Journal of the Linnean Society* **99**:377–386.

Sheffield E, Wolf PG, Rumsey FJ, Robson DJ, Ranker TA, Challinor SM. 1993. Spatial distribution and reproductive behaviour of a triploid braken (*Pteridium aquilinum*) clone in Britain. *Annals of Botany* **72**:231–237.

Shepherd LD, De Lange PJ, Perrie LR. 2009. Multiple colonizations of a remote oceanic archipelago by one species: how common is long-distance dispersal? *Journal of Biogeography* **36**:1972–1977.

Sigel EM, Windham MD, Pryer KM. 2014. Evidence for reciprocal origins in *Polypodium hesperium* (Polypodiaceae): a fern model system for investigating how multiple origins shape allopolyploid genomes. *American Journal of Botany* **101**:1476–1485.

Small RL, Lickey EB, Shaw J, Hauk WD. 2005. Amplification of non-coding chloroplast DNA for phylogenetic studies in lycophytes and monilophytes with a comparative example of relative phylogenetic utility from Ophioglossaceae. *Molecular Phylogenetics and Evolution* **36**:509–522.

Snell HM, Stone PA, Snell HL. 1996. A summary of geographical characteristics of the Galapagos Islands. *Journal of Biogeography* **23**:619–624.

Soltis DE, Soltis PS. 1987. Polyploidy and breeding systems in homosporous pteridophyta: a reevaluation. *The American Naturalist* **130**:219–232.

Soltis PS, Soltis DE. 1991. Multiple origins of the allotetraploid *Tragopogon mirus* (Compositae): rDNA evidence. *Systematic Botany* **16**:407–413.

Soltis PS, Soltis DE, Holsinger KE. 1988. Estimates of intragametophytic selfing and interpopulational gene flow in homosporous ferns. *American Journal of Botany* **75**:1765–1770.

Tan MK, Thomson JA. 1990. Variation of genome size in *Pteridium*. In: Thomson JA, Smith RT, eds. *Bracken 89: bracken biology and management*. Sydney: The Australian Institute of Agricultural Science.

Thomson JA. 2000. Morphological and genomic diversity in the genus *Pteridium* (Dennstaedtiaceae). *Annals of Botany* **85**:77–99.

Thomson JA. 2012. Taxonomic status of diploid southern hemisphere brackens (*Pteridium*: Dennstaedtiaceae). *Telopea* **14**:43–48.

Thomson JA, Alonso-Amelot ME. 2002. Clarification of the taxonomic status and relationships of *Pteridium caudatum* (Dennstaedtiaceae) in Central and South America. *Botanical Journal of the Linnean Society* **140**:237–248.

Thomson JA, Martin AB. 1996. Gnarled trichomes: an understudied character in *Pteridium*. *American Fern Journal* **86**:36–51.

Thomson JA, Mickel JT, Mehltreter K. 2008. Taxonomic status and relationships of bracken ferns (*Pteridium*: Dennstaedtiaceae) of Laurasian affinity in Central and North America. *Botanical Journal of the Linnean Society* **157**:1–17.

Tindale MD, Roy SK. 2002. A cytotaxonomic survey of the Pteridophyta of Australia. *Australian Systematic Botany* **15**:839–937.

Tryon RM. 1941. A revision of the genus *Pteridium*. *Rhodora* **43**:1–67.

Tryon RM, Tryon AF. 1982. *Ferns and allied plants with special reference to Tropical America*. New York: Springer.

Tye A, Francisco-Ortega J. 2011. Origins and evolution of Galapagos endemic vascular plants. In: Bramwell D, Caujapé-Castells J, eds. *The biology of island floras*. Cambridge: Cambridge University Press.

White WM, Mcbirney AR, Duncan RA. 1993. Petrology and geochemistry of the Galápagos islands: portrait of a pathological mantle plume. *Journal of Geophysical Reserch* **98**:19533–19563.

Wright SD, Yong CG, Wichman SR, Dawson JW, Gardner RC. 2001. Stepping stones to Hawaii: a trans-equatorial dispersal pathway for *Metrosideros* (Myrtaceae) inferred from nrDNA (ITS+ETS). *Journal of Biogeography* **28**:769–774.

Yokoya K, Roberts AV, Mottley J, Lewis R, Brandham PE. 2000. Nuclear DNA amounts in roses. *Annals of Botany* **85**:557–561.

Zhou S, Dong W, Chen X, Zhang X, Wen J, Schneider H. 2014. How many species of bracken (*Pteridium*) are there? Assessing the Chinese brackens using molecular evidence. *Taxon* **63**: 509–521.

15

Vertebrate seed dispersers maintain the composition of tropical forest seedbanks

E. M. Wandrag[1,2]*, A. E. Dunham[1], R. H. Miller[3] and H. S. Rogers[1,4]

[1] Biosciences at Rice, Rice University, Houston, TX 77005, USA
[2] Institute for Applied Ecology, University of Canberra, Bruce, ACT 2617, Australia
[3] College of Natural and Applied Sciences, University of Guam, Mangilao, GU 96923, USA
[4] Department of Ecology, Evolution, and Organismal Biology, Iowa State University, Ames, IA 50011 USA

Associate Editor: Anna Traveset

Abstract. The accumulation of seeds in the soil (the seedbank) can set the template for the early regeneration of habitats following disturbance. Seed dispersal is an important factor determining the pattern of seed rain, which affects the interactions those seeds experience. For this reason, seed dispersal should play an important role in structuring forest seedbanks, yet we know little about how that happens. Using the functional extirpation of frugivorous vertebrates from the island of Guam, together with two nearby islands (Saipan and Rota) that each support relatively intact disperser assemblages, we aimed to identify the role of vertebrate dispersers in structuring forest seedbanks. We sampled the seedbank on Guam where dispersers are absent, and compared this with the seedbank on Saipan and Rota where they are present. Almost twice as many species found in the seedbank on Guam, when compared with Saipan and Rota, had a conspecific adult within 2 m. This indicates a strong role of vertebrate dispersal in determining the identity of seeds in the seedbank. In addition, on Guam, a greater proportion of samples contained no seeds and overall species richness was lower than on Saipan. Differences in seed abundance and richness between Guam and Rota were less clear, as seedbanks on Rota also contained fewer species than Saipan, possibly due to increased post-dispersal seed predation. Our findings suggest that vertebrate seed dispersers can have a strong influence on the species composition of seedbanks. Regardless of post-dispersal processes, without dispersal, seedbanks no longer serve to increase the species pool of recruits during regeneration.

Keywords: Bird loss; community ecology; island ecology; mutualisms; plant recruitment; tropical forest ecology.

Introduction

Seeds present in or on the soil (the soil seedbank) provide the template for plant recruitment and can be important for the regeneration of habitats immediately following disturbance (Chandrashekara and Ramakrishnan 1993; Grombone-Guaratini and Rodrigues 2002). Because the spatial pattern of seed deposition can influence the interactions that seeds are involved in, processes that affect those patterns could have important consequences for

* Corresponding author's e-mail address: elizabethwandrag@gmail.com

seed fate (Bigwood and Inouye 1988), with implications for plant population and community dynamics (Beckman and Rogers 2013). Frugivorous vertebrates are a dominant mechanism of seed dispersal in many ecosystems, especially tropical forests where between 70 and 94 % of tree species are estimated to rely on vertebrates for the dispersal of their seeds (Jordano 2000). However, while we increasingly understand how seed dispersers might shape patterns of seed distribution at the plant species level (e.g. Caughlin et al. 2015), we know little about how they may alter the distribution and local community structure of seeds in the seedbank.

For species that rely on vertebrate dispersers for the dispersal of their seeds, there are three key ways in which those dispersers might influence the structure and composition of the seedbank. First, dispersal moves seeds away from parent plants and should decrease the probability that seeds in the seedbank are in close proximity to a conspecific adult. This is important because natural enemies such as fungal pathogens and seed predators often concentrate close to parent plants (Wright 2002; Comita et al. 2014) and dispersal can thus reduce distance-dependent seed mortality associated with proximity to conspecifics (Janzen 1970; Connell 1971). Second, by redistributing seeds within the landscape, seed dispersal may alter the spatial aggregation of seeds, leaving fewer 'seed gaps' on the forest floor (e.g. seedless areas under non-fruiting trees) and reducing areas of high local seed density (i.e. under fruiting trees). A disadvantage of seedless patches within the seedbank would be a reduction in the availability of seeds for seedling regeneration following disturbance, and an increase in density-dependent mortality associated with clustered seed deposition patterns (Russo 2005). Third, by increasing the movement of seeds within the forest, vertebrate seed dispersers expand the available species pool for any given area and should thereby increase the local species richness of seeds present in the soil seedbank.

Understanding the role of vertebrate-mediated seed dispersal in structuring forest communities is increasingly important because vertebrate populations continue to decline from forests around the world (e.g. Savidge 1987; Terborgh et al. 2008). If the loss of vertebrate dispersers has ramifications for forest seedbanks, then this could have implications for forest regeneration and persistence. Nevertheless, few studies have attempted to identify the role of vertebrate dispersers in structuring forest seedbanks. One reason for this is that it is difficult to manipulate entire vertebrate assemblages at scales large enough to meaningfully identify their role at the community level. Loss of vertebrate species from forests by hunting or other anthropogenic pressures may provide

one way in which to examine the role of vertebrate seed dispersal in structuring seedling communities (Terborgh et al. 2008; Effiom et al. 2013; Harrison et al. 2013). However, such forests are rarely completely free of vertebrate dispersers, and none of these studies have examined the seedbank.

We take advantage of an 'accidental experiment' (HilleRisLambers et al. 2013) provided by the functional extirpation of frugivorous vertebrates from the island of Guam to test predictions about the role of native vertebrate dispersers in determining both the distribution and local species composition of seeds of tree species in tropical forest seedbanks. The introduction of the brown tree snake, Boiga irregularis (Colubridae), to Guam in the 1940s resulted in the extinction of native bird species, including four of the five frugivorous birds that were previously present (Savidge 1987) **[see Supporting Information—Table S1]**. The fifth frugivorous bird experienced extreme range contractions and is now functionally extinct from Guam's forests. In contrast, the nearby islands of Saipan (Camp et al. 2009) and Rota (Camp et al. 2014), which both support similar limestone forest to Guam, have a more intact disperser assemblage, though some populations in Rota are declining (Camp et al. 2014). In addition to frugivorous birds, there is one species of frugivorous bat native to the Mariana Islands (Pteropus mariannus Desmarest). However, this bat species is also now functionally extinct from forests on both Guam and Saipan, and present only on Rota in reduced abundance. These native dispersers have not been replaced by non-native species, with the potential exception of feral pigs (Sus scrofa, present on Guam and Rota but excluded from plots used in this study). Since ~85 % of the tree species present in the Marianas have seeds adapted for dispersal by vertebrate frugivores (H. S. Rogers, unpubl. data), this system provides a rare opportunity to examine the consequences of the functional extirpation of native vertebrate dispersers for the seedbank.

We examined the distribution and composition of seeds in the seedbanks of forests with native frugivorous vertebrates (Saipan and Rota) when compared with those without (Guam). We predicted that if vertebrate seed dispersers are important for maintaining forest seedbanks, then we would see differences in the distribution and composition of the seedbank on Guam relative to Saipan and Rota. We hypothesized that the presence of frugivorous vertebrates on Saipan and Rota would be associated with greater seed movement that would result in (i) a greater proportion of species in the seedbank that lack nearby conspecific adults, (ii) a more regular distribution of seeds (i.e. fewer sites that are either devoid of seeds or have high seed densities) and (iii) higher species richness per seedbank sample.

Methods

Study area

The islands of Guam, Saipan and Rota are located within the Mariana Island chain, in the Western Pacific (Fig. 1). They have a mean annual temperature of ~27 °C, with little seasonal variation. All islands experience frequent typhoons, which can cause considerable damage to vegetation (Kerr 2000), and pronounced wet and dry seasons.

We conducted our study within limestone forest, which overlies karst formed by uplifted coral plateaus. The forest is moist, broadleaved and evergreen (Mueller-Dombois and Fosberg 1998) and characterized by species such as *Aglaia mariannensis* (Meliaceae); *Artocarpus mariannensis* (Moraceae); *Cynometra ramiflora* (Leguminosae); *Elaeocarpus joga* (Elaeocarpaceae); *Ficus prolixa* (Moraceae); *Meiogyne cylindrocarpa* (Annonaceae, previously known as *Guamia mariannae*); *Ochrosia mariannensis* and *O. oppositifolia* (Apocynaceae, previously known as *Neisosperma oppositifolia*); *Pandanus dubius* and *P. tectorius* (Pandanaceae); *Pisonia grandis* (Nyctaginaceae) and *Premna serratifolia* (Lamiaceae, previously known as *Premna obtusifolia*). A survey of this forest type in Saipan in 1992 recorded 27 tree species occupying the canopy, with a further 22 in the understorey (Craig 1992). The forest has a particularly short canopy with most trees <11 m tall (Donnegan et al. 2011), likely as a result of the frequent typhoons.

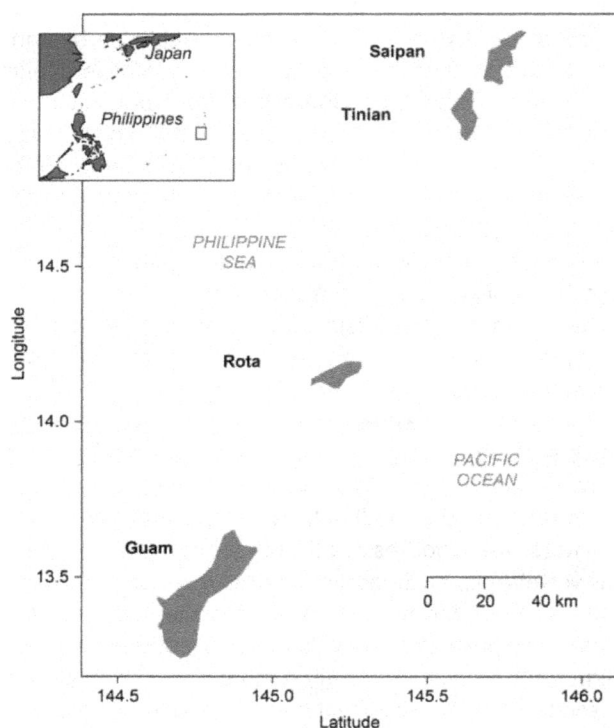

Figure 1. Map showing the location of Guam, Rota and Saipan.

Soil sampling

We collected soil samples between December 2013 and January 2014 to identify relative differences in the distribution and composition of the seedbank between islands. Although seedbank composition may vary seasonally (Dalling et al. 1997), we were primarily interested in identifying relative differences between islands at a single time point. Peak fruiting occurs between May and August in these islands, so we expect samples from December and January to be dominated by seeds waiting for an opportunity to germinate from the seedbank rather than recently fallen seeds.

We sampled within 44 plots spread across 11 forest sites on 3 islands. These sites were established between 2008 and 2009 as part of a long-term forest research project. There are five sites on Guam and three on each of Saipan and Rota. Within each of these sites, four plots ranging in size between 8 and 12 m² were demarcated for a separate experiment. This gave a total of 20 plots sampled on Guam and 12 each on Saipan and Rota. Plots were at least 20 m apart and centred on at least one of three common forest species (*A. mariannensis*, *C. ramiflora* or *M. cylindrocarpa*). All three species fruit during the peak fruiting season, with some low-level fruiting throughout the year. *Aglaia mariannensis* and *M. cylindrocarpa* are fleshy fruited. We chose to centre the sites on three of the most abundant tree species in the forest because we expected these species to have widespread seed rain due to their high abundance, and thus, differences in the seedbank around these species are more likely to reflect differences in dispersal of other species rather than differences in the canopy above each sample.

Plots were fenced during the peak fruiting season between 4 and 6 months before sampling to exclude invasive deer (*Cervus mariannus*) and feral pigs. Deer are thought to be primarily browsers in this system, so are unlikely to affect seed density. Pigs can act as both seed predators and seed dispersers (Sanguinetti and Kitzberger 2010; O'Connor and Kelly 2012), and there is some evidence they may do so in this system (A. Gawel, pers. comm.). Deer and pigs are present on Guam and Rota, but absent or at low densities on Saipan.

We took between 3 and 12 samples from each plot, depending on the heterogeneity of the substrate. Because we expected that the seedbank might vary based on the substrate at a particular microsite, we sampled separately from each of the four primary substrate types: soil, rocky soil, loose karst and solid karst **[see Supporting Information—Text]**. Within each plot, we took three soil samples from each substrate type that comprised at least 20 % of the forest floor. Each sample was separated by at least 1.5 m from a previous sample of the same substrate, but not necessarily a previous sample from a different substrate. If a site contained only 1 substrate, we took up to 3 samples, whereas if it contained all 4 substrates, we took up to 12 samples depending on the availability of each substrate.

We sampled using 0.15 × 0.15 m quadrats because the use of soil cores was not possible on the karst substrate, and soil, when present, rarely exceeded 6 cm in depth. Litter and soil samples were combined for each sample. For bare or loose karst areas, we searched within each quadrat for 5 min, using tweezers to extract seeds where necessary. Where moveable rocks were present on loose karst, we lifted rocks where necessary/possible and searched the area underneath, up to a maximum depth of 6 cm. For rocky soil and soil, we used a trowel to collect all soil and leaf litter up to a maximum depth of 6 cm, or less if bedrock was reached.

In processing samples, all visible seeds were removed from both soil and litter. The remaining soil was sieved to break up any large lumps and searched again. We considered whole, intact seeds as viable, and counted only those seeds. The seeds of herbaceous species and vines were not included, and we focussed only on seeds of

tree species. The smallest tree seed within these forests is *Pipturus argenteus* (Urticaceae), measuring 0.64 mm², which is visible by eye. For the genera *Ficus* and *Eugenia*, which both have more than one species present in these forests, seeds were assigned to genus level only. Only one primarily abiotically dispersed tree species was found in our seed samples (*Leucaena leucocephala*); however, the seeds of this species have previously been reported as dispersed by rodents, birds and cattle (Pacific Island Ecosystems at Risk 2012) (Table 1).

Distance to conspecific

We tested whether seeds were more likely to have a reproductively mature conspecific (i.e. conspecific adult) nearby in the absence of seed dispersers. We assumed that it would be unlikely for most seeds to arrive via gravity dispersal from a parent tree that is >2 m away, given the low stature of the forest canopy. To identify nearby adult conspecifics likely providing seeds via gravity, we surveyed all adult trees with a canopy overlapping each plot or within a distance of 2 m from the edge of the plot from which the sample was taken. These data

Table 1. Dispersal syndrome used in analyses for each species of seed recorded in the seedbank on each of the three islands, mean size of seeds and the island on which each species was recorded in the seedbank. The total number of seeds recorded and total number samples taken on each island are given. [1]A bird dispersal syndrome was assigned based on whether fruits of the species have previously been recorded as eaten by birds or based on the presence or absence of a fruity pulp and the size of a seed: where pulp was present and seeds were small enough to be consumed or carried by the largest vertebrate frugivore that once occurred on the island a species considered to be adapted for vertebrate dispersal (as in Caves et al. 2013). [2]E. Fricke, unpublished data, unless otherwise stated. [3]Although showing no adaptations for vertebrate seed dispersal, seeds of this species are reported to have been dispersed by rodents, birds and cattle (Pacific Island Ecosystems at Risk 2012). [4]http://pages.bangor.ac.uk/~afs101/iwpt/web-sp7.htm. [5]Wiles and Fujita (1992).

Species	Dispersal syndrome[1]	Approx. seed area (mm²)[2]	Island		
			Guam	Rota	Saipan
Aglaia mariannensis	Bird/bat	182.4	15	2	3
Aidia cochinchinensis	Bird	3.3	0	0	1
Eugenia spp.	Bird	96.7	27	0	1
Ficus spp.	Bird/bat	0.9	0	2	0
Guamia mariannae	Bird	89.3	9	0	2
Leucaena leucocephala	Gravity/wind[3]	21.0[4]	0	1	194
Macaranga thompsonii	Bird	10.6	187	6	0
Melanolepis multiglandulosa	Bird	18.3	1	1	20
Morinda citrifolia	Bird	35.3	79	0	0
Ochrosia mariannensis	Bat[5]	10.1	15	0	2
Ochrosia oppositifolia	Bat[5]	354.8	44	0	0
Carica papaya	Bird/bat	20.0	0	14	10
Premna obtusifolia	Bird/bat	9.2	7	3	67
Psychotria mariana	Bird	29.2	0	11	161
	Total number of seeds (total number of samples)		384 (130)	40 (56)	461 (68)

were also used to calculate the species richness of the surrounding canopy.

Analysis

To account for the nestedness of our design, we fitted linear mixed models to our data. We predicted that if frugivorous vertebrates are an important determinant of the distribution and composition of seeds present in forest seedbanks, then we would see differences between Guam and the two islands with dispersers in each of the variables we examined. We, therefore, assessed whether island was an important predictor of variation in each of our response variables, with site included as a random effect in each model. Additional fixed and random effects were assessed or included where relevant, and we detail those for each response variable below. To determine whether the inclusion of seeds from the only potentially wind-dispersed species in the study, *L. leucocephala*, affected each of the qualitative results, we ran analyses both with and without *L. leucocephala*.

We fitted models in R (R Development Core Team 2015) using the package lme4 (Bates and Maechler 2014). We assessed the significance of island as a fixed effect by comparing a model that included it with one that included only an intercept term using likelihood ratio tests. For models fitted using a normal distribution, we assessed the significance of differences between islands using Satterthwaite's approximation for degrees of freedom within the package lmerTest (Kuznetsova *et al.* 2014), with models re-levelled to enable pairwise comparisons between islands. For models fitted using a binomial or Poisson distribution, we assessed these differences using the Wald Z test in lme4.

First, we tested the hypothesis that greater seed movement in the presence of frugivores would be associated with a lower proportion of species in the seedbank with a conspecific adult within 2 m. We recorded whether or not (one or zero, respectively) a species recorded in the seedbank at each plot had a conspecific neighbour within 2 m, i.e. we had one value per species per plot. Because multiple species were often recorded within each plot, we included plot as an additional random effect in the model, nesting it within site. By recording the presence of a conspecific neighbour at the level of the species rather than seed, this measure is independent of seed number, which can vary among species. Instead, this measure reflects the proportion of species found in the seedbank that could only have arrived in the seedbank through dispersal. We specified a binomial error distribution.

Second, we tested whether frugivore absence would be associated with a patchy distribution of seeds by examining two response variables. Because we obtained between 3 and 12 samples per plot (depending on substrate), we examined the within-plot variation in the number of seeds per sample. For each plot, we calculated the coefficient of variation (CV) of the total number of seeds per sample. No seeds were recorded in 3 of the 44 plots. We excluded these plots from the analysis as we were specifically interested in testing for within-plot variation in seed density. Data were normally distributed and no transformations were made. In addition, we determined the proportion of samples that contained zero seeds on each island. Here, we specified a binomial error distribution. As multiple samples were taken per plot, we included 'plot' as an additional random effect, nesting it within site. We also examined the potential for substrate (which was recorded at the sample level) to influence the proportion of samples that contained seeds by including substrate as a fixed effect in the model and comparing this model with one that did not include substrate as a fixed effect using a likelihood ratio test.

Finally, to examine whether frugivores would increase the small-scale species richness of seeds in the seedbank, we quantified the mean number of species per sample. We examined the potential for the species richness of the surrounding canopy to influence seedbank richness by including the number of adult tree species recorded within 2 m of each plot as a fixed effect in the model, and comparing this model with one that did not include the number of adult tree species as a fixed effect using a likelihood ratio test. Because adult tree species richness was calculated at the plot level, we modelled mean number of species per sample by summing together the number of species recorded in each sample at each plot to give one value per plot (such that if a species was recorded in two samples it would count twice) and offsetting this by the number of samples taken at each plot. We specified a Poisson distribution.

We calculated marginal R^2 values $(R_m{}^2)$ and conditional R^2 values $(R_c{}^2)$ for the final model used in each case (Nakagawa and Schielzeth 2013) using the MuMIn R package (Barton 2014). Marginal R^2 values are those due to fixed effects only, while conditional R^2 values are those due to fixed plus random effects.

Results

Overall, the number of seeds we recorded in samples was low and variable (Table 1). We recorded few seeds of the focal tree species in seedbank samples. As expected in our system, the majority of species we recorded are primarily dispersed by vertebrates, with the exception of one species common on Saipan, *L. leucocephala*.

Proportion of seeds with a conspecific within 2 m

Island was a significant predictor of the proportion of species in the seedbank with a conspecific adult present within 2 m ($\chi^2 = 13.86$, df = 2, $P < 0.001$, $R_m{}^2 = 0.29$,

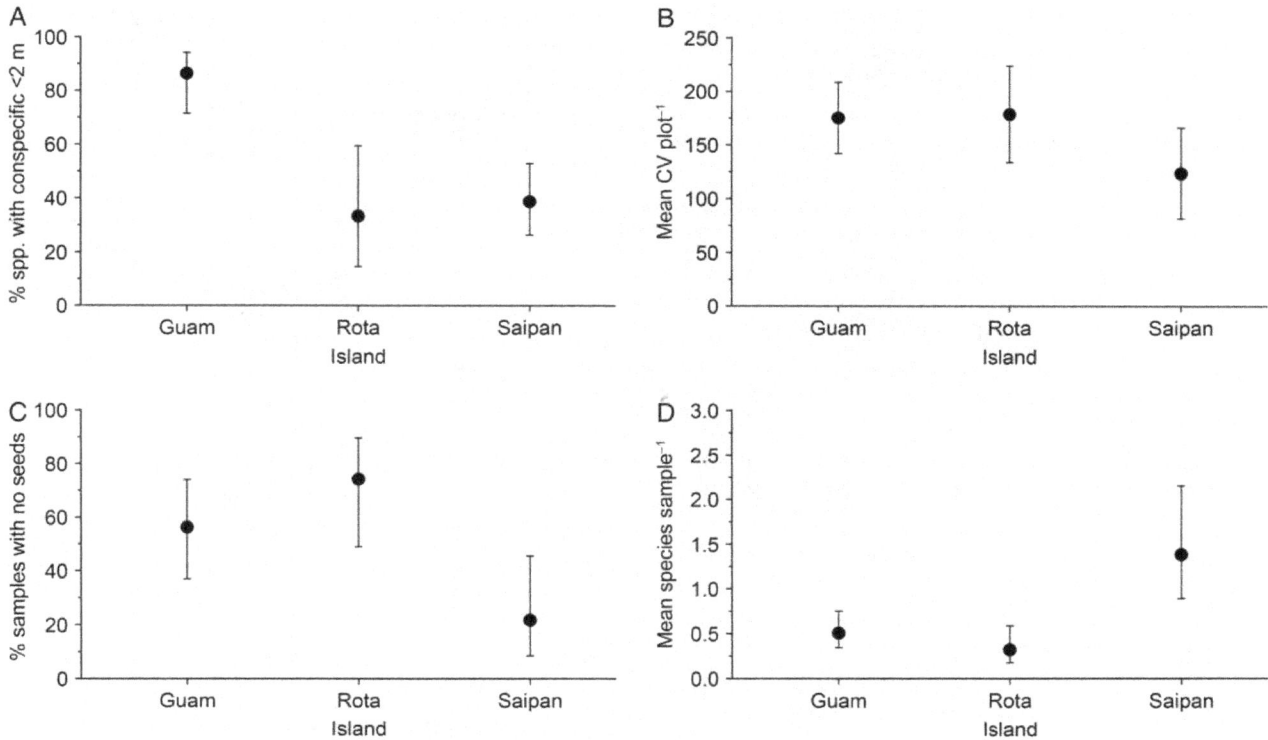

Figure 2. The per cent of species found in the seedbank at each plot that had an adult conspecific within 2 m on Guam where dispersers are functionally absent relative to Rota and Saipan where they are present (A), mean CV in seed density per seedbank sample at each plot (B), the per cent of seedbank samples that lacked any seeds (C) and the mean number of species per seedbank sample (D). Bars represent 95 % confidence intervals.

$R_c^2 = 0.29$). Species sampled within Guam's seedbank were more likely to have conspecific adults present within 2 m than on both Rota ($z = -3.50, P < 0.001$) and Saipan ($z = -4.11, P < 0.001$). While 86.5 % of species in the seedbank had a conspecific adult nearby on Guam, this was only true for 33.3 and 38.8 % of species in the seedbank on Rota and Saipan, respectively (Fig. 2A).

Spatial distribution of seeds within the seedbank

There were no differences between islands in the between-sample variation in seed density ($\chi^2 = 4.69, df = 2, P > 0.05, R_c^2 = 0.20 R_c^2 = 0.20$). Although the CV was lower for samples on Saipan when compared with Guam (Fig. 2B), this was not significant ($t = -2.19, P = 0.05$).

Substrate was not an important predictor of the per cent of samples that contained zero seeds ($\chi^2 = 6.23, df = 3, P = > 0.05$) and was excluded from the final model. There was a significant influence of island on the per cent of samples that contained zero seeds ($\chi^2 = 7.36, df = 2, P = 0.03, R_m^2 = 0.14, R_c^2 = 0.35$). While more than half of samples on both Guam (56.3 %) and Rota (74.2 %) were devoid of seeds, less than a quarter (21.7 %) of samples on Saipan lacked seeds (Fig. 2C). This difference was significant when comparing Saipan with both Guam ($z = -2.26, P = 0.02$) and Rota ($z = -2.99$,

$P = 0.003$). However, if we excluded seeds of *L. leucocephala*, the only species we recorded that is primarily wind or gravity dispersed and was found predominantly on Saipan, the difference between Guam and Saipan was no longer significant ($z = -1.87, P = 0.06$) **[see Supporting Information—Figure S1]**.

Species richness of the seedbank

A total of 14 species were recorded in the seedbank across the three islands (Table 1). The species richness of the surrounding canopy was not a significant predictor of the species richness of the seedbank ($\chi^2 = 0.42, df = 1, P > 0.05$) and was not included in the final model. The mean number of species per sample was low overall, but again varied between the three islands ($\chi^2 = 10.55, df = 2, P < 0.01, R_m^2 = 0.22, R_c^2 = 0.28$). Saipan had greater species richness than both Guam ($z = 3.36, P < 0.001$) and Rota ($z = 3.85, P < 0.001$) with 1.39 species per sample, compared with only 0.32 species per sample on Rota and 0.51 species per sample on Guam.

Discussion

The soil seedbank is an important source of regeneration following disturbance, but our understanding of the processes that determine the distribution and composition of seeds in the seedbank at the community level remains limited. We demonstrate a role of vertebrate frugivores in building and maintaining forest seedbanks. On the island of Guam, where most vertebrate frugivores have been absent for about 30 years, >80 % of seeds in the seedbank were found to have a conspecific adult neighbour. This was in contrast to the seedbank sampled on the nearby islands of Saipan and Rota (which support vertebrate frugivores) where a majority of seeds on the forest floor had no conspecific neighbour and thus were likely dispersed there from >2 m away. In addition, a greater proportion of samples lacked seeds, and the species richness of seedbank samples was lower on Guam relative to Saipan, as predicted if dispersers influence the spatial pattern and diversity of seedbanks. However, we did not find differences in the variability of seed densities or per sample species richness between Guam and Rota, indicating that the presence of dispersers alone is not sufficient to explain these patterns. We hypothesize that high rates of post-dispersal predation may be responsible for the reduced density and richness of seeds in the seedbank on Rota. Since seedbanks strongly reflect the surrounding trees when dispersers are lost from a system, patterns of forest regeneration are unlikely to be maintained by recruitment from either persistent or transient seedbanks.

Species in the seedbank escape conspecific adults where dispersers are present

Seed dispersal is considered important for moving seeds away from parent trees. Here, we demonstrate the magnitude of that effect for seeds in seedbanks: while 60–70 % of seeds found in the seedbank on islands with native vertebrate frugivores are likely the result of dispersal, in the absence of dispersers, >80 % of seeds are likely from nearby adult trees. This pattern in the seedbank is mirrored by seedling communities in other defaunated sites around the world, where the seedlings closely reflect the identity of the nearby adults (Terborgh et al. 2008; Harrison et al. 2013). The potential impact of recent declines in some frugivorous bird species on Rota (Camp et al. 2014) was not evident in our study, as a similar per cent of seeds in the seedbanks on Rota as on Saipan likely arrived through vertebrate dispersal (i.e. did not have a conspecific adult within 2 m). The failure of seeds present in the seedbanks on Guam to escape their parent plants could have implications for the role of the seedbank in forest regeneration because seeds landing in close proximity to conspecifics are predicted to experience an increase in distance-dependent mortality (Dalling et al. 1998; Kotanen 2007).

Seed dispersal results in a more even distribution of seeds

Seed dispersal is not only important for moving seeds away from parent plants but also for reducing density-dependent mortality by redistributing seeds within the landscape (Comita et al. 2014). Without dispersal, seeds should fall in higher densities underneath parent trees and fail to reach sites away from parent trees, leading to greater variation in seed density across the landscape. However, we found no evidence that seed density per sample was more variable on Guam than on other islands, although more samples lacked seeds on Guam than on Saipan.

We have identified four possible reasons for the increase in seedless areas on Guam relative to Saipan. First, we expect that more seeds would experience density- or distance-dependent mortality when not being moved away from conspecifics. High distance-dependent mortality associated with proximity to parent trees may explain why few seeds of the three target tree species were recorded in our sample. An alternative explanation is that seeds from the focal species may not persist in the seedbank, as is common with larger-seeded species (Hopkins and Graham 1983). Second, if seed longevity in the seedbanks is low across all species, which might be expected given the shallow soils and moist conditions of these islands, the seedbank in the absence of dispersal would only contain seeds if it is within close proximity to a tree that has recently reproduced. Third, the pattern could be driven by the presence of a non-native wind-dispersed tree species, Leucaena, which is more common on Saipan. However, while the difference between Saipan and Guam is no longer significant when Leucaena seeds are omitted, the trend that Saipan has fewer samples without seeds remains. Finally, post-dispersal process such as seed predation may vary across islands, which we discuss in more detail below.

Counter to our predictions, the seedbank on Rota more closely resembled that on Guam in terms of the proportion of samples that lacked seeds, and overall seed abundance (Table 1). Since there appears to be adequate dispersal on Rota, with >65 % of seeds coming from trees >2 m away, we hypothesize that post-dispersal seed predators are responsible. The two most likely post-dispersal predators on Rota are the Malayan black rat (Rattus rattus diardii Jentink) and the Cuban slug. Rat densities are similar between Rota and Saipan and higher on both those islands than on Guam (Wiewel et al. 2009), indicating that seed predation by rats is unlikely the reason for the varying seed densities between Saipan and Rota. Instead, it is possible that the Cuban slug (Veronicella cubensis L. Pfeiffer), a seed predator that is considered a major pest on Rota but is only present in low abundance on Guam and Saipan (Robinson and Hollingsworth 2004; H. S. Rogers, unpubl. data), is responsible for this change. The limited seedbank present in Rota could, therefore, indicate the potential for post-dispersal seed predation to decrease the role of the seedbank in forest regeneration. If so, further work will be needed to tease apart the relative contribution of seed dispersal and seed predation to the spatial distribution of seedbanks.

Seed dispersal promotes the species richness of the seedbank

Although the movement of seeds away from parent trees and reduction of distance- and density-dependent mortality are generally considered key mechanisms through which the species richness of forests is maintained (Harms *et al.* 2000; Mangan *et al.* 2010), the importance of biotic seed dispersal in structuring forest communities is still often overlooked. One reason for this is that when dispersers are lost from a system, the impacts on adult tree communities may not be seen for several decades (Terborgh 2013) or even centuries (Kelly *et al.* 2004). However, recent evidence demonstrates an impact of disperser loss on the species composition of the seedling community (Terborgh *et al.* 2008; Effiom *et al.* 2013; Harrison *et al.* 2013), a finding that highlights the importance of biotic seed dispersal in structuring forest communities over even short timescales. Our finding that the seedbank has fewer species not present in the surrounding canopy and lower species richness in the absence of vertebrate frugivores suggests that biotic seed dispersal has an important role in determining the species composition of soil seedbanks. Since the early regeneration of forests following disturbance is expected to start with the seedbank (Chandrashekara and Ramakrishnan 1993; Grombone-Guaratini and Rodrigues 2002), we would expect any influence on the species composition of the seedbank to translate to the regenerating seedling community.

Conclusions

The seedbank is thought to be important for storing seeds until the conditions are right for germination, and for facilitating early regeneration after disturbance (Grime 1989; Grombone-Guaratini and Rodrigues 2002). In that way, seedbanks may have a role in buffering forests against short-term losses in seed input (Thompson 2000). By comparing the seedbanks on islands with and without seed dispersers, we demonstrate a role of seed dispersal in structuring forest seedbanks. We show that seed dispersal is important for moving seeds away from adult conspecifics and maintaining the species richness of the seedbank. These findings not only highlight the

importance of dispersers for building and maintaining forest seedbanks, but suggest another mechanism by which current global declines in vertebrate assemblages could have important implications for forest persistence.

Sources of Funding

Funding was provided by a United States National Science Foundation award DEB-1258148 to H.S.R., A.E.D., R.H.M. and H.S.R. was also supported by a Huxley Fellowship from the Department of BioSciences at Rice University.

Contributions by the Authors

E.M.W., H.S.R., A.E.D. and R.H.M. conceived the idea. E.M.W. and H.S.R. collected the data and carried out the analyses. E.M.W. led the writing with assistance from H.S.R., A.E.D. and R.H.M.

Acknowledgements

We thank Anthony Castro, Lauren Thompson, Micah Freedman, Alexandra Kerr and Kyle Ngiratregd for field assistance and data entry. We thank the Guam Department of Agriculture, the CNMI Division of Fish and Wildlife, Guam National Wildlife Refuge at Ritidian Point, the US Fish and Wildlife Service and the Blas family for permission and support to work on their lands.

Supporting Information

The following additional information is available in the online version of this article –

Table S1. Lists bird species present or previously present on Guam, Rota and Saipan.

Text. Description of the four substrate types present in the forests of each island.

Figure S1. Results of each analysis when *Leucaena leucocephala*, the only species predominantly dispersed by wind or gravity, is excluded.

Literature Cited

Barton K. 2014. *Package 'MuMIn': multi-model inference.* R package version 1.13.4. https://cran.r-project.org/web/packages/MuMIn/index.html (8 September 2015).

Bates DM, Maechler M. 2014. *Linear mixed-effects models using S4 classes.* R package version 1.1-6. https://cran.r-project.org/web/packages/lmerTest/lmerTest.pdf (29 July 2015).

Beckman NG, Rogers HS. 2013. Consequences of seed dispersal for plant recruitment in tropical forests: interactions within the seedscape. *Biotropica* **45**:666–681.

Bigwood DW, Inouye DW. 1988. Spatial pattern analysis of seed banks: an improved method and optimized sampling. *Ecology* **69**:497–507.

Camp RJ, Pratt TK, Marshall AP, Amidon F, Williams LL. 2009. Recent status and trends of the land bird avifauna on Saipan, Mariana Islands, with emphasis on the endangered Nightingale Reed-warbler *Acrocephalus luscinia*. *Bird Conservation International* **19**:323–337.

Camp RJ, Brinck KW, Gorreson MP, Amidon FA, Radley PM, Berkowitz PS, Banko PC. 2014. *Status of forest birds on Rota, Mariana Islands*. Hilo, Hawai'i: Hawai'i Cooperative Studies Unit.

Caughlin TT, Ferguson JM, Lichstein JW, Zuidema PA, Bunyavejchewin S, Levey DJ. 2015. Loss of animal seed dispersal increases extinction risk in a tropical tree species due to pervasive negative density dependence across life stages. *Proceedings of the Royal Society B: Biological Sciences* **282**:20142095.

Caves EM, Jennings SB, HilleRisLambers J, Tewksbury JJ, Rogers HS. 2013. Natural experiment demonstrates that bird loss leads to cessation of dispersal of native seeds from intact to degraded forests. *PLoS ONE* **8**:e65618.

Chandrashekara UM, Ramakrishnan PS. 1993. Germinable soil seed bank dynamics during the gap phase of a humid tropical forest in the Western Ghats of Kerala, India. *Journal of Tropical Ecology* **9**:455–467.

Comita LS, Queenborough SA, Murphy SJ, Eck JL, Xu K, Krishnadas M, Beckman N, Zhu Y. 2014. Testing predictions of the Janzen-Connell hypothesis: a meta-analysis of experimental evidence for distance- and density-dependent seed and seedling survival. *Journal of Ecology* **102**:845–856.

Connell JH. 1971. On the role of natural enemies in preventing competitive exclusion in some marine animals and in rain forest trees. *Dynamics of Populations* **298**:312.

Craig R. 1992. Ecological characteristics of a native limestone forest on Saipan Mariana Islands. *Micronesica* **25**:85–97.

Dalling JW, Swaine MD, Garwood NC. 1997. Soil seed bank community dynamics in seasonally moist lowland tropical forest, Panama. *Journal of Tropical Ecology* **13**:659–680.

Dalling JW, Swaine MD, Garwood NC. 1998. Dispersal patterns and seed bank dynamics of pioneer trees in moist tropical forest. *Ecology* **79**:564–578.

Donnegan JA, Butler SL, Kuegler O, Hiserote BA. 2011. *Commonwealth of the Northern Mariana Islands' forest resources, 2004*. Portland, OR: US Department of Agriculture, Forest Service, Pacific Northwest Research Station.

Effiom EO, Nuñez-Iturri G, Smith HG, Ottosson U, Olsson O. 2013. Bushmeat hunting changes regeneration of African rainforests. *Proceedings of the Royal Society B: Biological Sciences* **280**: 20130246.

Grime J. 1989. Seed banks in ecological perspective. In: Leck MA, Parker VT, Simpson RL, eds. *Ecology of soil seed banks*. San Diego: Academic Press.

Grombone-Guaratini MT, Rodrigues RR. 2002. Seed bank and seed rain in a seasonal semi-deciduous forest in south-eastern Brazil. *Journal of Tropical Ecology* **18**:759–774.

Harms KE, Wright SJ, Calderón O, Hernández A, Herre EA. 2000. Pervasive density-dependent recruitment enhances seedling diversity in a tropical forest. *Nature* **404**:493–495.

Harrison RD, Tan S, Plotkin JB, Slik F, Detto M, Brenes T, Itoh A, Davies SJ. 2013. Consequences of defaunation for a tropical tree community. *Ecology Letters* **16**:687–694.

HilleRisLambers J, Ettinger AK, Ford KR, Haak DC, Horwith M, Miner BE, Rogers HS, Sheldon KS, Tewksbury JJ, Waters SM, Yang S. 2013. Accidental experiments: ecological and evolutionary insights and opportunities derived from global change. *Oikos* **122**:1649–1661.

Hopkins MS, Graham AW. 1983. The species composition of soil seed banks beneath lowland tropical rainforests in North Queensland, Australia. *Biotropica* **15**:90–99.

Janzen DH. 1970. Herbivores and the number of tree species in tropical forests. *The American Naturalist* **104**:501–528.

Jordano P. 2000. Fruits and frugivory. In: Fenner M, ed. *Seeds: the ecology of regeneration in plant communities*. Wallingford, UK: CABI Publishing.

Kelly D, Ladley JJ, Robertson AW. 2004. Is dispersal easier than pollination? Two tests in New Zealand Loranthaceae. *New Zealand Journal of Botany* **42**:89–103.

Kerr AM. 2000. Defoliation of an island (Guam, Mariana Archipelago, Western Pacific Ocean) following a saltspray-laden 'dry' typhoon. *Journal of Tropical Ecology* **16**:895–901.

Kotanen PM. 2007. Effects of fungal seed pathogens under conspecific and heterospecific trees in a temperate forest. *Canadian Journal of Botany* **85**:918–925.

Kuznetsova A, Brockhoff P, Christensen R. 2014. *lmerTest: tests for random and fixed effects for linear mixed effect models (lmer objects of lme4 package)*. R package 2.0-11. https://cran.r-project. org/web/packages/lmerTest/lmerTest.pdf (21 September 2015).

Mangan SA, Schnitzer SA, Herre EA, Mack KML, Valencia MC, Sanchez EI, Bever JD. 2010. Negative plant–soil feedback predicts tree-species relative abundance in a tropical forest. *Nature* **466**:752–755.

Mueller-Dombois D, Fosberg FR. 1998. *Vegetation of the tropical Pacific islands*. New York: Springer.

Nakagawa S, Schielzeth H. 2013. A general and simple method for obtaining R^2 from generalized linear mixed-effects models. *Methods in Ecology and Evolution* **4**:133–142.

O'Connor S-J, Kelly D. 2012. Seed dispersal of matai (*Prumnopitys taxifolia*) by feral pigs (*Sus scrofa*). *New Zealand Journal of Ecology* **36**:228–231.

Pacific Island Ecosystems at Risk. 2012. *Leucaena leucocephala (Lam.) de Wit, Fabaceae (Leguminosae): plant threats to Pacific ecosystems*. www.hear.org/pier/species/leucaena_leucocephala. htm (September 2015).

R Development Core Team. 2015. *R: a language and environment for statistical computing*. Version 3.2.1. Vienna, Austria: R Foundation for Statistical Computing.

Robinson D, Hollingsworth R. 2004. Survey of slug and snail pests on subsistence and garden crops in the islands of the American Pacific: Guam and the Northern Mariana Islands. Part 1. The leatherleaf slugs (Family Veronicellidae). Commonwealth of the Northern Mariana Islands' forest resources. USDA. http://www.botany.hawaii.edu/ basch/uhnpscesu/pdfs/sam/Robinson2006AS.pdf (8 June 2015).

Russo SE. 2005. Linking seed fate to natural dispersal patterns: factors affecting predation and scatter-hoarding of *Virola calophylla* seeds in Peru. *Journal of Tropical Ecology* **21**: 243–253.

Sanguinetti J, Kitzberger T. 2010. Factors controlling seed predation by rodents and non-native *Sus scrofa* in *Araucaria araucana* forests: potential effects on seedling establishment. *Biological Invasions* **12**:689–706.

Savidge JA. 1987. Extinction of an island forest avifauna by an introduced snake. *Ecology* **68**:660–668.

Terborgh J. 2013. Using Janzen–Connell to predict the consequences of defaunation and other disturbances of tropical forests. *Biological Conservation* **163**:7–12.

Terborgh J, Nuñez-Iturri G, Pitman NCA, Valverde FHC, Alvarez P, Swamy V, Pringle EG, Paine CET. 2008. Tree recruitment in an empty forest. *Ecology* **89**:1757–1768.

Thompson K. 2000. The functional ecology of soil seed banks. In: Fenner M, ed. *Seeds: the ecology of regeneration in plant communities*. Wallingford, UK: CABI Publishing.

Wiewel AS, Yackel Adams AA, Rodda GH. 2009. Distribution, density, and biomass of introduced small mammals in the Southern Mariana Islands. *Pacific Science* **63**:205–222.

Wiles GJ, Fujita MS. 1992. Food plants and economic importance of flying foxes on Pacific islands. In: Wilson DE, Graham GL, eds. *Pacific island flying foxes: Proceedings of an international conservation conference*. US Fish and Wildlife Service Biological Report 90 (23). WA, USA: US Fish and Wildlife Service, 24–35.

Wright JS. 2002. Plant diversity in tropical forests: a review of mechanisms of species coexistence. *Oecologia* **130**:1–14.

Permissions

List of Contributors

Manuel Nogales
Island Ecology and Evolution Research Group (CSIC-IPNA), C/Astrofísico Francisco Sánchez n8 3, 38206, La Laguna, Tenerife, Canary Islands, Spain

Tomás A. Carlo
Department of Biology, Pennsylvania State University, 208 Mueller Laboratory, University Park, PA 16802, USA

Aarón González-Castro
Island Ecology and Evolution Research Group (CSIC-IPNA), C/Astrofísico Francisco Sánchez n8 3, 38206, La Laguna, Tenerife, Canary Islands, Spain
Department of Biology, Pennsylvania State University, 208 Mueller Laboratory, University Park, PA 16802, USA
Instituto de Ciencia Innovación Tecnología y Saberes Universidad Nacional de Chimborazo, Avenida Antonio Joséde Sucre, Riobamba, Ecuador

Suann Yang
Department of Biology, Pennsylvania State University, 208 Mueller Laboratory, University Park, PA 16802, USA
Biology Department, Presbyterian College, 503 South Broad Street, Clinton, SC 29325, USA

Inger Greve Alsos
Tromsø Museum, University of Tromsø, NO-9037 Tromsø, Norway

Dorothee Ehrich
Department of Arctic and Marine Biology, Faculty of Biosciences, Fisheries and Economics, University of Tromsø, NO-9037 Tromsø, Norway

Pernille Bronken Eidesen
The University Centre in Svalbard, NO- 9171 Longyearbyen, Norway

Heidi Solstad
Museum of Natural History and Archaeology, Norwegian University of Science and Technology, NO-7491 Trondheim, Norway

Kristine Bakke Westergaard
Norwegian Institute for Nature Research, Sluppen, NO-7485 Trondheim, Norway

Peter Schönswetter
Institute of Botany, University of Innsbruck, Sternwartestraße 15, A-6020 Innsbruck, Austria

Andreas Tribsch
Department of Organismic Biology, University of Salzburg, Hellbrunnerstraße 34, A-5020 Salzburg, Austria

Siri Birkeland
The University Centre in Svalbard, NO- 9171 Longyearbyen, Norway
Centre for Ecological and Evolutionary Synthesis, Department of Biosciences, University of Oslo, Blindern, NO-0316 Oslo, Norway

Reidar Elven and Christian Brochmann
National Centre for Biosystematics, Natural History Museum, University of Oslo, Blindern, NO-0318 Oslo, Norway

Filipa Monteiro
Biosystems and Integrative Sciences Institute (BioISI), Faculty of Sciences, University of Lisbon, Campo Grande 1749-016, Lisbon, Portugal

Maria M. Romeiras
Biosystems and Integrative Sciences Institute (BioISI), Faculty of Sciences, University of Lisbon, Campo Grande 1749-016, Lisbon, Portugal
Tropical Research Institute (IICT/JBT), Trav. Conde da Ribeira 9, 1300-142 Lisbon, Portugal

M. Cristina Duarte
Tropical Research Institute (IICT/JBT), Trav. Conde da Ribeira 9, 1300-142 Lisbon, Portugal
Centre in Biodiversity and Genetic Resources (CIBIO/InBIO), University of Porto, Campus Agrário de Vairão, 4485-661 Vairão, Portugal

Hanno Schaefer
Technische Universitaet Muenchen, Biodiversitaet der Pflanzen, D-85354 Freising, Germany

Mark Carine
Plants Division, Department of Life Sciences, Natural History Museum, Cromwell Road, London SW7 5BD, UK

Aaron B. Shiels
USDA, National Wildlife Research Center, Hawai'i Field Station, Hilo, HI 96721, USA

Donald R. Drake
Department of Botany, University of Hawai'i, 3190
Maile Way, Honolulu, HI 96822, USA

Luís Silva, Elisabete Furtado Dias and Mónica Moura
InBIO, Rede de Investigação em Biodiversidade,
Laboratório Associado, CIBO, Centro de Investigação
em Biodiversidade e Recursos Genéticos, Polo-Ac͵ores,
Departamento de Biologia, Universidade dos Açores,
9501-801 Ponta Delgada, Açores, Portugal

Julie Sardos
Bioversity-France, Parc Scientifique Agropolis II, 34397
Montpellier Cedex 5, France

Eduardo Brito Azevedo
Research Center for Climate, Meteorology and
Global Change (CMMG - CITA-A), Departamento de
Ciências Agrárias, Universidade dos Açores, Angra do
Heroísmo, Açores, Portugal

Hanno Schaefer
Plant Biodiversity Research, Technische Universita¨t
Mu¨nchen, D-85354 Freising, Germany

Anna Traveset
Laboratorio Internacional de Cambio Global (LINC-
Global), Institut Mediterrani d'Estudis Avançats
(CSIC-UIB), C/Miquel Marqués 21, 07190-Esporles,
Mallorca, Balearic Islands, Spain

Susana Chamorro
Laboratorio Internacional de Cambio Global (LINC-
Global), Institut Mediterrani d'Estudis Avançats
(CSIC-UIB), C/Miquel Marqués 21, 07190-Esporles,
Mallorca, Balearic Islands, Spain
Universidad Internacional SEK, Facultad de
Ciencias Ambientales, Calle Alberto Einstein y 5ta
transversal, Quito, Ecuador

Jens M. Olesen
Department of Bioscience, Aarhus University, Ny
Munkegade 114, DK-8000 Aarhus C, Denmark

Ruben Heleno
Centre for Functional Ecology, Department of Life
Sciences, University of Coimbra, Calçada Martim de
Freitas, 3000-456 Coimbra, Portugal

Liba Pejchar
Department of Fish, Wildlife and Conservation
Biology, Colorado State University, Fort Collins, CO
80523, USA

Pablo Vargas
Real Jardín Botánico de Madrid (RJB-CSIC), 28014
Madrid, Spain

Manuel Nogales
Island Ecology and Evolution Research Group (IPNA-
CSIC), 38206 La Laguna, Tenerife, Canary Islands, Spain

Yurena Arjona
Real Jardín Botánico de Madrid (RJB-CSIC), 28014
Madrid, Spain
Island Ecology and Evolution Research Group (IPNA-
CSIC), 38206 La Laguna, Tenerife, Canary Islands,
Spain

Ruben H. Heleno
Centre for Functional Ecology, Department of Life
Sciences, University of Coimbra, 3000-456 Coimbra,
Portugal

Philippe Birnbaum
CIRAD, UMR 51 AMAP, 34398 Montpellier, France
Laboratory of Applied Botany and Plant Ecology,
Institut Agronomique néo-Calédonien (IAC),
Diversitébiologique et fonctionnelle des écosystèmes
terrestes, 98848 Noumea, New Caledonia

Thomas Ibanez, Hervé Vandrot and Elodie Blanchard
Laboratory of Applied Botany and Plant Ecology,
Institut Agronomique néo-Calédonien (IAC),
Diversitébiologique et fonctionnelle des écosyste`mes
terrestes, 98848 Noumea, New Caledonia

Robin Pouteau
Laboratory of Applied Botany and Plant Ecology,
Institut Agronomique néo-Calédonien (IAC),
Diversitébiologique et fonctionnelle des écosystèmes
terrestes, 98848 Noumea, New Caledonia
Laboratory of Applied Botany and Plant Ecology,
Institut de Recherche pour le Développement (IRD),
UMR 123 AMAP, 98848 Noumea, New Caledonia

Vanessa Hequet and Tanguy Jaffré
Laboratory of Applied Botany and Plant Ecology,
Institut de Recherche pour le Développement (IRD),
UMR 123 AMAP, 98848 Noumea, New Caledonia

Kim R. McConkey
School of Biological Sciences, Victoria University of
Wellington, Wellington, New Zealand
School of Natural Sciences and Engineering,
National Institute of Advanced Studies, Indian
Institute of Science Campus, Bangalore, India

Donald R. Drake
Department of Botany, University of Hawai'i at
Manoa, 3190 Maile Way, Honolulu, HI 96822, USA

Janice M. Lord
Botany Department, University of Otago, Dunedin
9054, New Zealand

Koji Takayama
Museum of Natural and Environmental History, Shizuoka, Oya 5762, Suruga-ku, Shizuoka-shi, Shizuoka 422-8017, Japan

Patricio López-Sepúlveda, Marcelo Baeza and Eduardo Ruiz
Departamento de Botánica, Universidad de Concepción, Casilla 160-C, Concepción, Chile

Josef Greimler and Gudrun Kohl
Department of Botany and Biodiversity Research, University of Vienna, Rennweg 14, A-1030 Vienna, Austria

Daniel J. Crawford
Department of Ecology and Evolutionary Biology and the Biodiversity Institute, University of Kansas, Lawrence, KS 60045, USA

Patricio Peñailillo
Instituto de Ciencias Biológicas, Universidad de Talca, 2 Norte 685, Talca, Chile

Karin Tremetsberger
Institute of Botany, Department of Integrative Biology and Biodiversity Research, University of Natural Resources and Life Sciences, Gregor Mendel Straße 33, A-1180 Vienna, Austria

Alejandro Gatica
Bioma Consultores S.A., Mariano Sanchez Fontecilla No. 396, Las Condes, Santiago, Chile

Luis Letelier
Universidad Bernardo O'Higgins, Centro de Investigaciones en Recursos Naturales y Sustentabilidad, General Gana 1702, Santiago, Chile

Patricio Novoa
Jardín Botánico de Viña del Mar, Corporación Nacional Forestal, Camino El Olivar 305, Viña del Mar, Chile

Johannes Novak
Institute for Applied Botany and Pharmacognosy, University of Veterinary Medicine, Veterinärplatz 1, A-1210 Vienna, Austria

Tod F. Stuessy
Department of Botany and Biodiversity Research, University of Vienna, Rennweg 14, A-1030 Vienna, Austria
Herbarium, Department of Evolution, Ecology, and Organismal Biology, The Ohio State University, 1315 Kinnear Road, Columbus, OH 43212, USA

Matt S. McGlone
Landcare Research, Lincoln 7640, New Zealand

Janet M. Wilmshurst
Landcare Research, Lincoln 7640, New Zealand
School of Environment, The University of Auckland, Private Bag 92019, Auckland 1142, New Zealand

Chris S.M. Turney
School of Biological, Earth and Environmental Sciences, University of New South Wales, NSW 2052, Australia

Carol A. Rowe and Martin P. Schilling
Department of Biology, Utah State University, Logan, UT 84322, USA

Paul G. Wolf
Department of Biology, Utah State University, Logan, UT 84322, USA
Ecology Center, Utah State University, Logan, UT 84322, USA

Joshua P. Der
Department of Biological Science, California State University, Fullerton, CA 92834, USA

Clayton J. Visger
Department of Biology, University of Florida, Gainesville, FL 32611, USA

John A. Thomson
National Herbarium of NSW, Royal Botanic Gardens and Domain Trust, Mrs Macquaries Road, Sydney, NSW 2000, Australia

A. E. Dunham
Biosciences at Rice, Rice University, Houston, TX 77005, USA

E. M. Wandrag
Biosciences at Rice, Rice University, Houston, TX 77005, USA
Institute for Applied Ecology, University of Canberra, Bruce, ACT 2617, Australia

R. H. Miller
College of Natural and Applied Sciences, University of Guam, Mangilao, GU 96923, USA

H. S. Rogers
Biosciences at Rice, Rice University, Houston, TX 77005, USA
Department of Ecology, Evolution, and Organismal Biology, Iowa State University, Ames, IA 50011 USA

Index